T0362140

ELSEVIER OCEAN ENGINEERING BOOK SERIES

VOLUME 1

PRACTICAL SHIP DESIGN

Elsevier Science Internet Home Page

http://www.elsevier.nl (Europe)
http://www.elsevier.com (America)
http://www.elsevier.co.jp (Asia)

Consult the home page for full catalogue information on all books, journals and electronic products and services.

Elsevier Titles of Related Interest

MOAN & BERGE
ISSC '97, 13th International Ship and Offshore Structures Congress
(3 Volume Set)
ISBN: 008 042829 0

VUGTS
BOSS '97, Behaviour of Offshore Structures (3 Volume Set)
ISBN: 008 041916 X

GUEDES-SOARES
Advances in Safety and Reliability
(3 Volume Set)
ISBN: 008 042835 5

FUKUMOTO
Structural Stability Design
ISBN: 008 042263 2

BJORHOVDE, COLSON & ZANDONINI
Connections in Steel Structures III
ISBN: 008 042821 5

CHAN & TENG
ICASS '96, Advances in Steel Structures (2 Volume Set)
ISBN: 008 042830 4

FRANGOPOL, COROTIS & RACKWITZ
Reliability and Optimization of Structural Systems
ISBN: 008 042826 6

GODOY
Thin-Walled Structures with Structural Imperfections:
Analysis and Behavior
ISBN: 008 042266 7

Related Journals

Free specimen copy gladly sent on request: Elsevier Science Ltd, The Boulevard, Langford Lane, Kidlington, Oxford OX5 1GB, UK

Applied Ocean Research
Coastal Engineering
Computers and Structures
Engineering Failure Analysis
Engineering Structures
Finite Elements in Analysis
Journal of Constructional Steel
  Research
Marine Structures
Ocean Engineering
Structural Safety
Thin-Walled Structures

ELSEVIER OCEAN ENGINEERING BOOK SERIES

VOLUME 1

PRACTICAL SHIP DESIGN

DAVID G.M. WATSON

BSc, CEng, FRINA, FIMarE, FIES, FRSA, DEng
SCOTLAND

"Life is the art of drawing sufficient conclusions
from insufficient premises"

Samuel Butler

OCEAN ENGINEERING SERIES EDITORS

R. Bhattacharyya & M.E. McCormick

ELSEVIER
Amsterdam – London – New York – Oxford – Shannon – Paris – Tokyo

ELSEVIER SCIENCE Ltd
The Boulevard, Langford Lane
Kidlington, Oxford OX5 1GB, UK

First edition 1998 (hardcover)
Second impression 2002
First impression 2002 (softcover)

Library of Congress Cataloging in Publication Data
Watson, D. G. M. (David G. M.)
    Practical ship design / D.G.M. Watson.
       p.   cm. -- (Ocean engineering series)
    ISBN 0-08-042999-8 (hardcover)
    1. Naval architecture.   2. Ships--Design and construction.
 I. Title.   II. Series: Ocean engineering series (Elsevier)
    VM156.W38   1998
    623.8'1--dc21   98-13994
                                                            CIP

British Library Cataloguing in Publication Data
A catalogue record from the British Library has been applied for.

ISBN: 0-08-042999-8 (hardcover)
ISBN: 0-08-044054-1 (softcover)

♾ The paper used in this publication meets the requirements of ANSI/NISO Z39.48-1992 (Permanence of Paper).

Transferred to digital printing 2005.

Foreword

We have in this book a distillation of the wisdom and knowledge acquired from the lifetime's work of a successful ship designer. Shining through it comes the author's obvious concern to hand on the fruits of his long and wide experience for the benefit of others. The reader cannot fail to be impressed by the scope of the subject as presented, and by the meticulous care taken to cover every aspect of ship design. The book deals with merchant ships and naval ships. It is not often that both of these are dealt with by one author, let alone handled with such authority. There is coverage of cargo ships and passenger ships, right on to tugs and dredgers and other service craft. There is concept design, leading on through detail design, to the study of the effect of regulations, the preparation of specifications, and on to matters of cost and economics. There is structural design and hydrodynamic design. No aspect of design has been left out.

Because the Author's span of working years closely paralleled my own, I can appreciate how all the changes in ship design and operation during those exciting years have been enjoyed by him; and I can only admire the way he has recorded, not just the outcome of these changes as they affect ship design, but also the reasoning behind the changes. It is the latter that means so much to the seriously enquiring reader. No doubt we all feel that our own little sector of personal history is the most significant ever, but I think the Author would agree that more has happened on the maritime scene in the years since World War 2, a period which covered our working lives, than in all the preceding centuries. It was so because of the ever growing demand for commercial activities at sea. Matched against that demand has been the greater ability to meet the design requirements, and that has been made possible by having more knowledge and better means of handling it.

The Author is genuinely competent to write on Practical Ship Design because of his long history in the actual business. Today there are very many people ready to discourse on design. "Design" has become a subject in its own right. But the number of people who have actually designed ships is small. In his presentation he has taken, from a position of great authority, a pleasing stance of humility. His views are not handed down from lofty professional eminence, but are offered in a real and sympathetic attempt to help fellow designers. There are endearing admissions of difficulties he has experienced in summarising and presenting concise data. There are biographical references to his own design problems along the way. Explanation of and reference to the underlying principles is laid alongside his design data and his recommendations and advice.

The question must have been asked at the inception of writing as to whom the book was directed and whom it would benefit. Now that the book is written, it is easier to answer. It is likely to interest almost everybody involved, not just in design, but in ships generally. Students, especially those needing a counterbalance to pure theory, young designers, designers working in isolation (the days of large shipbuilding design teams have gone, at least in UK), designers who are faced with unusual types of vessel, will all benefit from the vast store of design data and the conclusions and recommendations. But because of the style of writing and the more discursive approach, the Author has produced a text book that is not only interesting, but provides an educational experience for the interested or the curious, even those not directly involved in the ship design process.

It has become less common to find senior personnel in industry who have been able or allowed to pursue their own professional interests throughout their career. The pressure towards general management and the attraction of wider responsibilities is all too pervasive. The author has been fortunate and successful in following his technological career to a conclusion where it has brought to fruition an opus magnum of great potential benefit to many others. A man of the sea in every respect, he has demonstrated his depth of knowledge, his dedication in keeping such detailed records, and his sharp memories of the why and wherefore. As I read the book my own mind goes to a myriad of events in my design career where it is obvious that both of us suffered the same doubts, fears and uncertainties about the same topics; but perhaps we had our little successes too. I can see the problem he has faced as he comes to specific paragraphs and subjects where I am sure he would have wanted to lay out the pros and cons at much greater length, but then he would have had continually to accept that he was writing a book and not presenting and discussing a series of learned papers such as he has participated in so often in his career. What I also see as I read, is his native caution that will be transmitted to the reader. It will remind the designer that he will seldom have all the information or facts to hand; but he will still have to make his decisions. I can hear coming through, the Author's Scottish sense of humour, dryish and pawky, which I have known so well over the years. and underlying all, the principled beliefs from which has come the dedication behind the recording, documenting, evaluating and presenting of such a mass of design knowledge.

Today the shipyards where the Author and I worked have gone, as have the great UK shipowning companies we served. David Watson's book on Practical Ship Design based on his vast experience reminds us of all that was commendable and successful in our long shipbuilding history. It may well be one of the most lasting and valuable products to be carried forward from it.

Marshall Meek, CBE RDI FEng
Past President, Royal Institution of Naval Architects 1990–93
Master, Faculty of Royal Designers for Industry 1997–

Preface

During the years spent writing this book, I have become very conscious of how rapidly ships are developing and how fast the rules governing their construction and operation are changing.

I have also come to appreciate that although I can justifiably claim to have a particularly wide experience of ship design covering many ship types, my knowledge of specific subjects lacks the depth that an expert whose interest is limited to one or two subjects can be expected to have in his speciality.

Reading and re-reading IMO and DTI rules, appropriate parts of LLoyds rules, books on naval architecture and many technical papers and magazines has helped considerably, but the simplification and drastic condensation of some very lengthy and complex rules to the precis form which is the essence of this book was not an easy task and it seemed quite likely that there might be some errors.

I therefore asked a number of friends and former colleagues to read a chapter or two on subjects in which they had particular expertise. Suggestions from these readers have undoubtedly eliminated a number of errors and have also resulted in many improvements and I would like to acknowledge my very sincere thanks to all of them.

Former colleagues from YARD who helped in this way include Jack Bowes, Richard Benson, George Davison, John Jack, Andrew Kerr, Andrew Macgregor, Graeme Mackie, Richard Simpson and Bob Tait.

Friends who brought their expertise to bear include Alan Armstrong, a Director of Denholm Ship Management; Dr. Christopher Grigson, a distinguished hydrodynamicist and author of many technical papers on ship resistance and propulsion; David Moor, formerly Superintendent of Vickers Experimental Tank and the author of many technical papers which I have used extensively both in design work and in writing this book; John Sadden, Chief Naval Architect of Yarrow Shipbuilders. A number of others whose help was greatly appreciated asked to remain anonymous.

Even with this editing of individual chapters there remained a very real possibility that there might be discrepancies between the chapters or that the balance between them might not be altogether satisfactory — the former a thing that years of specification writing has elevated to a capital sin in my mind. It was therefore both a relief and a great pleasure when Allan Gilfillan, my colleague for many years and a very distinguished naval architect, agreed to read the whole book in draft. This was a major task involving very considerable time and effort and resulted in a number of further corrections and improvements for which I would wish to express my most sincere thanks. Finally, two distinguished naval architects, who acted as referees for a possible publisher, suggested a number of helpful changes which have now been incorporated to the benefit of the final text.

Having acknowledged all this most generously given assistance, I would wish to absolve all the helpers of responsibility for any residual errors —indeed I would rather like to "cover" myself also against any consequences which may stem from any remaining errors (of which there must surely be some) by recommending that users of this book should make their own checks on all data, approximate formulae and rules given before they use them "in anger". This is a policy that I can claim to have followed throughout my career and it is one which has saved me from making a number of errors that might have resulted from the use of data and/or formulae which investigations showed to be less accurate than was claimed by their protagonists !

A book or this sort would be a poor thing if it did not draw on many sources and I would acknowledge that some parts of it almost take the form of an anthology and would express my thanks to the authors of many technical papers from which either text or illustrations have been drawn.

I have tried to obtain permission for all these quotations and think I have acknowledged the source in every case but ask forgiveness if any permissions have not been obtained or if any acknowledgements have been accidentally omitted.

It would be very helpful if any reader spotting an error would inform the publishers so that a correction can be made should there be a later edition — one of the joys of word processing being the ease with which such changes can be made.

David G.M. Watson

Publisher's Note

Since the publication of this book the author, in recognition of his life's work in ship design, much of which is recorded in this book, was given an honorary degree by the University of Glasgow and is now Dr. Watson, D.Eng.

Contents

"Life is the art of drawing sufficient conclusions from insufficient premises."

Samuel Butler

Chapter 1. Introduction, Methods and Data

Chapter 2. Setting Design Requirements

Chapter 3. The Design Equations

Chapter 4. Weight-Based Designs

Chapter 5. Volume, Area and Dimension-Based Designs

Chapter 6. Powering I

Chapter 7. Powering II

Chapter 8. Design of Lines

Chapter 9. Machinery Selection

Chapter 10. Structural Design

Chapter 13. Other Statutory Rules

Chapter 14. Special Factors Influencing Warship Design

Chapter 17. Specification and Tender Package

Chapter 18. Cost Estimating

Chapter 19. Operational Economics

Chapter 20. Conversions

Chapter 1

Introduction, Methods and Data

1.1 INTRODUCTION

1.1.1 Design and naval architecture

There are many excellent books on naval architecture. Most of the recent ones have been written by authors of considerable ability and handle admirably the highly mathematical treatment that is demanded nowadays by the naval architecture of advanced ship types.

The last chapter of most of these books is generally entitled "ship design", but unfortunately, in the author's opinion, these chapters rarely show the same mastery of their subject that the other chapters of these books do, possibly because most of the authors have an academic background and few have worked for any significant time as designers.

There is, in fact, a surprising dearth of books which specialise in ship design. Presumably this is partly because practitioners in this field —whether they work for shipyards, for shipping companies or consultancies — are usually too busy exercising their skills to find time to write and partly because they, or the firms they work for, are reluctant to give away what they consider to be commercially valuable secrets.

This book's thesis is that ship design although based on the science of naval architecture involves something more. In the author's view naval architecture consists of a number of quite distinct subjects which are generally taught and dealt with in almost complete isolation from one another — structural strength, trim and stability, and resistance and propulsion being three such subjects. Design, on the other hand (or it may be more correct to say 'initial design') requires the designer to keep the essentials of all these separate subjects of naval architecture and indeed a number of other factors in his mind at one time so that he can synthesise a ship concept which both in its main dimensions and in its general arrangement satisfies or comes close to satisfying all these requirements.

If he can do this successfully from the start of a project, he will greatly reduce the time and effort required to produce a design. If he fails to do so the design is likely to require major changes as it is developed and detailed calculations can be made.

Once an initial design has been completed, each facet of it must of course be tested using the appropriate rigorous scientific naval architectural methods, but in the author's view it is ponderous and time wasting to apply these methods whilst the initial design is still being developed, although it must be admitted that the use of computers has opened the door to the possibility of making detailed calculations much earlier in the design process than used to be possible.

To give one example of the way in which thinking ahead can greatly reduce design effort: the development of an outline design in which the stability is satisfactory, or nearly so, need not necessitate detailed stability calculations at the initial design stage (when the all-important weights are in any case likely to have a considerable margin of error) but can instead be reasonably assured by choosing a ratio of beam/depth which experience has shown will result in satisfactory stability.

Similar thinking can ensure that a design, almost from its inception, is such that no really nasty surprises in strength or powering will be found when it is subjected to the detailed scientific examination that comes at a later stage.

1.1.2 Reader's background knowledge

This book makes no attempt to teach scientific naval architecture and it is assumed that professional naval architects will bring a well developed background know-ledge of naval architecture to their reading and use of the book.

Ships are, however, a fascinating subject and reading this book should not be too difficult for anyone interested in learning how they are designed. Lay readers will want to skip those parts that invoke terms with which they are unfamiliar, but should still find much that is intelligible to them and be able to see why ship designers find their profession so absorbingly interesting.

1.1.3 Scope in terms of ship types

This book covers the design of a wide range of monohull displacement ship types, but this needs to be set in the context of the even wider range of marine vehicles shown in Fig. 1.1. These range from surface skimming vessels, through displacement ships and semi-submersibles, whose main buoyancy is well under the water surface, to wholly submerged submarines.

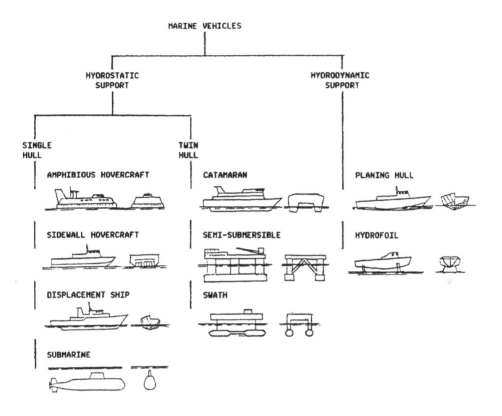

Fig. 1.1. Marine vehicle types.

The extremes of amphibious hovercraft and submarines have unique capabilities; the former has an ability to travel over land or ice as well as over the sea; the latter an ability to travel under ice flows and to remain invisible.

Many other types of marine vehicles share their market place, to a greater or less extent, with the choice between them being determined by the required speed and carrying capacity together with the wind and sea conditions in which they are required to operate. The building and operational costs which these factors entail for the alternative types of vessel determines the "winner".

Apart from some discussion in Chapter 2 — in which the importance of setting objectives in broad terms which admit unconventional solutions and a brief treatment later in that chapter of planing and multi-hull vessels — this book is devoted to monohull displacement ships. The great majority of ships sailing the seas today are monohull displacement ships, with this solution having been shown

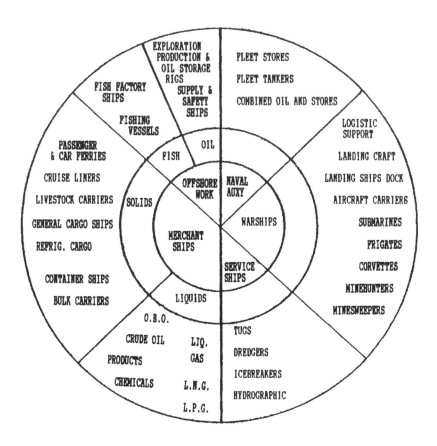

Fig. 1.2. Main ship types and their purposes.

to provide the most economical answer to the majority of design requirements. Some challenges to this supremacy may, however, be on the way: wave-piercing catamarans are becoming competitive for passenger ships and the excellent sea-keeping ability of the SWATH type of vessel enables a ship of this configuration to be smaller than a competing monohull so that this type may become economically competitive for a service in which minimum motions in a seaway are a prime need — aircraft carriers and some research ships being distinct possibilities.

Monohull displacement ships can be divided into many categories, some of the principal divisions by use being shown in Fig. 1.2. From a design point of view there is, however, an alternative classification according to which design requirements are most critical in the determination of the main dimensions of the ship (a subject discussed in Chapter 2).

1.1.4 Transfer of technology between ship types

It is perhaps obvious that a design for a particular type of ship can most readily be prepared by a naval architect who has recently designed a successful ship of that type. From such a background of experience, a competent design can be confidently expected, but there must be a probability that the new design will closely follow the trends of recent designs and is unlikely to include much innovative thought.

On the other hand, a naval architect experienced in designing a wide variety of ships, but lacking detailed up-to-date knowledge of the particular type, will have a harder task as he will have to start by studying magazine articles etc. about recently built ships of the type to acquire the necessary background knowledge. Once he has gained this background he may, however, go on to produce a more innovative design, possibly bringing into play ideas used in other ship types which can be adapted to the ship type on which he is working.

Unfortunately for naval architects, the tendency today is for shipyards, and to a lesser extent shipping companies, to specialise in one (or at most a very few) ship types, reducing the range of experience which used to be common in the versatile shipyards of some decades ago.

1.1.5 The author's design experience

The author was lucky to have the opportunity of gaining a particularly wide range of experience and would like to use this book to share this with his readers. The shipyard in which he spent the first half of his career built passenger liners, cross-channel passenger, car and train ferries, refrigerated and general cargo ships, bulk carriers, oil tankers, many dredger types, logistic support landing ships, frigates and destroyers, and he was deeply involved in the design of all of these except the warships. As consultants, the firm was also involved in the design of some of the earliest stern trawlers and fish factory ships, and of the first generation container ships.

In the second half of his career, the author joined a major firm of consultants which under his direction designed another wide assortment of merchant ships and warships. The merchant ships included cargo liners, container ships, bulk carriers, sewage-disposal ships, fishery research vessels, hydrographic and oceanographic research ships, fishing boats.

The warships and naval auxiliary vessels included aircraft and helicopter carriers, frigates, corvettes, mine hunters, landing ship docks, logistic support ships, fleet auxiliary combined oil tanker and store ships.

Some of these ships feature in Chapter 16, in which the general arrangements of a number of ship types are examined. Other ships featuring in this chapter have been selected as representing good recent practice.

1.1.6 The general layout of the book

The next two sections of this chapter deal in a general way with design methods and design data respectively. The section on design methods starts by discussing the place of some of the "back of the envelope" type calculations outlined in later chapters and then goes on to describe computer methods and how these can speed up and increase the accuracy of design work.

The section on data outlines the importance of data to a naval architect and the need to store this in an easily accessible format. The sources drawn on in the writing of this book are given together with suggestions of other sources that designers will find useful.

Chapter 2 starts by dealing with the very important subject of setting the design requirements. For merchant ships this task will often be carried out by the commercial side of a shipping company; for warships by naval staff; for specialist ships by the scientists or others involved in the specialism. The naval architect has, however, a great deal to contribute to this task and should be fully consulted. If he is not so consulted he should have no inhibitions about questioning the design requirements with which he is eventually faced. The chapter then introduces the design spirals for merchant ships and warships, compares these and goes on to suggest how to establish which criteria are most critical in seeking a solution which meets the requirements.

Chapters 3, 4 and 5 draw quite largely on the R.I.N.A. paper "Some ship design methods" which the author wrote in 1976 in collaboration with A.W. Gilfillan, to whom he is indebted for permission to draw on this joint work. Most of what was written in 1976 seems to have stood the test of time very well, but some updating has of course been necessary and there has been some expansion of a text which was originally limited by R.I.N.A. publication guidelines.

Chapter 3 gives the fundamental design equations for both weight and volume-based designs. This includes data on the dimensional relationships applicable to a variety of ship types. Data on the deadweight/displacement ratio and the cargo capacity/hull volume ratio are given, again for a variety of ship types.

Chapter 4 deals with weight-based designs describing both approximate and detailed methods for calculating steel-weight, outfit weight and machinery weight.

Chapter 5 deals with volume-based designs describing how to calculate the volume required to accommodate all the space requirements of a passenger ship and how to translate a space requirement to appropriate ship dimensions.

Chapters 6 and 7 which deal with powering, Chapter 8 which follows on to the closely related subject of the ship lines, and Chapter 9 which deals with machinery selection all draw on the author's Parsons Memorial paper "Designing ships for fuel economy" published by R.I.N.A. in 1981.

The treatment of powering in Chapters 6 and 7 kept expanding under the influence of the author's advisers. Interestingly one of these favoured the newer treatment of this subject as more scientific whereas the other felt that there was much more useful data available in the earlier Froude format and believed that with appropriate "fiddle factors" use of this data can still give satisfactory answers. The chapters have tried to keep a balance between these two approaches.

Readers may feel with some justification that the treatment of powering falls short of the full treatment they would like to have as the subject of propeller efficiency has been omitted for the very good reason that the author can claim no expertise in this science (or is it a black art?). He has instead always used the shortcut to the quasi-propulsive efficiency which is given in Chapter 7, having found this to be remarkably accurate.

Chapter 8 deals not only with the design of lines to minimise powering, but looks at the qualities that the lines must have to ensure good sea-keeping, good manoeuvrability and good stability for given dimensions.

Chapter 9, in its treatment of machinery selection, starts with a statement of the criteria against which main engines are chosen and goes on to consider which of these are important for different ship types and which types of machinery best meet them.

Chapter 10 deals with the factors influencing structural design. Although no detailed structural calculation methods are given, the chapter gives a lot of advice on how to design both the general arrangement and the structure itself for economy in steel-weight and in fabrication costs, whilst avoiding many of the pitfalls of fatigue, brittle fracture, vibration, corrosion that can be the consequence of less then satisfactory structural design.

Chapters 11, 12 and 13 deal with the main statutory rules for merchant ships, the need to ensure compliance with which forms a prominent part of the work undertaken in the later design spirals.

Chapter 11 has freeboard and subdivision as its subject and gives a full treatment of the new probabilistic rules for the subdivision and damaged stability of cargo ships. The corresponding rules for passenger ships are not dealt with in the same detail as it is expected that they will be brought into line with the cargo ship rules within a relatively short time.

Chapter 12 deals with stability and trim and after dealing with the statutory rules for these subjects for merchant ships outlines the treatment that these are given in warship design and operation.

Chapter 13 deals with some of the remaining subjects for which there are statutory rules for merchant ships, such as fire protection, life-saving, marine pollution and tonnage.

Chapter 14 deals with some of the special requirements which are involved in the design of a warship.

Chapters 15 and 16 bear a considerable responsibility for this book being written as it was the author's view that the arrangement aspects of design were badly neglected, both in textbooks on naval architecture and in teaching in universities and technical colleges, that provided much of the original motivation.

Both the task of creating the general arrangement of a ship and the work involved in drawing detailed arrangements of each part of it seem to be regarded by many lecturers at universities and technical colleges and, to only a slightly lesser extent, by some designers themselves as simple tasks which can be left to draughtsmen. This attitude is compounded by the fact that draughtsmen are given only limited instruction in much of the skills of their trade and are largely left to learn for themselves by studying the plans of "the last ship". Whilst studying the plans of ships should be a "must" for all designers, this study ought to go well beyond knowing "what" was done towards a clear understanding of "why" it was done.

Designers should have an ability to appreciate when good reasoning about a multitude of factors has led to a good arrangement and, even more importantly, an ability to see the faults in other arrangements. These abilities, which can and should be taught, deal with as fascinating a subject as anything in ship design.

Chapter 17 goes back a long way in the author's career to when he wrote a standard specification for the Clydeside shipyard of Alexander Stephen & Sons. This specification was intended both to ease the task of writing ship specifications and to lay down standards to be followed where owner's specifications lacked detail.

Chapter 18 dates back to the same period of his career, but needed substantial updating in later years to deal with the exceptionally difficult problem that faces a consultant when his client wants an estimate of the price of a ship which may be built, not in the adjoining shipyard (which shipyard estimators find difficult enough) but in Japan or Korea.

Chapter 19 is, of course, closely related to Chapter 2 and might have adjoined it in the book. The author cannot claim any specialised knowledge of this subject but feels strongly that a book of this sort would be incomplete unless it addressed the subject of operational economics, which is both the test of whether a good merchant ship has been built and the starting point for the design of a new merchant ship.

Chapter 20 deals with some solutions to the design problems that arise in major conversion work whether this is undertaken to enable a ship to operate in a new role or to rectify design errors.

In the hope of easing reference to the bibliography, this has been divided into six sections, with each section covering a group of chapters whose subjects are related.

It has to be admitted that the bibliography is far from complete, but it is hoped that the references given will lead to other relevant bibliographies.

There are such a mass of symbols and abbreviations in use in naval architecture and ship design that it was thought best to define these in close context to the formulae in which they are used.

The author has tried to write this book in as plain English as possible as he has a strong dislike of some of the modern words with which a number of today's technical papers seem to be inflated. In keeping with this policy, the book tries to describe practical ship design methods and not to "elaborate the systems methodology of the design of marine artifacts"!

One of the aims of this book is to help naval architects to co-operate closely and harmoniously with marine engineers and other specialists whose expertise is required in ship design and construction, which seems preferable to indulging in "synergetic integration".

The author has tried to follow the one acronym he really likes, "AAEFTR" — "all acronyms explained first time round" and hopes that his readers will appreciate this.

1.2 DESIGN CALCULATION METHODS

1.2.1 General discussion

It is perhaps worth emphasising that this book deals primarily with initial design, although it also looks forward to the detailed design development which follows once the initial design is accepted. Whilst many of the methods given were originally used on slide rules or calculators, they can equally well be developed into computer programs and in a number of cases this development has been outlined in the book. The author hopes that readers will write their own computer programs using some of the other methods suggested.

To bring himself up-to-date with modern computer methods the author wrote to several firms who offer computer-aided ship design programs and received a number of helpful replies. In general these showed that there are a number of good programs covering what the author would call "design development".

Although these programs, which are discussed in §1.2.3, can also contribute to initial design, none of them seem to deal with the first, and arguably the most important, step in initial design — the determination of suitable main dimensions, block coefficient and the arrangement concept. That this should be the case is not altogether surprising as these aspects of initial design (at any rate as the author has practised it) do not require any considerable mathematical treatment but do need quite a bit of lateral thinking, which is not easy to program.

As soon as these initial design steps have been completed CAD (computer-aided design) can come into its own with many of the programs discussed in §1.2.4

being used to speed the process. If used at this early stage, some of the input is likely to be tentative but one great joy of computer methods is the ease with which calculations can be updated as better information becomes available. CAD methods will generally be used when drawing the arrangement plan, the lines plan and other plans, speeding this work enormously and greatly improving its quality. The fact that scales can be changed during this process, with what started life as a 1/200 initial plan being transformed first to 1/100 and then to 1/50 enabling more and more detail to be added, is a tremendous advantage. The scale change ability means that even at a very early stage large-scale detailing can be used to investigate areas of difficulty.

It is interesting to recall that initial design was one of the first subjects to be attacked in the early days of computers. The method used at that time involved the processing of multiple designs on a batch basis with the aim of identifying an optimum solution from the resulting mass of designs. The process was not very successful for two main reasons: the design processes used some algorithms of doubtful accuracy and the criteria used to identify the optimum answer were unsatisfactory.

An unfortunate side effect of the use of such computerised ship design programs was the tendency this had of deflecting the designer from innovative thinking about the design. No-one today would dispute that the biggest contribution to the efficient transport of general cargo in recent years came, not from all the effort put into optimising ship design using advanced computer methods, but from the lateral thinking of the trucker who came up with the idea of making mixed general cargo into an easily handled bulk cargo by putting it in standardised containers.

1.2.2 Expert systems

The modern approach to the use of computers for initial design makes use of an expert system technique which almost certainly encourages innovative thinking and is a trend which the author welcomes unreservedly. In the past when one of the stalwarts of his design team was approaching retirement or leaving to take another job, it had always to be a top priority to ensure that the departing expertise was passed on to another member of the team. This was never easy but the advent of computerised expert systems means that it is now possible to accumulate know-how in an organised format as a standard routine.

Even more important, however, is the ability that an expert system has to make use of the expertise of a large number of specialists enabling these to be consulted at their convenience without any need for them to be present at the time when a design is being prepared — usually in a hurry and when they may have other commitments.

Whilst it is still early days in the development of expert systems, a 1990 R.I.N.A. paper "The application of an expert system to ship concept design investigations" by Welsh, Buxton and Hills gives a good idea both of the way to set up such a system and of the advantages that can stem from its use. In the paper, the design of a container ship was taken as an example and it is interesting to note the wealth of information about containers that was assembled from shipping company experts, going well beyond what a naval architect could expect to know and which could in principle have major design implications. It was salutary to note that the knowledge base contained around 7000 lines of statements.

The author thinks he would have enjoyed writing, or contributing to the writing of an expert system but unfortunately has never had the opportunity to do so. He believes that this book contains data and formulae that can provide a substantial contribution to ship design expert systems for a variety of ship types.

1.2.3 Use of spreadsheets

Although the use of programs in which all the data is input by the user, and all the formulae involved are standard naval architecture (like most of the programs discussed in §1.2.4.) can be unreservedly recommended, it may be wise to have some reservations about programs which use stored data and algorithms that are not known and approved by the user.

Spreadsheets present an easy way of using a computer for a wide variety of tasks without having to write or purchase special-purpose programs. One advantage of using a spreadsheet is that a user will generally write the program himself and will therefore know exactly what formulae are included in his calculations and what confidence can be given to the answers.

In writing this book, the author made considerable use of spreadsheets, a number of which are included in the text.

1. In the next section of this chapter spreadsheets have been used to marshal data on dimensional relationships, deadweight/displacement ratios and capacity/volume relationships, all of which are required in Chapter 3.

2. In Chapter 4 a spreadsheet format is used for a calculation sheet for initial design.

3. In Chapters 6 and 7 spreadsheets are used for various aspects of powering: to convert © F or © ITTC to C_t ITTC57 or C_t ITTC78; to calculate $(1 + K)$ using Holtrop and Mennen's formula; to calculate $(1 + K)$ using a modified Prohaska method.

4. In Chapter 12 a sheet for dredger spill-out calculations is given.

In addition, formats have been suggested elsewhere which it is hoped will assist readers in developing programs for a number of other design calculations.

An example showing how to develop a special-purpose spreadsheet has been included because the author has found most handbooks on computer programs singularly difficult to follow (a criticism not particularly directed at the Lotus 123 system which the author used, as this handbook is better than most) and wondered whether writing programming instructions which would be readily intelligible to a computer novice was as difficult as these books made it seem. At the end of this section readers will be able to judge whether he has succeeded!

Incidentally, almost the only computer books which the author has read that he would exempt from this criticism are the entertaining series with the general title "Computing for Dummies" which deal with a number of well known computer programs and are a delight with their clarity and wit.

Table 1.1 originally set up to produce data for Chapter 3 is used here to illustrate the development of a spreadsheet. If readers can follow these instructions and create their own versions of this spreadsheet, they should have little difficulty in writing other spreadsheet programs to speed up their ship design calculations.

Some of the instructions which follow are applicable to all spreadsheets but some are specific to Lotus 123.

– Before starting to use the computer, prepare a draft of what you intend to do; assess the width which each column needs to be, the number of decimal places which are appropriate and collect all the equations it is intended to use.

– At the MS/DOS prompt A> put the Lotus 123 system disc in drive A and enter 123.

– After an interval, an outline spreadsheet generally as shown in Fig. 1.3 will appear, but without the menu which is discussed later. The spreadsheet has a top line with alphabetic labels to identify the columns, and the first column gives numbers to identify the lines. Neither of these identifications appear when the spreadsheet is printed unless the print screen key is used as has been done for the illustration. As this alphanumeric index can be useful for descriptive purposes it has been repeated in the third line and second column respectively of the example.

– The next step is to title the columns. Position the cursor and type the required title — this will first appear in the top right of the screen and will transfer to its correct position when "enter" is pressed.

– If it is desired to centre the title within a column the word should be preceded by ^. If the word is required to start at the left of the column it should be preceded by '.

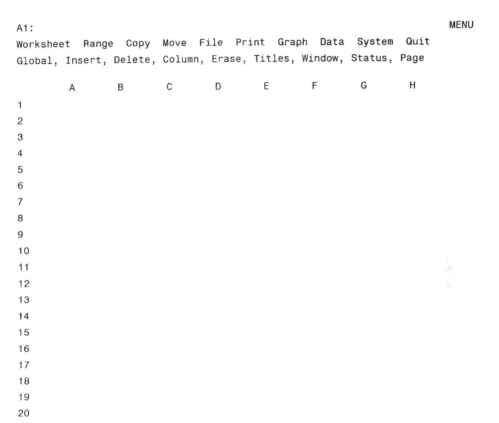

A1:   MENU

Worksheet Range Copy Move File Print Graph Data System Quit

Global, Insert, Delete, Column, Erase, Titles, Window, Status, Page

	A	B	C	D	E	F	G	H
1								
2								
3								
4								
5								
6								
7								
8								
9								
10								
11								
12								
13								
14								
15								
16								
17								
18								
19								
20								

Fig. 1.3. Lotus 123 Spreadsheet starting point. This is a print screen copy. The first menu is shown.

- To draw a horizontal line type '- - - - - - - - - and press "enter". The line will appear in the one column only but can be copied to as many other columns as required using the copy procedure described later. To draw a vertical line type '| and press enter. The line will appear on one line only but can be copied for as many lines as desired by the use of the copy procedure.

- It is now time to introduce the Lotus 123 menu system which is obtained by the use of the / command. This produces a first menu which is activated by either moving the curser to the selected menu item and pressing "enter" or by pressing the initial letter of a menu item.

- For copy, press "copy" on the first menu.

- The question "from?" will then be asked.

- Reply by typing in the address or addresses that are to be copied: typically, B2 if only what is in B2 is to be copied or D2..M2 if everything in line 2 from column D to column M inclusive is to be copied to another line, or B2..B30 if everything in column B from line 2 to line 30 inclusive is to be copied to another column.

- After the reply to this question has been entered, a second question "to?" will then appear. The answer to this may be a single address or two addresses indicating a line or a column. When this is entered the required copy will appear on the spreadsheet.

- After setting up the outline of the table in this way, the next step should be to adjust each of the columns to the required width, from the standard 9-digit setting. At the top of each column in turn use / to obtain the menu. On the first menu enter "worksheet", on the second menu enter "column", then "set-width". Then give the required number and on pressing "enter", the column will change to the required size. For a vertical line using | the column width should be set to 1.

- The next step should be to set the number of decimal places to be used for the figures which will ultimately appear in each column. Position the curser at the first line in each column in which figures are to be entered. Use / to obtain the menu. Then enter "range", "format", "fixed" and the number of decimal places. The questions "from" and "to" will then be posed in succession. Type the answers and press "enter". The decimal places and column widths will appear typically as (F2) and [W5] respectively on the sheet.

Names and symbols together with numbers not intended for use in any calculation purpose should be preceded by either ^ or '.

Numbers to be used in calculations should not have any prefix.

Formulae to be used should be entered with the prefix +. They should be entered in the first address to which they apply and the copy procedure used to apply the same formula to the data on successive lines.

The symbols + and − are used for addition and subtraction; the symbols * and / for multiplication and division.

In a complex equation an alternative to using / for division is to give the divisor an exponent of (−1) and use the multiplication sign *. This sometimes seems to work better. The sign for an exponent is ^.

The only mathematical function used in this particular table is "@ATAN" which is tan (−1) in radians, but there are a large number of other mathematical functions available in the Lotus system, all of which should be preceded by the symbol "@".

Column

A	[W4] ^ ref no.
B	[W5} ^ ship type T,B or R.
C	[W17] ' Ship Name
D	(F2)[W7] data L
E	(F2)[W6] data B
F	(F2)[W6] data D
G	(F3)[W7] data T
H8	(F3)[W6] +0.7+).125*@ATAN (0.25*(23-100N8)
H9	/ copy from H8 to H9..H26
I8	(F0)[W7] + 1.03*D8*E8*G8*H8
I9	/ copy I8 to I9..I26
J	(F0)[W7] data deadweight
K8	(F3)[W7] +J8/I8
K9	/ copy K8 to K9..K26
L	(F2)[W7] data speed
M8	(F2)[W6] +D8^0.5
M9	/ copy M8 to M9..M26
N8	(F3)[W6] +0.165*L8/M8
N9	/ copy N8 to N9..N26
O8	(F2)[W5] +D8/E8
O9	/ copy O8 to O9..O26
P8	(F2)[W5] +E8/F8
P9	/ copy P8 to P9..P26
Q8	(F3)[W6] + G8/F8
Q9	/ copy Q8 to Q9..Q26

One slightly disconcerting feature of the Lotus 123 and probably most, if not all, other spreadsheet programs is that when an equation has no data to work on, an entry of either 0 when the calculation is a multiplication or ERR if a division is included appears in the answer address. Fortunately these entries disappear when data is supplied.

After a little practice, a spreadsheet of this sort can be very quickly produced and once available both eases the completion of tables of this sort and ensures an accuracy which the author must admit he had not achieved in the hand calculations. which preceded its use.

A completed version of this spreadsheet is given as Table 1.1.

Table 1.1 (*continued opposite*)

Deadweight / displacement ratio based on L.R. data or from *Significant Ships* where this does not give the displacement

A	B	C	D	E	F	G	H	I	J
Ref	Type	Name	*L*	*B*	*D*	*T*	C_b	Disp (tonnes)	Dwt (tonnes)
8	T	Arabiyah	242.19	43.28	23.80	15.62	0.845	142529	121109
9	T	Australia Sky	171.00	26.80	16.40	10.67	0.804	40492	33239
10	T	Columbia *	313.00	56.60	28.60	19.44	0.849	301000	258076
11	T	Golar Coleen	256.01	46.21	23.83	17.02	0.842	174566	152385
12	T	Golar Cordelia	315.02	57.21	30.41	22.03	0.849	347078	304622
13	T	Olympic Serenity	222.12	42.00	20.30	14.20	0.835	113948	96733
14	T	Achilles	215.02	32.20	18.30	13.20	0.829	78031	63800
15	B	Amelia	250.02	43.01	23.42	17.20	0.838	159562	135000
16	B	An Ping	185.00	28.40	15.80	11.00	0.818	48699	38987
17	B	Angel Feather *	172.00	30.50	15.80	11.23	0.823	49939	42448
18	B	Clarita	260.00	43.00	24.11	17.62	0.842	170912	152065
19	B	Dahlia	300.00	50.00	25.70	18.02	0.852	237264	207063
20	B	Yeoman Burn	235.00	32.20	20.10	14.00	0.834	90949	77500
21	R	Arctic Ocean	140.01	22.01	13.21	9.35	0.559	16600	10600
22	R	Del Monte Cons.	147.56	23.51	12.76	9.00	0.639	20545	12700
23	R	Del Monte Harr.	130.03	22.31	12.76	9.20	0.578	15882	10000
24	R	Hokkaido Rex	140.00	20.60	12.80	9.42	0.610	17075	11622
25	R	African Reefer	136.66	23.60	15.40	10.02	0.632	21047	14572
26	R	Kowhai	136.00	18.50	10.65	7.30	0.631	11935	7168

* Indicates double hull construction.
T = tanker, B = bulk carrier, R = refrig. cargo.

1.2.4 Design development calculations

Whilst most firms undertaking ship design will already possess a suite of programs covering all the main design calculations that they have either written themselves or purchased, it may be helpful to mention some of the software available for sale.

There are quite a number of firms offering software for both CAD and CAM (computer-aided manufacture) which for shipyards is the natural next step. In

Table 1.1

continuation

A	B	C	K	L	M	N	O	P	Q	R
Ref	Type	Name	$\dfrac{\text{Dwt}}{\text{Disp}}$	Speed (knots)	$(L)^{0.5}$	F_n	L/B	B/D	T/D	L/D
8	T	Arabiyah	0.850	13.00	15.56	0.138	5.60	1.82	0.656	10.17
9	T	Australia Sky	0.821	14.75	13.08	0.186	6.38	1.63	0.651	10.42
10	T	Columbia *	0.857	14.00	17.69	0.131	5.53	1.98	0.680	10.94
11	T	Golar Coleen	0.873	14.00	16.00	0.144	5.54	1.94	0.714	10.74
12	T	Golar Cordelia	0.878	14.00	17.75	0.130	5.51	1.88	0.724	10.35
13	T	Olympic Serenity	0.849	14.00	14.90	0.155	5.29	2.07	0.700	10.94
14	T	Achilles	0.818	14.50	14.66	0.163	6.68	1.76	0.721	11.74
15	B	Amelia	0.846	14.50	15.81	0.151	5.81	1.84	0.734	10.67
16	B	An Ping	0.801	14.40	13.60	0.175	6.51	1.80	0.696	11.70
17	B	Angel Feather *	0.850	13.50	13.11	0.170	5.64	1.93	0.711	10.88
18	B	Clarita	0.890	14.00	16.12	0.143	6.05	1.78	0.731	10.78
19	B	Dahlia	0.873	12.75	17.32	0.121	6.00	1.95	0.701	11.67
20	B	Yeoman Burn	0.852	14.61	15.33	0.157	7.30	1.60	0.697	11.69
21	R	Arctic Ocean	0.639	22.50	11.83	0.314	6.36	1.67	0.708	10.59
22	R	Del Monte Cons.	0.618	18.50	12.15	0.251	6.28	1.84	0.705	11.56
23	R	Del Monte Harr.	0.630	20.00	11.40	0.289	5.83	1.75	0.721	10.19
24	R	Hokkaido Rex	0.681	19.00	11.83	0.265	6.80	1.61	0.736	10.93
25	R	African Reefer	0.692	18.00	11.69	0.254	5.79	1.53	0.651	8.87
26	R	Kowhai	0.601	18.00	11.66	0.255	7.35	1.74	0.685	12.76

* Indicates double hull construction.
T = tanker, B = bulk carrier, R = refrig. cargo.

alphabetical order some of the better known names in this business are:
- Autoships (Coast Design Canada);
- Kockums Computer Systems (Sweden, but with a strong U.K. presence as it incorporates the former B.M.T. ICONS);
- Macsurf-Island Computer (U.K.);
- Senermar (Spain);
- Wolfson Institute (U.K.).

Macsurf and Wolfson concentrate on CAD, whilst Senermar and Kockums continue through to CAM. Whilst CAM is beyond the scope of this book it is worth mentioning that an integration of CAD and CAM can have big benefits to a shipyard, both because it results in a big reduction in the input task and because it should reduce errors since all users will know the wide use to which the data has been or will be put.

The range of CAD programs offered by each of the companies differs, but most of the items in the following list are available from at least two of the companies mentioned above.

1. Lines plan production — either from program data based on main dimensions, block coefficient and LCB or from the digitising of an approximate lines plan; fairing of lines; production of offsets.

2. Deck, bulkhead and other outlines based on lines plan data to feed into CAD draughting.

3. Shell development for plate ordering.

4. Weights and centres — can be used in the make up of a lightship weight or in developing a stability condition.

5. Space — capacity and centres.

6. Hydrostatics and cross curves of stability.

7. Stability conditions.

8. Damaged stability deterministic and probabilistic.

9. Grain stability.

10. Floodable length curves.

11. Longitudinal strength — bending moment and shear force.

12. Freeboard and tonnage.

13. Ship motions — roll, pitch, heave, yaw, sway for a variety of sea spectra.

14. Manoeuvring.

15. Powering — resistance and effective horsepower.

16. Powering — propulsion factors, quasi propulsive coefficient and propeller design.

Some of these items, or sometimes a particular firm's approach to them, merit a few comments:

(1) Lines

At least one firm offers lines, hydrostatics etc. for asymmetric forms. The parameters used for the design of lines seem to be based on block coefficient and LCB only and do not appear to include consideration of the features required to ensure good sea-keeping and manoeuvrability or high KM which are discussed in Chapter 8. The alternative approach via the digitising of an approximate lines plan enables these features to be built in by the designer.

The system of co-ordinates used by one (and possibly some others) of the systems examined involves the use of a load waterline or similar datum which facilitates the definition of both the ship's bottom and the superstructures with negative and positive Z values. Longitudinal or X values are positive forward, whilst Y values are positive to starboard.

(2) and (3) Outline structural plans

The use of these outlines along with computerised area measurement holds out hope of speeding and improving the accuracy of steel-weight calculations.

(4) Capacities and centres

The ability to update these for minor changes on a regular basis enables the design to be updated regularly and goes a long way to ensuring the avoidance of the nasty surprise which not infrequently occurred in the days when such updates were made manually at too lengthy intervals.

(13) and (14) Ship motions

In the past these calculations were only rarely made and then only for ships designed for a special-purpose role. Ship motions were investigated for ships involved in work at sea, either of a defence or scientific nature or in support of oil exploration or production.

Manoeuvring was only investigated for ships requiring a special capability. With the programs now available both of these important ship characteristics can be assessed early in the design process almost as a routine matter.

(15) and (16) Resistance and effective horsepower

Each of the firms offering this software provides options as to the data and calculation methods used for normal displacement ships and in addition offer the use of special methods and data for particular ship types. For normal displacement ships, the methods on offer include:

- Taylor
- Series 60
- BSRA methodical series
- Guldhammer and Harvald
- Holtrop and Mennen

For small displacement ships, tugs, etc.:
- Van Oortmerssen

For stern trawlers:
- Dankwardt

For fast craft NPL round bilge craft analysis:
- Davidson Regression for round bilge and hard chine craft
- Savitsky Planing prediction for planing craft.

Some of the programs have limits on hull parameters and Froude number.

In general, the ITTC'78 performance prediction method is used, but the user has the option of modifying the hull factor if he wishes. A number of these methods are discussed in Chapter 6.

(16) Propulsion factors and propeller characteristics

Again most programs allow a choice of the data and methods used for these calculations.

Propulsion characteristics:
- Series 60
- BSRA Methodical series
- Van Oortmerssen
- Holtrop and Mennen

Propeller characteristics:
- Wageningen B-series
- Gawn–Burrill series.

1.3 SHIP DESIGN DATA

1.3.1 Collecting data and making use of it

A task to which all naval architects should devote a lot of time and attention throughout their career is the accumulation of data of all sorts relevant to their work. From the start they should set wide bounds to their interests. Even a naval architect who plans to spend his life designing one type of ship should collect data on any other type of ship to which he may have access. Even if his concern is entirely technical, he should note cost data and shipyard management techniques against a possible change in the direction of his career.

Storing a mass of data in a way which enables the holder to find what he wants quickly was always a matter of some considerable difficulty in the notebook era, although even then it was well worth doing. Nowadays, those who have their own personal computer can store an immense amount of data in small bulk provided they can make time to input it.

Programs such as DBASE and ACCESS can be used to store vast amounts of information and with a well devised retrieval system, access is almost immediate.

As well as collecting data it is important to organise its storage in a way that enables it to be put to advantage — a way that facilitates interpolation against scientifically chosen parameters. Quite frequently, there will be a need to extrapolate beyond the field covered by the existing data and the use of a well chosen parameter may make this possible, although the dangers involved must be recognised.

Data should always be dated. Unfortunately much of it is liable to become out of date as new developments occur and if it has to used after the lapse of some years it is helpful to know its provenance so that the necessary corrections can be made to it.

Data to which a naval architect may want to refer can take many forms, but the principal items which he should collect are:
- data on ship dimensions and dimensional ratios (see Chapter 3);
- data on ship's lines, with block coefficient and LCB position, etc. (see Chapter 8);
- data on powering (see Chapters 6 and 7);
- data on general arrangements (see Chapters 15 and 16);
- data on steel-weight, outfit weights and machinery weights (see Chapter 4);
- data on areas and volumes (see Chapter 5);
- data on the many rules applicable to ships (see Chapters 11, 12 and 13);
- data on the many different items that make up outfit with notes on their capabilities, weights and the services they require (see Chapters 4 and 17);
- cost data of all sorts (see Chapters 18 and 19).

1.3.2 Data sources

Pride of place amongst data sources must go to data on ships which have been designed, built, or owned by the companies for whom the data collector works.

The calculations, plans, specifications, tank test results, trial trip reports, material orders, man-hours and cost data, all or part of which will be available in these companies provide a comprehensive data base covering all the items mentioned in the last paragraph. The great point about this data is that its provenance means that it can be used with complete confidence, whereas it is always wise to have some reservations about data obtained from articles in the technical press and similar sources. This should not be interpreted as an attack on the technical press which the author believes does a very good job and is generally accurate, but reflects the fact that with the best will, errors do occur.

The snag with in-house data is that it is unlikely to be very large, except in a few very exceptional companies and even in these will probably be limited in its diversity. Naval architects therefore find it necessary to use other sources but when they do so should try to reduce the effect of any errors that there may be in this data by using (and comparing) several sources and amassing such a quantity of data that the effect of errors can be minimised by "fairing".

Some particularly useful sources for the data items suggested are discussed below.

1.3.3 Ship dimensions

1.3.3.1 Data from Lloyd's Register

Lloyd's Register is a prime source of data on the dimensions of almost all types of ship but suffers from the disadvantage that it is arranged in alphabetical order of ship's name with old and new ships and all types of ship mixed up.

When writing Chapter 3 the author wanted to use data from recent ships and saved a lot of work by using Lloyd's "May up-date" — the last update before the publication of a new volume of the register which contains particulars of all the ships registered for the first time in the year in question.

Unfortunately his wish to have data on the ship's design deadweight was frustrated because the deadweight quoted by Lloyds is that at the full draft. In addition, Lloyds do not quote the load displacement so if this is required it must be synthesised. In the spreadsheet already discussed in §1.2.3, this was done using the service speed quoted to calculate the Froude Number and this in turn to estimate the block coefficient using the mean line on the graph (reproduced as Fig. 3.12), which represented the practice of a large number of naval architects at the time it was produced and seems to remain close to current practice. The block coefficient is then used to calculate the displacement, but it should be noted that the method assumes that the quoted speed is related to the quoted deadweight which may not be correct and this could introduce some error.

Table 1.1 records information for tankers, bulk carriers, container ships and refrigerated cargo ships, the only types for which the 1988–89 and 1989–90 volumes seemed to have an adequate statistical sample, and it came as quite a surprise to note that general cargo ships had almost disappeared as a category except for small vessels with deadweights of up to about 5000 tonnes. The ships included in each sample were selected to cover the range of deadweight but in other respects the sample was arbitrary, although it has to be admitted that there can be a considerable temptation in work of this sort to omit data points that do not seem to plot well with the majority of the data.

Whether such a divergence is due to error in the input data or whether it means that the theory is not as exact as one would like, must trouble anyone working in this way. Caution clearly needs to be exercised when deducing design information from this or similar sources, but provided an adequate number of samples are used and extreme values are discarded or at least are not allowed to influence conclusions to any significant extent, useful lessons can be learnt.

1.3.3.2. Data from Significant Ships

In the past, designers obtained a lot of information from articles in the technical press. These articles, based on data provided by the shipbuilders concerned, gave much useful design information including plans. In recent years, fewer such articles have appeared and it was for this reason that the *Register* was used. Since writing the first draft of this section however it has been a pleasure to see this information gap largely filled by R.I.N.A's *Significant Ships* series.

The author of *Significant Ships* plainly wanted to quote displacements and lightship weights as well as deadweights. In some cases the shipbuilders appear to have been coy, but it is most satisfactory to see that this data is given for almost 50% of the designs shown. Knowing the displacement provides a most important key to any design as it enables the block coefficient to be calculated and if the deadweight is known it gives the lightweight. It is perhaps understandable that this information used to be kept confidential but the freer interchange of it that *Significant Ships* seems to be achieving will certainly be a great help to designers.

Table 1.2 uses data from *Significant Ships* of 1990 and 1991. It will be noted that because *Significant Ships* gives displacements there are no columns for F_n as there was in Table 1.1, but the columns for C_b are retained for use as a check on the displacements quoted.

The deadweight/displacement ratios calculated in Tables 1.1 and 1.2 are plotted in Fig. 3.3, whilst the dimensional ratios are given in Fig. 3.8.

Cargo capacity is also given in *Significant Ships* and a form for tabulating and using this data in the way discussed in §3.2 is given as Table 1.3, which has been completed for the same ships used in Table 1.1 enabling the information on the block coefficient at the load draft given in that figure to be used to estimate the block coefficient at the hull depth and thus the total hull volume. The data from this investigation is plotted in Fig. 3.5.

To use *Significant Ships* effectively in design, it is desirable to have a form of index which identifies the location in the volumes, of which there are now six, of ships of the type and size which may be suitable as guidance for a particular design and this is given in Chapter 16, §16.9.

Table 1.2 (*continued opposite*)

Deadweight/displacement ratio, based on data from *Significant Ships* 1990/1991

A	B	C	D	E	F	G	H	I	J
Ref	Type	Name	L	B	D	T_d	T_s	Dwt (d) (tonnes)	Dwt (s) (tonnes)
8	T	Argo Electra	315.00	57.20	30.40		20.80		285000
9	T	B.P.Admiral	169.00	30.80	17.00	10.00	11.52	33000	41100
10	C	Cap Polonio	188.20	32.30	18.80	10.00	12.00	22263	33205
11	C	CGM Provence	166.96	27.50	14.30		10.52		26288
12	B	China Pride	215.00	32.20	18.00	12.50	13.11	61687	65655
13	R	Del Monte Pride	147.50	23.50	12.75	6.70	9.10	6300	12700
14	M	Hornbay	141.50	23.00	13.90	7.30	8.70	5900	9429
15	R	Ice Star	84.27	15.10	7.75		5.30		3187
16	T	Jahre Traveller	260.00	44.50	24.20	15.60	16.60	131000	142000
17	C	Katherine Sif	120.70	22.70	11.30	7.60	8.60	9766	12112
18	C	Nordlight	145.20	22.86	11.20	7.65	8.62	11420	14190
19	M	Serenity	146.85	23.05	13.40		10.09		17175
20	T	Zafra *	218.70	32.24	21.60	11.58	16.00	54000	84000
21	T	Bunga Siantan	133.00	22.40	11.80		9.10		16294
22	T	Fandango	173.00	32.20	17.80		12.25		46087
23	B	Front Driver *	275.00	45.00	25.90		18.50		169178
24	C	Hanover Express	281.60	32.25	21.40	12.50	13.52	55590	67780
25	B	Solidarnosc *	224.60	32.24	19.00	12.50	14.10	63000	74000
26	C	Vladivostok	225.25	32.20	18.80	11.00	12.00	40250	47120
27	B	Western Bridge	239.00	38.00	21.50		15.02		96275

*Indicates double hull construction.   Suffix d = design draft, s = summer draft.

1.3.3.3 Warship dimensions and data from Jane's Fighting Ships

As a source of warship data, *Jane's Fighting Ships* must take pride of place. Like *Lloyds Register*, this volume is published every year and therefore provides up-to-date information. It is well illustrated with line drawings of ship profiles together with photographs of almost every ship in the world's navies and sometimes several different views in the case of the more important ships. It gives fairly extensive

Table 1.2

continuation

A	K	L	M	N	O	P	Q	R	S	T	U
Ref	Disp (d) (tonnes)	Disp (s) (tonnes)	C_b (d)	C_b (s)	$\dfrac{Dwt(d)}{Disp}$	$\dfrac{Dwt(s)}{Disp}$	L/B	B/D	$\dfrac{T_s}{D}$	L/D	Speed
8		319600		0.828		0.892	5.51	1.88	0.684	10.36	14.00
9	40866	48966	0.762	0.793	0.81	0.839	5.49	1.81	0.678	9.94	14.00
10	35545	46487	0.568	0.619	0.63	0.714	5.83	1.72	0.638	10.01	18.50
11		33690		0.677		0.780	6.07	1.92	0.736	11.67	18.60
12	73310	77278	0.822	0.827	0.84	0.850	6.68	1.79	0.728	11.94	14.91
13	13320	19720	0.557	0.607	0.47	0.644	6.28	1.84	0.714	11.56	20.00
14	13300	16829	0.544	0.577	7.30	8.700	6.15	1.65	0.626	10.17	20.00
15		5043		0.726		0.632	5.58	1.95	0.684	10.87	13.30
16	152000	163000	0.818	0.824	0.86	0.871	5.84	1.84	0.686	10.74	14.00
17	14234	16580	0.664	0.683	0.69	0.731	5.32	2.01	0.761	10.68	17.20
18	17995	20765	0.688	0.705	0.63	0.683	6.35	2.04	0.770	12.96	17.00
19		24668		0.701		0.696	6.37	1.72	0.753	10.95	16.20
20	69200	98900	0.823	0.851	0.78	0.849	6.78	1.49	0.741	10.12	14.40
21		21739		0.779		0.750	5.94	1.90	0.771	11.27	13.50
22		56583		0.805		0.815	5.37	1.81	0.688	9.71	14.50
23		192651		0.817		0.878	6.11	1.74	0.714	10.61	14.73
24	76330	88520	0.653	0.700	0.73	0.766	8.73	1.51	0.632	13.15	23.80
25	76375	87575	0.819	0.833	0.82	0.845	6.97	1.70	0.742	11.82	13.80
26	54735	61605	0.666	0.687	0.74	0.765	7.00	1.71	0.638	11.98	22.00
27		115473		0.822		0.834	6.29	1.77	0.699	11.11	15.00

*Indicates double hull construction.   Suffix d = design draft, s = summer draft.

data on the armament, machinery, complement, etc., but its statement on the main dimensions, possibly because of the security constraints placed on such data by navies, does not give the figures that a naval architect would like to have.

The length LOA and the length LWL always used to be given, but some recent entries only give LOA. The breadth is given but for ships with flared sides this is usually the maximum beam and not that on the LWL which is what a naval architect is mainly interested in. For some inexplicable reason the depth is never

Table 1.3 (*continued opposite*)

Cargo capacity/total hull volume

A	B	C	D	E	F	G	H	I
Ref	Type	Name	L	B	D	T	C_b	$d(C_b)$ from T to D
8	T	Arabiyah	242.19	43.28	23.80	15.621	0.845	0.027
9	T	Australia Sky	171.00	26.80	16.40	10.674	0.803	0.035
10	T	Columbia	313.00	56.60	28.60	19.436	0.848	0.024
11	T	Golar Coleen	256.01	46.21	23.83	17.017	0.841	0.021
12	T	Golar Cordelia	315.02	57.21	30.41	22.026	0.848	0.019
13	T	Olympic Serenity	222.12	42.00	20.30	14.200	0.835	0.024
14	B	Achilles	215.02	32.20	18.30	13.201	0.828	0.022
15	B	Amelia	250.02	43.01	23.42	17.201	0.837	0.020
18	B	An Ping	185.00	28.40	15.80	11.000	0.818	0.026
17	B	Angel Feather	172.00	30.50	15.80	11.228	0.822	0.024
18	B	Clarita	260.00	43.00	24.11	17.620	0.842	0.019
19	B	Dahlia	300.00	50.00	25.70	18.019	0.852	0.021
20	R	Arctic Ocean	140.01	22.01	13.21	9.350	0.559	0.061
21	R	Del Monte Cons.	147.53	23.51	12.76	9.002	0.638	0.050
22	R	Del Monte Harv.	130.31	22.31	12.76	9.202	0.577	0.055
23	R	Hokkaido Rex	140.00	20.60	12.80	9.417	0.610	0.047
24	R	African Reefer	136.66	23.60	15.40	10.015	0.632	0.066
25	R	Kowhai	136.00	18.50	10.65	7.300	0.630	0.057
26	C	Cap Polonio	188.20	32.30	18.80	12.000	0.619	0.072
27	C	CGM Provence	166.96	27.50	14.30	10.520	0.677	0.039
28	C	Hannover Express	281.60	32.25	21.40	13.520	0.700	0.058

T = tanker, B = bulk carrier, L = liquid, G = general, R = refrig., C = container.

given. In many cases, and particularly for recent ships, the draft quoted is that to the propeller tip, which, whilst of interest from an operational point of view since it determines the depth of the water in which the ship can navigate, is again of little use to a naval architect. The draft moulded amidships, which is what a naval architect wants to know, is only occasionally given.

Table 1.3

continuation

A	B	C	J	K	L	M	N
Ref	Type	Name	$C_b(D)$ at D	V_h Total hull	V_c Cargo	Cap. type	$\dfrac{V_c}{V_h}$
8	T	Arabiyah	0.872	217552	150048	L	0.690	
9	T	Australia Sky	0.838	62999	40386	L	0.641	
10	T	Columbia	0.872	441762	306300	L	0.693	
11	T	Golar Coleen	0.862	243072	163505	L	0.673	
12	T	Golar Cordelia	0.867	475323	350000	L	0.736	
13	T	Olympic Serenity	0.859	162606	114580	L	0.705	
14	B	Achilles	0.850	107716	80428	G	0.747	
15	B	Amelia	0.857	215740	160699	G	0.745	
18	B	An Ping	0.844	70102	50082	G	0.714	
17	B	Angel Feather	0.846	70136	52125	G	0.743	
18	B	Clarita	0.861	232190	169176	G	0.729	
19	B	Dahlia	0.873	336553	236359	G	0.702	
20	R	Arctic Ocean	0.620	25226	13875	R	0.550	
21	R	Del Monte Cons.	0.688	30465	16424	R	0.539	
22	R	Del Monte Harv.	0.632	23427	10477	R	0.447	
23	R	Hokkaido Rex	0.657	24242	13734	R	0.567	
24	R	African Reefer	0.698	34666	18244	R	0.526	
25	R	Kowhai	0.687	18398	9432	R	0.513	
26	C	Cap Polonio	0.691	78965	38923	C	0.493	1011 TEU
27	C	CGM Provence	0.716	46990	27258	C	0.580	708 TEU
28	C	Hannover Express	0.758	147370	87857	C	0.596	2282 TEU

T = tanker, B = bulk carrier, L = liquid, G = general, R = refrig., C = container.

Two displacements are usually quoted, one of these being the so-called standard displacement which has little technical value but the other is the full load displacement. Attempting to correlate the full load displacement with the dimensions quoted assuming a block coefficient appropriate to the stated speed makes it clear that quite a few of the figures quoted are not to be trusted. Whether this is due to

Table 1.4 (*continued opposite*)

Data sheet for destroyers, frigates, and corvettes. Data from *Jane's Fighting Ships* . O = Ownership; B = build

A	B	C	D	E	F	G	H	I	J
Ref	Ship type	Name	Country O/B	Year	LOA	LWL	B (LWL)	D	Super-structure
8	F	Niteroi	Brazil/UK	1976	129.2	122.0	13.5	9.1	68%
9	F	Halifax	Canada	1990	134.1	124.5	16.4	11.1	0%
10	C	Niels Juel	Denmark	1980	84.0	80.5	10.3	6.3	59%
11	F	Lafayette	France	1994	119.0	110.0	13.8	7.3	69%
12	D	Geo. Leygoues	France	1979	139.0	132.8	14.0	9.5	0%
13	F	Bremen	Germany	1979	130.0	121.8	14.5	9.2	0%
14	D	Animoso	Italy	1992	147.7	137.0	15.0	9.8	0%
15	F	Maestrale	Italy	1982	122.7	115.0	12.9	8.9	22%
16	F	Lupo	Italy	1977	113.2	107.0	11.3	8.0	20%
17	F	Dat Assawari	Libya/UK	1973	101.5	95.4	11.7	7.3	50%
18	F	Abukama	Japan	1989	109.0	102.0	13.4	7.7	0%
19	F	Karel Dorman	Netherlands	1990	122.3	115.5	14.4	9.0	0%
20	F	Kortenauer	Netherlands	1978	130.5	122.0	14.6	9.0	0%
21	D	Sovremenny	USSR	1980	156.0	144.4	17.3	10.1	42%
22	F	Krivak I	USSR	1981	123.5	114.9	14.0	9.5	86%
23	F	Type 22 batch 1	UK	1979	131.2	125.0	14.8	9.8	68%
24	F	Type 22 batch 2	UK	1984	146.5	136.5	14.8	9.8	79%
25	F	Type 23 #	UK	1989	133.0	123.0	15.3#	8.9	0%
26	D	Spruance	USA	1975	171.7	161.6	16.8	12.8	0%
27	F	Oliver H Perry	USA	1977	138.1	135.6	13.7	9.7	0%

Where Cb figures calculatedfrom displacements appear incorrect by up to 10% these are marked *; incorrect by more than 10% these are marked **.
Max. beam = 16.1 m.

deliberate misinformation in the interests of security or coyness about weight overruns it is impossible to say. It is also impossible to be sure precisely which figures are wrong, but it seems likely that the displacements are in most cases fairly correct and it is the drafts which are incorrect — in some cases by up to a metre.

Table 1.4

continuation

K	L	M	N	O	P	Q	R	S	T	U	V
T	T screw	L/B	B/D	T/D	L/D	C_b	Δ	Crew	Speed	Machy. type	SHP (metric)
4.20	5.5	9.04	1.48	0.46	13.40	0.533*	3800	209	30.0	CODOG	56000
4.60		7.59	1.48	0.41	11.21	0.491	4750	225	29.0	CODOG	50000
3.10		7.82	1.63	0.49	12.77	0.499	1320	98	28.0	CODOG	18400
4.00		7.97	1.89	0.55	15.06	0.512	3200	156	25.0	CODAD	20000
4.50	5.7	9.49	1.47	0.47	13.97	0.484	4170	216	30.0	CODOG	46200
4.15	6.5	8.40	1.58	0.45	13.31	0.477	3600	225	30.0	CODOG	51600
5.00		9.13	1.53	0.51	13.97	0.477	5045	400	31.5	CODOG	55000
4.10	8.4	8.91	1.45	0.46	12.92	0.485	3040	232	32.0	CODOG	50000
3.70		9.47	1.41	0.46	13.37	0.543*	2500	185	35.0	CODOG	50000
3.40		8.15	1.60	0.47	13.06	0.455	1780	130	37.5	CODOG	N/A
3.80		7.61	1.74	0.49	13.24	0.467	2500	115	27.0	CODOG	48000
4.30		8.02	1.60	0.48	12.83	0.451	3320	141	29.0	CODOG	27000
4.30		8.36	1.62	0.48	13.55	0.460	3630	176	30.0	COGOG	50000
6.00	6.5	8.35	1.71	0.59	14.29	0.473	7300	320	32.0	ST TUR	110000
5.00		8.21	1.47	0.53	12.09	0.471	3900	180	32.0	COGAG	69000
4.80	6.0	8.45	1.51	0.49	12.75	0.481	4400	222	30.0	COGOG	54600
4.80	6.4	9.22	1.51	0.49	13.92	0.481	4800	222	30.0	COGOG	54600
4.30	5.5	8.04	1.72	0.48	13.82	0.504**	4200	177	28.0	CODLAG	38400
5.80	8.8	9.62	1.31	0.45	12.62	0.496	8040	319	33.0	G.T.	80000
4.50	7.5	9.90	1.42	0.47	14.05	0.476	4100	206	29.0	G.T.	41000

Where Cb figures calculatedfrom displacements appear incorrect by up to 10% these are marked *; incorrect by more than 10% these are marked **.
Max. beam = 16.1 m.

A proforma for extracting dimensions and similar data from *Jane's* has been completed as Table 1.4 with data on modern frigates and corvettes. Column 9 in this table is included in an attempt to make sense of the big differences in *B/D* and *T/D* which result from the extremes of short and very lengthy superstructures.

After quite a lot of thought, only ships whose displacements and drafts appear to be in reasonable accord with one another as shown by a block coefficient calculation based on the data given, have been included. This unfortunately necessitates the omission of a number of quite important ships that would otherwise have been included.

Some of the designs shown in *Jane's* such as the British types 22 and 42, were built in two batches of different lengths. The dimensional ratios *L/B* and *L/D* of the first batches could be assumed to represent the naval architect's intent whilst the later ships are *ad-hoc* modifications. On the other hand, it is just possible that the lengths of the first batches were squeezed below their designer's wishes by economic constraints and the later versions are nearer to the designer's preferred figures!

On the basis that approximate information is better than none, some of the missing information for the ships for which data is given has been obtained by scaling from the small-scale profiles.

Ten countries are represented as designers in Table 1.4 and it is interesting to note that the dimension ratios calculated are remarkably similar and no national trends can be identified.

Occasionally more and/or better dimensional and other information is given in technical papers and in the present context of dimensions and their ratios, mention must be made of a 1992 R.I.N.A. paper "On the variety of monohull warship geometry" by W.J. Van Griethuysen, from which Table 1.5, which covers most types of warships, is abstracted. The paper makes the point, well illustrated by the figures, that different types of warships have quite distinctly different form characteristics.

Table 1.5

Summary of warship dimension ratios (for volumetric Froude number, see §6.3, ix)

Type of warship	V (10^3 m^3)	$L/V^{1/3}$	L/B	L/D	B/D	B/T	Volumetric Froude no.
World War II battleship	40.6	7	7	14	1.8	2.5–3	0.8
Destroyer	2.3	8–9	10	16	1.8	3–3.5	1.5
Minehunter	0.5	5–6.5	5–6	8	1.4	3.2–4	0.8
Corvette	1.2	7–8	7–8	11	1.5	3.5	1.3
Frigate	3.5	7–8.5	8–9.5	13	1.5	2.8–3.2	1.2
Cruiser	7.1	7–8.5	8–10	12	1.4	2.5–3.2	1.1
Aircraft carrier	13.9	6–7.5	6–8	9	1.3	3.3–4.1	0.8

1.3.3.4 Computerised dimension data

Fairplay Information Systems now offer an information system which is designed for personal computers and can hold almost 100 different data items about the ships entered in the system. These include most of the items normally given in *Lloyd's Register* but also include such items as the new building price and the sale price if the ship has changed hands. There are also a number of additional details relating to the ship's outfit and capability.

Whether this system is sufficiently developed to be of immediate use may be doubtful, but it will be surprising if it does not build up in the course of a few years to become a most useful tool for the designer.

1.3.4 Data on lines and powering

Data on lines and powering should be kept together. An integrated package of the lines plan, related tank test and trial trip results is particularly valuable, but designers are unlikely to acquire many such items. Technical papers published by R.I.N.A., S.N.A.M.E., etc., provide the next best available information but in these days of photocopying extracts of anything that looks useful should be filed away.

1.3.5 General arrangement plans

Significant Ships and its sister publication on smaller vessels is now a most useful source of data on a wide variety of ship types and the *Naval Architect* and its sister publications are another important source. In addition, each of the principal ship-building countries has one or more technical magazines which give useful data on ships built in their respective countries. Although language may be a barrier to the detail, the plans will be clear and an occasional glance at these publications can provide useful information.

1.3.6 Outfit and machinery data

Keeping abreast of developments in outfit and machinery requires a great deal of reading of the technical press and manufacturers' catalogues. A computerised data base for these items which gives performance characteristics, services requirements, weight and cost would be high on the author's list of priorities if he was still a practising designer.

1.3.7 Weight data

Good weight data is vitally important to a naval architect. Almost all naval architectural calculations depend on weights and their distribution but accurate weight calculations require a well advanced general arrangement plan, a lines plan, structural plans, specification and equipment lists, much of which will not be complete until the design is well advanced. During the early stages of a design, the uncertainties surrounding the weight estimate are almost always more significant than those associated with hydrostatics or hydrodynamics of the vessel. Far more ship design problems arise through bad weight estimation than from errors in other much more difficult calculations.

There are two ways of minimising this problem: one is the collection of good weight data and its intelligent use, and the other is the frequent iteration of the weight estimate as better data becomes available.

1.3.8 Cost data

Cost data and such related information as man-hours/tonne for steelwork etc. should be zealously sought as it is particularly difficult to obtain since *Shipyards* — almost the sole source of this data — severely restrict its circulation for obvious commercial reasons. There is the further complication that this data gets out of date particularly quickly for a number of reasons such as improvements in productivity, changes in currency exchange rates etc., so always date this data!

1.3.9 Data on rules

Before starting to design a ship type with which he is not familiar, a naval architect has in recent years had the daunting task of identifying the rapidly growing number of rules that will apply and of familiarising himself with the more significant of these. Fortunately, help is now at hand because *Lloyd's Register* now offers a computerised solution in "Rulefinder", which is accessible on PCs and covers virtually all the rules that need to be considered.

1.3.10 Making use of data

One of the best ways of storing data and of interpolating between or extrapolating beyond available information can often be to graph it. Vital to the success of graphing is the choice of a suitable base parameter, which must be a measure that can readily be obtained or calculated at the stage in the design process at which the ordinate value is required and one against which the ordinate scales on a scientific basis.

Possibly because the modern generation can turn so readily to and achieve so much so quickly by the use of computers, they appear to have some reluctance to use graphical methods which are seen as old fashioned, but the author hopes that the extensive use of graphical methods in this book may show how useful they can be.

Chapter 2

Setting Design Requirements

2.1 STATING OBJECTIVES IN BROAD TERMS

It is most important that the objectives which a new design is to meet should be stated in a way that does not rule out any possible solution. It is only too easy when setting requirements to have a particular type of design in mind and write terms of reference in a way that leads to a solution along these lines but excludes some other equally good or better answer. Objectives should be set at their most desirable level even if their attainment seems unlikely or impossible. This will stretch designers and may cause them to come up with novel ideas that are ahead of any current solution.

When setting objectives, it is wise to differentiate between qualities which are essential and those which are only desirable and can be modified if the price of their attainment is too expensive or turn out to be to the detriment of a higher rated goal. It is debatable whether such relaxations in the statement of requirements should be exposed to the designers at the outset or only be released to them when the impossibility, or excessive cost of meeting the ideal requirements becomes apparent. Concealing them will force maximum effort and good lateral thinking; on the other hand, if this leads to the designers feeling that they have not been fully trusted they may not give of their best.

A merchant ship's requirements will usually originate in a transportation study which examines the economic background to the projected service. The requirements for a warship will have been based on consideration of possible threats and will usually have been preceded by many strategic studies.

2.2 DIMENSIONAL CONSTRAINTS

Dimensional constraints may impose a limit on length, breadth, draft and air draft, or two or more of these.

A constraint on length may be set by the dimensions of canal locks or docks. It may also be set by a need to be able to turn the ship in a narrow waterway. In either case the necessity of the limit set should be thoroughly questioned if it appears likely to limit the ship's length to less than that which would be desirable if there was no such constraint.

If the limit is set by a dock or canal, question whether the use of these is essential or so desirable that this limit must be accepted, or whether rerouting could avoid the canal, or the choice of another port avoid the constraint set by the dock. A limit set by turning ability can be considerably eased by fitting a high-performance manoeuvring device such as a bow thruster. The constrained length will usually be the overall length but in some cases the constraint may apply at the waterline or at some definite height above the waterline at which the ship will be floating.

A limit on breadth is usually set by canal or dock lock gates, but the breadth of vehicle ferries is sometimes limited by the dimensions and position of shore ramps giving vehicles access to bow or stern doors. The outreach of other shore-based cargo-handling devices such as grain elevators or coal hoists can limit the desirable distance of the offshore hatch side from the dockside and thereby limit the breadth of the ship. In general, breadth limits apply to the maximum breadth measured over fenders (if fitted) and usually must be maintained to at least quay height above the waterline at low tide. Above this it may be possible for the breadth to be increased if flaring the ship's sides is desirable (see Chapter 8) or if overhanging decks, lifeboats, dredger suction pipes or any similar item are a feature of the design. Any overhanging features demand a most careful survey of the places where the ship may berth to ensure that there is no possibility of contact with objects on the quay.

A draft limit is usually set by the depth of water at low spring tides in the ports (or their approaches) to which the ship is intended to trade.

For very large tankers the depth of the ocean itself must be considered. The impact of a draft limit can be minimised if the ship's routing, fuelling and storing are so arranged that fuel and stores are at a minimum and the ship is on level keel when it is passing through the shallow water that sets the limit.

The last of the dimensional constraints is that of air draft. This is the vertical distance from the waterline to the highest point of the ship's structure and denotes the ship's ability to pass under a bridge spanning the seaway which forms part of the projected route. Where necessary, air drafts can be greatly reduced by equipping the ship with folding or telescopic masts and funnels. Other measures which can be taken to reduce the effect of an air draft limit are to arrange that transit under the bridge takes place at low tide and/or to load or ballast the ship to the deepest

Table 2.1

Three important dimensional restraints

	Max. length (m)	Max. beam (m)			Max. draft (m)	Air draft (m)
Panama Canal*	289.56	32.31			12.04 TFW	57.91 m
Suez	No restriction	74.0	and		11.0	
		or 48.0	and		17.7	
St. Laurence	225.5	23.8			8.0	35.5

*294.13 m for passenger and container ships. Reduced dimensions apply to some special types of vessel. Addresses of these authorities are given in Chapter 19.

permissible mean draft, in association with a trim that maximises the draft at the fore and aft position of the highest point of the ship. Three of the most important dimensional restraints are given in Table 2.1.

2.3 ENVIRONMENTAL CONDITIONS

The wind and sea states in which a ship is required to operate are major factors to be considered in its design, although for the majority of merchant ships these states are not mentioned in the specification, their place in that document being taken by a Classification Societies notation "100A1" or similar, denoting an ability to trade worldwide in ice-free waters. For ships intended to trade only in sheltered waters reduced strength and other requirements are permitted by classification societies, IMO rules and those of national authorities. If it is intended to take advantage of these concessions and the reduced associated costs, the trading limits within which the ship will operate must be clearly defined.

For the other categories of ships shown in Fig. 1.2, namely warships and naval auxiliaries, service ships and floating production vessels, it is essential to define the wind and sea states in which the ships are required to carry out various tasks.

Three sets of conditions are usually defined for warships. The first defines wind and sea states in which the ship should be able to maintain its full service speed and operate all its equipment at maximum efficiency. The second is a more severe set of conditions in which a reduced speed is acceptable, but the ship must still be able to operate its helicopter and other weapon systems, possibly with some reduction in performance. The third is a still more severe set of conditions in which survivability is the requirement.

For an oil production vessel there will be generally be two sets of conditions: one in which production must be able to continue and another in which production will be shut down and the safety of the vessel and crew becomes the design criterion. For vessels which are usually stationed in one location, the likely weather is more easily defined than for a mobile ship. For most of the main production areas records have been kept from which projections of wind speed and direction, wave length and height can be made. These predictions are commonly made on the basis of the likelihood of occurrence in a particular period, for instance the ten-year storm or the hundred-year storm, although it must be said that the latter has shown an alarming ability in recent years to turn up rather early in its century!

Where vessels are required to operate in ice, it is necessary to define whether the ice is first-year or multi-year ice as well as stating the thickness. Lloyds have five notations for first-year ice ranging in thickness from 0.4 to 1.0 m. For multi-year ice they have another four notations for ice thicknesses ranging from 1.0 to 3.0 m.

For ships operating in Arctic or Antarctic waters the formation of ice on deck can be a severe problem both because of its effect on the operation of deck machinery and because of the loss of stability caused by the accretion of top-weight. Inclusion in the statement of requirements of a need for ice accretion to be considered can lead to both a reduction in the amount of gear arranged on the open decks on which the ice can form and to the provision of systems that will speed ice clearance.

A statement on the maximum and minimum air and sea temperatures in which the ship will operate is required. This information has several uses and these need to be remembered when the figures are being stated. The design of the accom-modation heating system requires a knowledge of the ambient cold air temperature which will be encountered in winter. The design of the air-conditioning system requires knowledge of the ambient hot air temperature and associated relative humidity which will be met in summer. In both cases the desired inside temp-eratures and relative humidities must also be stated.

Air and sea temperatures also affect the power output and the fuel efficiency of both diesel engines and gas turbines and it is desirable to give values of these temperatures to be used as the basis of powering and of fuel consumption. It is worth noting that these may or may not be the same as the temperatures on which the air conditioning is to be based.

2.4 MERCHANT SHIP REQUIREMENTS AND TRANSPORTATION STUDIES

The main requirements that need to be set for a merchant ship are:
- the type and quantity of cargo to be carried;
- the service speed, the voyage route and distances.

The type of cargo and how it is to be stowed on the ship and handled on and off the ship determines the ship type, whilst the quantity of cargo to be carried is obviously the main determiner of the ship's size.

The type of cargo is generally the starting point, although some transportation studies commence at a more fundamental level by looking at a country's or region's economic forecasts to identify the cargo-carrying capacity that will be needed in the future.

Even if consideration is limited to a particular cargo, a wide-ranging economic study will usually be necessary to assess the quantity that will need to be carried in the future and the loading and discharge ports which will best meet the trade.

At the next level it is necessary to assess what competition there will be and the optimum shipment size. Consideration must be given to whether the ship should cater for one cargo only or should be so designed that it can carry more than one cargo at the same time or a different cargo or cargoes on different voyages. The possibility of there being a suitable return cargo for the "homeward" leg of the voyage avoiding a ballast leg must be investigated as this can appreciably improve the economics.

In a simple case it may be possible to link the number of ships, their carrying capacity and the number of voyages per year directly to the quantity of cargo requiring transport. The aim may either be to ensure that the transport need can be met or that the trade is a profitable one, or more usually both.

The number of voyages per year is clearly a function of time spent at sea and that spent in port loading and discharging. The sea time is set by the voyage distance and the ship's speed; the port time by the cargo-handling arrangements provided.

In real life things are much more complicated and shipping companies require more than a little luck as well as very sophisticated calculations to lay down requirements that will result in ships which will operate profitably.

Once a preliminary decision has been taken to fix the ship's cargo capacity, the next step should take the form of a sensitivity study to optimise such things as the speed and cargo-handling methods. When these have been optimised, further refinement of the cargo capacity may follow.

For heavy cargoes the cargo capacity should be stated as the cargo deadweight; for light cargoes the cargo capacity should be stated as the cubic capacity; for intermediate cargoes both deadweight and cubic capacity should be stated. For container ships, vehicle ferries and passenger ships the capacity should be a stated in numbers divided into appropriate categories such as forty- and twenty-foot containers, goods vehicles and cars, first- and second-class passengers, etc.

As well as the route between the cargo loading and discharge ports, the transportation study should consider where the ship should load fuel. A short addition to the voyage route may enable the ship to divert to a port where fuel can be shipped at a lower price than prevails in the cargo terminal ports. If fuel can be obtained cheaply at more than one port in a round voyage it may be wise to use this option, thereby reducing the average displacement and hence the fuel consumption per mile.

2.5 REQUIREMENTS OF SERVICE SHIPS AND OFFSHORE WORKING VESSELS

Each of the ship types shown in the service ship and offshore working sectors of Fig 1.2 has its own quite different requirements.

2.5.1 Tugs

The requirement that overrides all others for a tug is the bollard pull. This is determined by the size and type of ship the tug is designed to assist, the number of tugs that will share the work, and the currents, tides and winds in which towage may take place.

Tugs for long-distance offshore towage must be very seaworthy vessels and have ample fuel and stores capacity.

The turning capability should be tightly specified. All tugs, but particularly harbour tugs, must be highly manoeuvrable, not least to avoid the danger of capsize that can arise if the tow rope pull goes round to the beam and by the same token tugs must have very good stability.

One way of providing a high manoeuvring capability is by fitting one or more Voith–Schneider propellers or steerable thrusters in lieu of conventional propellers. If these are positioned near the bow rather than at the stern they operate to pull the tug rather than push it and, as a result, the danger of a sideways pull is almost completely eliminated.

A required free-running speed may be specified, but this is usually of lesser importance and in practice is often determined by the power installed to give the bollard pull. However, if a relatively high free running speed would be advantageous, this should be stated, as a variable pitch propeller can significantly improve free running performance.

2.5.2 Dredgers

The leading requirements for a dredger are the types of spoil that it must handle and the range of water depths in which it must be able to operate. Next in importance is the quantity of spoil that has to be removed per day and whether the removal of the spoil to the dumping ground is a function of the dredger itself or is carried out by a separate hopper vessel or by pipeline. If the ship is to have its own hopper, then the spoil deadweight together with its lowest likely specific gravity which is required to determine the hopper size must be stated together with the distance to the dump ground. If the spoil reaching the hopper is likely to be thixotropic, this should be stated as this has a major "free surface" effect on stability.

Hopper dredgers with bottom dump doors permitting the jettisoning of the spoil in an emergency and open hoppers from which the spoil will spill before the ship heels to an angle at which it starts to lose stability are, in certain circumstances, permitted to operate at reduced freeboards. The requirements should therefore state what type of freeboard is to be used.

Most hopper suction dredgers have diesel–electric machinery, which is usually sized to meet the high power demand of the dredge pumps. This provides the possibility of a relatively high sea speed, although most dredgers do have quite full lines and a high appendage resistance from the overside dredging gear even when this is in the stowed position. It is probably wise therefore to specify the minimum speed required for operational reasons whilst making it clear that advantage is to be taken of the installed power to maximise the actual speed.

The type of dredging gear to be fitted should usually be specified in the statement of requirements, although this may be left to the discretion of specialised dredger designers provided they are given a clear description of the tasks that the ship has to accomplish.

Where the spoil to be moved from the sea bottom is handleable by a suction pipe and there is room for the dredger to move under power, the trailing suction dredger is almost invariably the most suitable type.

Where the spoil is suitable for suction dredging but in some part of the area to be dredged there is no room for the ship to move, the addition of a bow suction pipe extending forward can provide the answer.

Where the bottom is so compacted or stony that suction pipes are inadequate it is necessary to use a bucket dredger.

As well as their use in deepening or maintaining depth in channels, hopper suction dredges are used to collect cargoes of sand and/or gravel from offshore deposits and bring these back to land for civil engineering developments. These vessels are usually fitted with special self-unloading features unless they go to a berth which has shore-based special unloading equipment. The unloading method must be specified.

2.5.3 Icebreakers

The leading requirement for an ice breaker is the thickness of ice through which it is required to clear a passage, the ice in this case being almost invariably multi-year ice (see §2.3).

The next requirement is the breadth of the passage needed by the ships that will follow it through the ice.

The required bollard pull must be stated and should be adequate to provide effective assistance to the vessels for which the icebreaker is making a passage, should any of these become trapped in the ice.

Fuel and stores must be provided for a lengthy endurance, and the accommodation must be to a high standard and for a large crew.

2.5.4 Research vessels — hydrographic, oceanographic and fishery

These are fairly small ships but often have to go to remote waters where help, if needed, will not come quickly. Some of their voyages will take them to the more stormy regions of the world. Good sea-keeping and good reliability must therefore come high in their requirements. Whilst a need for good sea-keeping is based on safety considerations, another almost as important consideration is the need for these ships to have limited motions to facilitate the work of their scientists who may have to carry out delicate tasks and are frequently unused to working in a violently mobile environment.

Some aspects of scientific work require a minimum of noise and vibration and limits should be set for these in consultation with the scientists who should, however, be given a clear idea of the costs that may be associated with meeting low signatures to dissuade them from over-specification.

An ability to operate at low speeds is usually required as is an ability to handle boats and scientific equipment over-side, over the stern or via an internal moon pool.

2.5.5 Fishing vessels

The main requirements that need to be set for fishing vessels relate to the type or types of fishing for which they are intended: bottom trawling for demersal species or mid water trawling for pelagic species, long lining, etc. The next most important requirements are the time that has to be spent at the fishing ground and its distance from the port at which the catch is to be landed. As most fishing vessels are small and they go to sea in all but the most stormy conditions, another requirement is very good seaworthiness.

2.5.6 Fish factory ships

Where the distance to the fishing grounds involves a lengthy voyage, it may be necessary to process the fish on board so that the product remains in good condition until it is landed. Once the decision to process on board is taken, the corollary can clearly be a much longer stay on the fishing grounds, a bigger catch per trip and a much larger ship. There are several different types of processing to suit different types of fish and different markets and it is necessary to specify the one required.

2.5.7 Oil production vessels

There is a wide range of vessels involved in oil production: exploration/drilling rigs, supply boats, safety vessels, oil storage and tanker loading vessels, and floating oil production vessels.

2.5.8 Exploration/drilling rigs

Most exploration work has been carried out by jack-up rigs inshore and by semi-submersibles offshore, although some monohull vessels have been built for this purpose. This has usually been done where there is an also a need to provide oil storage capacity.

2.5.9 Supply boats

The requirements for these will usually be stated in terms of a clear deck area on which the extraordinary range of spare parts and stores required by a drilling rig or a production platform can be stowed. In addition there will generally be a need for tanks for the carriage of liquid cargoes such as drilling mud. Supply boats are generally designed so that they can undertake towage tasks and a bollard pull must therefore be specified.

2.5.10 Safety vessels

In the past, many of these have been converted trawlers but conversions of this sort have not been able to provide some most desirable requirements which would be specified for a new build vessel. These should include having a length of ship side with a low freeboard for the recovery of people from the water and an ability to manoeuvre quickly and precisely at slow speeds. Although trawlers generally have good sea-keeping characteristics, this is a subject to which particular attention would be given on a custom built safety vessel.

2.5.11 Oil storage and tanker loading facilities

Most of those built so far have been conversions of existing, and generally redundant, tankers. With fewer of these available in recent years in a condition that justifies the expenditure of considerable sums on conversion the new building of this type of vessel is now becoming more attractive.

The most important requirement is the quantity of oil to be stored, together with the rate at which it will be delivered to the storage vessel by pipeline and the rate at which it is to be passed on to the shuttle tankers. If, as is usually the case, the storage vessel is to be moored to a single point mooring buoy that will also carry the pipeline through which the oil comes on board, details of this interface must be given. The intended position of the shuttle tanker whether this is to be astern or alongside must be specified. The sea and wind conditions, respectively, must be specified for accepting a shuttle tanker, terminating loading and disconnecting the shuttle tanker, and adopting survival tactics, which may mean abandoning the pipeline.

2.5.12 Oil production vessels

Most of the requirements stated for the storage vessel apply, with one major difference being that a production vessel will generally have a moon pool surmounted by the production derrick with a pipeline dropping directly to the sea bottom and the oil well beneath. This requires a definitive statement of the permissible movement of the ship about this fixed point, together with a specification of the worst wind and sea conditions in which production should be able to continue. As the oil reaching the ship is raw crude, it is necessary to specify what type of oil is envisaged and in particular the anticipated gas/oil ratio.

2.6 STAFF REQUIREMENTS FOR WARSHIPS AND NAVAL AUXILIARY VESSELS

2.6.1 Staff requirements — general

Setting the staff requirements for warships or naval auxiliary vessels is a very difficult task which may have to start with a political assessment of possible enemies and allies and an estimate of the naval assets each of these may have at some time in the future.

With "a week being a long time in politics" it is essential to consider a large number of alternative scenarios. Only after pondering these very carefully is it possible to move to consideration of the requirements for a particular ship or class of ships — noting that these will not enter service for several years and will then remain in service for a lengthy period.

2.6.2 Frigates and corvettes

One of the more significant differences between setting the requirements for a warship and those for a merchant ship, including service and offshore working ships, is that the requirements of all these can usually be set on an individual ship basis, whereas those of a warship may have to consider scenarios both for the ship operating on its own and with other vessels, either getting support from or giving support to these.

This greatly complicates the task of setting the requirements for a new ship, as does the related and continually debated question of whether a limited budget should be devoted to a small number of highly capable ships or be spread more thinly over a larger number of cheaper but less effective vessels.

The argument for spending money on minimising signatures, for example, goes thus: if the enemy can detect your ship, he is a long way towards sinking or disabling it; if on the other hand he cannot detect it but you know where he is, you are well on the way to winning. On the other hand, detection equipment is constantly improving so maybe today's expensive low signature ship will be as readily detected in the future as today's cheaper one so there is undoubtedly an argument for more ships.

Measures to minimise signatures of all types — underwater noise, airborne noise, infra-red, radar, magnetic, etc. — are very important staff requirements for most types of warship. The importance of the various signatures varies with the vessel's primary role, with seaborne noise being most important for submarines and anti-submarine frigates and the magnetic signature being so vital for a mine hunter that the whole construction of these ships is of non-magnetic materials.

The principal requirements of a warship are the combat systems with which the ship is to be fitted, i.e. the guns, missile systems, helicopters and/or aircraft plus the whole range of accompanying command and control systems. Backing up the offensive weapons are a range of self-defence weapons: anti-missile missiles and guns, chaff launchers, etc.

The next most important requirement is the operational service speed and in particular the speed that can be maintained in adverse weather. Another speed that must be specified is the economical cruising speed, which is often only about half the service speed.

The endurance of a warship is generally stated in terms of distance or of the number of days that can be spent at the economical speed plus a shorter distance or time at the maximum service speed. The endurance of most warships is low by merchant ship standards, reflecting the cost of providing space and deadweight in a warship and the availability of replenishment at sea which navies provide as a corollary.

Replenishment at sea can also supply replacement stores and ammunition, reducing the quantity that must be carried.

The requirements for shock and vulnerability are two important standards which must be specified for a warship.

2.6.3 Naval auxiliary vessels — general

The requirements for these vessels tend to combine most of those discussed under warships with others mentioned in the section on merchant ships. However, a few naval auxiliary types bring their own special problems. It is worth commenting that in the British Navy some of the ship types discussed in this section are naval manned and can therefore in some respects be regarded as warships, others are merchant seaman manned and a third category has both naval and merchant manning.

2.6.4 Landing craft and logistic support ships

The unusual feature of these ships is their beaching role. This requires the staff requirements to specify the deadweight that the ship has to be able to land on an open beach, as well as the greater deadweight that the ship should be able to carry in a transport role. This in turn requires a statement on the wading ability of the vehicles which may be landed, defining the maximum permissible forward draft, and a statement on the gradient of the beach, defining the draft aft.

2.6.5 Landing ship docks

The distinctive feature of these vessels is the dock incorporated in their stern section. The length and breadth of this must suit the number and dimensions of craft that are to be carried in it or use it. The L.S.D. must be able to be trimmed to provide the necessary depth of water in the dock to suit the draft of the user craft and the change of trim from the seagoing condition must be to a rapid timescale against the probability that enemy action is likely to be imminent.

2.6.7 Fuel and stores replenishment ships

The distinctive feature of these ships is the provision of rigs designed to achieve the transfer of stores or fuel, or both to warships at sea and the number and types of rig must be specified.

Until fairly recently, liquids were carried in fleet tankers and dry cargoes in fleet store ships. For a ship requiring both liquids and solids this meant two storing periods so the advent of a new ship type (AOR) which could provide both on a "one stop" basis was a major improvement. The staff requirements for such a ship have to take account of the fact that this is becoming an exceedingly valuable ship and needs a good self-defence capability.

2.7 ADVANCED MARINE VEHICLES

The first section of this chapter advocated that design objectives should be stated in broad terms and should demand the most desirable qualities. Although there is little point in setting objectives that cannot be achieved, the possible area of achievement can be greatly extended if the whole range of marine vehicles can be considered as contenders, although it must be said that the transport efficiencies of most of the alternatives to displacement ships are much lower than can be obtained from a conventional ship. Figure 2.1, abstracted from Eames 1980 RINA paper "Advances in Naval Architecture for Future Surface Warships", gives an approximate indication of where, in terms of speed as measured by volumetric Froude number, the use of these vessels is worth considering.

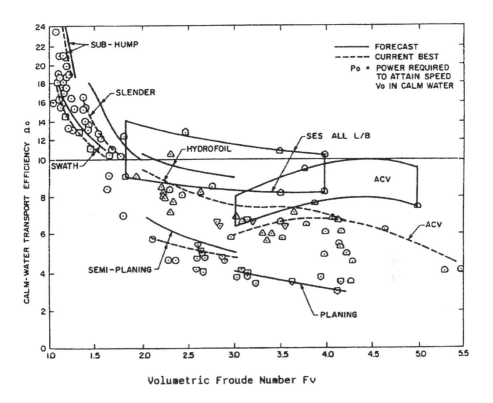

Fig. 2.1. Transport efficiency *n* versus volumetric Froude number.

Although the measure of efficiency used in Fig. 2.1 is quite a useful way of introducing the relative capabilities of the various marine vehicles, it does not tell the whole story. The numerator in the transport efficiency formula contains a displacement term, whereas a better measure of efficiency would involve the use of the cargo deadweight.

As most displacement ships have a deadweight/displacement ratio of between 0.7 and 0.9, and a figure of about 0.1 would apply to both hydrofoils and air cushion vehicles, changing to a deadweight criterion would introduce a further factor of 7 to 9 emphasising the efficiency of the displacement ship.

The range of volumetric Froude number (VFN) shown is conditioned by the fact that the paper was dealing with warships. If it was extended into the merchant ship area the displacement range would increase to about 500,000 tonnes whilst the speeds would drop to below 15 knots. The corresponding VFN for the biggest and slowest vessels would then drop to a fraction of unity and the transport efficiency would rise to the range 100–1000. However, even this is a long way from showing the true economic advantage of the displacement monohull as the efficiency comparison is limited to the propulsion aspects whereas the monohull has further very large advantages in its much smaller building and running costs.

Clearly, within a range of VFN from zero to about 1.3 there is no competition. Above this and extending to a VFN of about 2.0, a Swath configuration must be considered and this is particularly the case where the ship has to work at sea and especially if this makes minimum motions in a seaway desirable. Unfortunately, present day Swath designs require more power for the same speed than an equivalent monohull and this is inhibiting their development —possibly temporarily.

Developed since the graph was drawn in 1980 but now demanding consideration in this range and extending to a VFN of about 2.5, is the wave-piercing catamaran. Although one of these now holds the blue ribbon of the Atlantic, the sea-keeping ability of these vessels in other than moderate weather may still require more convincing demonstration before they win a wider acceptance. If this is achieved there may be a considerable place for this type of vessel as passenger and car ferries.

In the range of volumetric Froude number from 2.0 to 3.0 the hydrofoil shows to advantage; from a VFN of 4.0 to 5.0 the air cushion vehicle seems to have both the best record and the best potential. Between VFNs of 3.0 and 4.0 these two types compete.

Semi-planing and planing vessels do not show to advantage on the criterion used, but their cheapness of construction keep them very much in the picture for small fast pleasure boats, and of course the wave-piercing catamaran may also be a planing vessel, so maybe this type is developing.

2.8 THE DESIGN SPIRALS

Design spirals for merchant ships and warships are shown together in Fig. 2.2. The similarities and differences are both worth noting. The similarities predominate, although in some cases these are disguised by the use of different names such as the "total deadweight" of the merchant ship and the "variable weights" of the warship.

The differences start with the first spoke of the spirals, which denotes the most important feature of the two types of ship: cargo handling in its broadest sense for the merchant ship, and the weapons configuration in its totality for the warship. Most of the spokes thereafter are identical or nearly so until the penultimate one which is tonnage for the merchant ship and vulnerability and signature for the warship.

Feeding in to the merchant ship design all the way round the cycle are Classification Society rules, IMO and national rules, whilst warships are similarly guided by the relevant naval standards.

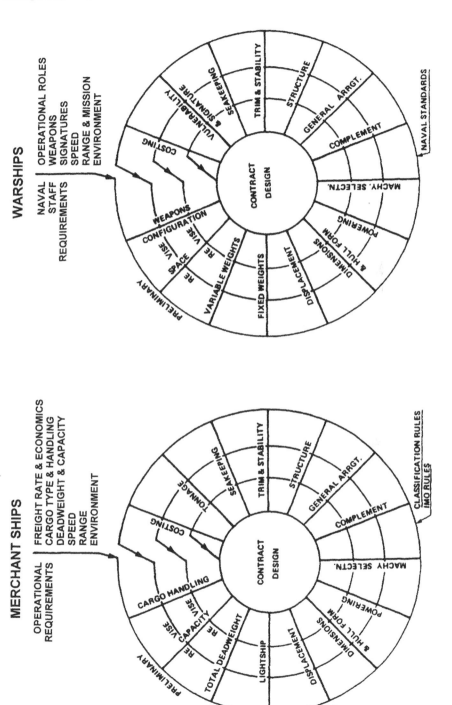

Fig. 2.2. Ship design spirals.

Design spirals should not be thought of as showing the exact order in which the different aspects of design should be tackled: this will depend on the type of ship being designed. For example the cargo handling of a Ro-Ro ship must be considered at an early stage in the design process, whereas most aspects of the cargo handling of a tanker can be dealt with quite late in the design process.

Some ship types require a near simultaneous assault on many of the features, or failing this an interactive approach becomes desirable or even necessary. If, however, the designer is able to recognise quickly which feature or features are the main driving force for the design he can speed the design process. In the following paragraphs an attempt is made to indicate the signs — some obvious and some less so — that help a designer to spot the driving criteria.

2.9 THE CRITICAL CRITERIA

It is very helpful while setting the design requirements, and even more when starting to convert these to a design, to be able to identify which of the requirements are likely to become the critical criteria. The following paragraphs consider which criteria are most likely to be critical for various ship types and why.

2.9.1 Ships for which weight is critical

Weight, coupled almost invariably with speed, is critical for the majority of ships, although there are some surprising exceptions. Because at the end of a design, a good designer will have brought weight and space into harmony with each other, many designers think that the type of ship that they design cannot be categorised as either weight or space critical.

However, at the start of a design when no holds are barred, one or other of these will usually be dominant. Weight is clearly the critical factor when the cargo to be carried is "heavy" in relation to the space provided for it. At one extreme, iron ore loaded in alternate holds, and therefore using less than half the available space, will take a bulk carrier down to its maximum draft even if this involves a B-60 freeboard (see Chapter 11).

The question of what is the critical cargo stowage rate — which decides whether a design is weight or space critical — is an interesting one. The answer depends on several factors such as the deadweight/displacement ratio, the proportion of the cargo deadweight to total deadweight, the type of freeboard, and the ratio of cargo capacity to the total volume below the upper deck. Some of these factors in turn involve the ship's speed, the power and type of machinery, the distance between fuelling ports, and whether any space below the upper deck is required for any other purpose such as passengers.

An approximate critical stowage rate can be synthesised as follows:

$$\text{Cargo S.G.} = \frac{\text{cargo dwt}}{\text{cargo vol}} = \frac{\dfrac{\text{cargo dwt}}{\text{total dwt}} \times \dfrac{\text{total dwt}}{\text{displt}}}{\dfrac{\text{cargo vol}}{\text{total vol}} \times \dfrac{\text{total vol}}{\text{displt}}}$$

If we make:

$$\frac{\text{Cargo dwt}}{\text{total dwt}} = K_1 \qquad \text{and} \qquad \frac{\text{total dwt}}{\text{displt}} = K_2$$

Draft/depth ratio $\qquad\qquad T/D = K_3$

$$\frac{\text{Cargo vol}}{\text{total vol}} = K_4 \qquad \text{and} \qquad \frac{C_b'(\text{at depth } D)}{C_b(\text{at draft } T)} = K_5$$

$$\frac{\text{total vol}}{\text{displt}} = \frac{C_b' \times L \times B \times D}{1.025 \times C_b \times L \times B \times T} = \frac{1}{1.025} \times \frac{C_b'}{C_b} \times \frac{D}{T} = \frac{1}{1.025} \times \frac{K_5}{K_3}$$

Cargo density (S.G.) $= \dfrac{1.025 K_1 K_2 K_3}{K_4 K_5}$ \hfill (2.1)

Putting some fairly arbitrary figures to this equation

$K_1 = 0.90;\ K_2 = 0.70$ (see Fig. 3.3)

$K_3 = 0.73$ (see Fig. 3.10)

$K_4 = 0.58$ (see Fig. 3.5) and $K_5 = 1.05$

Cargo density $= \dfrac{1.025 \times 0.90 \times 0.70 \times 0.73}{0.58 \times 1.05} = 0.77$ or 1.29 m^3/tonne

The ship in this example will be weight critical if the cargo that it is designed to carry has a cargo density of more than 0.77 or stows at less than 1.29 m^3/tonne and volume critical if the cargo is lighter.

Other requirements which may make, or help to make, weight the critical factor can be the need for particularly heavy construction or particularly heavy items of outfit; and/or the inclusion of a severe limitation on draft.

2.9.2 Ships for which volume, deck area, linear dimensions or stability are critical

Volume, coupled once again with speed, becomes the critical criterion when the cargo to be carried is light — with lightness being determined as shown above. Particular factors that may make volume critical can be the need to provide space within the normal cargo space area for some other need, e.g., for passenger accommodation or the fitting of some special machinery.

Modern tankers in which a large segregated ballast capacity has to be provided have moved from being the epitome of weight-based designs (which older ships with class A freeboard were), to being volume controlled.

Most modern warships are volume controlled, the principal dimensions being determined by the internal space required (with the length or the external deck area needed sometimes being an additional factor), rather than by the need to provide adequate buoyancy. This stems from the fact that most modern weapon systems and modern machinery are of low density whilst armour and other heavy items are now a thing of the past.

The requirement for these fairly small ships to achieve high speeds makes it desirable to keep the displacement as low as possible to minimise the power required. This, coupled to some extent with weight overruns in some construction, has led to the adoption of weight reduction techniques and the imposition of strict weight control measures becoming standard in warship design and building practice. The desirability of constraining weight for hydrodynamic reasons should not, however, be seen as making warships weight critical.

It may be noted that the fact that there are no freeboard rules for warships means that the load draft can be adjusted when the ship is complete to take the as-built weight into account. When this has been done the as-built data suggests that there is a balance between weight and volume, although it may not be that originally intended by the designer.

Ship types for which deck area is the most important criterion include car and train ferries and possibly aircraft carriers, although it can be argued that linear dimensions are equally important for these ship types and stability may in fact be the real determining factor in fixing the dimensions in some cases.

Linear dimensions are very important for a container ship whose length, breadth and depth should be tailored to maximise container numbers. On the other hand, if maximising the carriage of tiers of containers on deck is important stability becomes the ultimate criterion.

Stability is also the factor which determines how many superstructure decks can be fitted on a cruise liner and therefore becomes the critical criterion for the other ship dimensions as well (see §5.3.2).

2.9.4 Ships for which speed and/or seakeeping are critical

It has already been noted that speed is a joint criterion along with either weight or volume in the ship types discussed above. In some other ship types, however, speed and/or sea keeping can be critical by themselves. This applies to the smaller types of warship where a high speed/high Froude number is required particularly if, as is usually the case, there is a requirement for speed to be sustained in rough seas.

Even if the weight and space required could both be provided by a smaller ship, these ship types should be built with a hull whose length enables the speed to be obtained economically and the ship to meet the expected seas with acceptable motions.

Whilst in theory this means building a ship with surplus space, it may be noted that in practice uses for space are very quickly found. This, in fact, presents a problem because although the provision of unused space costs very little it has been found that very good project control is essential if the uses to which spare space are put do not to result in a cost overrun — with all the uses found being, of course, matters of high priority!

Other ships whose dimensions may be determined by speed and/or seakeeping include research vessels.

2.9.5 Ships for which tonnage is critical

There used to be several types of smaller cargo ships for which IMO or similar rules created significant commercial advantages — usually by a reduced manning requirement — for ships whose net tonnage was less than a critical number. Two such numbers being 499 tons and 1499 tons. Owners and shipbuilders specialising in these vessels become very expert in designing ships which met these criteria by the smallest of margins and found ways of providing quite extraordinarily large deadweights and cargo capacities within tonnage limits. Today these rules seem to be of reduced significance.

2.9.6 Other critical criteria

Some ships in which particular limitations or specialist requirements have an over-riding importance in determining the design are described in Chapter 16.

2.10 TRADE-OFFS BETWEEN OPERATING ECONOMY AND FIRST COST

Although these are not critical criteria in relation to the determination of the main dimensions in the way that previous paragraphs in this section have been, this seems an appropriate place to discuss what is in another sense a very critical criterion — the relative importance to be attached to operating economy and first cost, respectively.

The requirements for a ship rarely indicate the relative importance that the owner attaches to operating economy and first cost, although this should be an important design consideration.

Designs prepared by shipyards for a competitive tender are almost bound to give priority to features which will minimise the first cost, unless a clear indication is given that the potential customer will include an assessment of operating costs when evaluating tenders and give this due weight when comparing capital costs.

Designs prepared by a shipowner should include features which will reduce operational costs, with the criterion for any consequent increase in first cost being that this should be recoverable from the operational savings within an appropriate timescale.

If an owner is uncertain about the economic viability of any feature he should include it as an option in the tender specification with a request for an alternative price along with the main tender. Only in this way is it possible to obtain realistic prices that enable satisfactory trade-off calculations to be made.

Chapter 3

The Design Equations

3.1 THE WEIGHT EQUATIONS

The dimensions of a ship whose design is weight based are determined by the following equations.

$$\Delta = r \cdot C_b \cdot L \cdot B \cdot T \cdot (1+s) \tag{3.1}$$

and either

$$\Delta = W_d + W_l \tag{3.2}$$

or

$$\Delta = W_d / K_d \tag{3.3}$$

where

r = specific gravity of the liquid in which the ship is intended to float
= 1.000 for fresh water
= 1.025 generally for salt water
L = length BP or length WL in metres
B = breadth mld. in metres
T = load draft mld. in metres
C_b = moulded block coefficient at draft T on length L
Δ = full displacement in tonnes
s = displacement of shell, stern and appendages expressed as a fraction of the moulded displacement
W_d = full deadweight in tonnes
W_l = light ship weight in tonnes
K_d = deadweight/displacement ratio W_d / Δ

3.1.1 Comments on and finding a solution to eq. (3.1)

The length used in eq. (3.1) differs between merchant and warship practice. Whilst the use of LBP is general for merchant ships, warship designers use LWL.

There are arguments for both usages. The use of LBP is appropriate to single screw ships in which the AP is defined either as the after side of the rudder post or as the centre of the rudder stock if there is no rudder post and the stern is regarded as an appendage. The use of LWL is more appropriate for twin screw ships and in particular for those with twin rudders. For these ships there is no sensible "aft perpendicular" and the stern is very much an integral part of the hull lines.

Lloyds Register covers these cases by the statement that L is to be not less than 96% and need not be greater than 97% of LWL. With most warships being twin screw it is not surprising that LWL is generally used in warship design.

The difference between LBP and LWL is small but it is important to remember that values of F_n and C_b must be associated with the type of length on which they are based.

With the introduction of flared ship sides it is necessary for some ship types to designate that the breadth B is that at the load waterline.

It will be noted that finding a solution to this equation is a complex matter as there are three dimensions to evaluate plus the block coefficient which is a function of speed and length, as shown in Fig. 3.12.

In his 1962 paper, the author suggested a series of "best practice" relationships between the various ship dimensions all of which took the form $y = m_x + c$, and the use of a "three trial ships" method was suggested.

In this method dimensions, weights, powers, etc. were prepared for three ships spanning the likely size range. From a plot of the deadweight of each of these ships against length it was possible to read off the length which would give the required deadweight. Whilst this was a clean scientific method it did involve quite a lot of work.

In the 1975 paper, the dimensional relationships were reduced to simple ratios making it possible to alter eq. (3.1) to a cubic equation in L.

As a first step introduce dimension ratios to give

$$\Delta = r \cdot (1 + s) \cdot C_b \cdot L^3 \cdot (B/L) \cdot (B/L \cdot D/B \cdot T/D) \tag{3.4}$$

This can then be transformed to

$$L = \left(\frac{\Delta (L/B)^2 \cdot (B/D)}{r \cdot (1+s) \cdot C_b \cdot (T/D)} \right)^{1/3} \tag{3.5}$$

Values of the ratios L/B, B/D and T/D can be obtained from the graphs given later in this chapter. To solve the equation it is still necessary to make a first guess at C_b

but, even with the consequent need for successive approximation, a speedy solution is possible.

3.1.2 Choosing between the use of eqs. (3.2) or (3.3)

Historically, the required displacement for a merchant ship was generally arrived at by the use of eq. (3.3), but the increasing complication of ships, particularly in the period 1950–1970 when many general cargo ships had refrigerated chambers, cargo oil tanks and twelve passengers, made the selection of a correct value of K_d almost impossible and the author in his 1962 paper suggested that it was better to use eq. (3.2) and he continued to advocate this in his 1975 paper.

Since the 1950–1970 period, there seems however to have been a reversion to a number of fairly standardised types of ship and the possibility of deriving K_d values for particular ship types now seems worth investigating again. The sources of data for doing this have been explored in §1.3.

The eq. (3.2) method must still be preferred when designing an unusual ship type, although it requires much more work at this early stage in the design and can only be carried out satisfactorily and reasonably expeditiously if a good stock of up-to-date weight information is available and is listed against appropriate estimating parameters.

Even if eq. (3.2) is not used for the first design spiral, it is of course always completed at the next stage of the design and the calculation of lightweight and deadweight are discussed in Chapter 4.

3.1.3 Equation (3.3) and the deadweight displacement ratio

When using a deadweight/displacement ratio it is important to note that for many ship types more than one deadweight and corresponding displacement and draft may be quoted, each set having a distinctly different deadweight/displacement ratio.

In this section four main ship types are considered: bulk carriers, tankers, container ships and refrigerated cargo ships. Of these types bulk carriers are unique in having their design and full deadweights identical. This used also to be the case for tankers but as already noted these are now volume design ships and commonly have a design draft less than their full draft.

The same applies to container ships although in this case the full draft is generally a scantling draft which is less than the geometric freeboard draft.

Refrigerated cargo ships also have design and full drafts but in this case the full draft is usually the freeboard draft.

From a design point of view, the deadweight which matters is the design deadweight as it is at the displacement corresponding to this that the service speed is specified and therefore it is this displacement that determines the dimensions

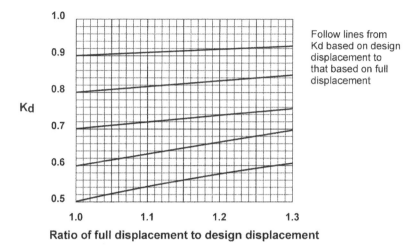

Fig. 3.1. K_d versus design displacement.

and block coefficient. It is an unfortunate fact that the design deadweight is not commonly quoted, although it is pleasing to see it being given for quite a number of ships in *Significant Ships*. The only deadweight given in *Lloyds Register* on the other hand is that at the full draft.

Because there are these alternative deadweights, displacements and drafts it is essential that great care is taken when plotting, and when subsequently using, K_d values.

Figure 3.1 explores the change in K_d value resulting from it being based on a displacement greater than the design displacement. It starts with the design K_d value and shows how this changes for a full draft at which the displacement is 10, 20 or 30% more than the design displacement.

Figure 3.2 starts at the other end with the K_d at the full draft and shows how the value of K_d at the design draft can be estimated if the percentage reduction in displacement can be estimated.

For ships such as large crude tankers which have a high K_d value, the difference in value from design to full load is not too significant. (An examination of a modest sample of these vessels suggests that the differences between the design and load displacements for these vessels is between 5% and 15%.)

For ships such as container ships and refrigerated cargo ships which have K_d values of about 0.70 and 0.60, respectively, the change in the values from full load to design can become very important.

In the end, a lack of data on design deadweight dictated the use of data relating to the full deadweight for all ship types and this is what is plotted in Fig. 3.3. The data used for this plot comes from Tables 1.1 and 1.2.

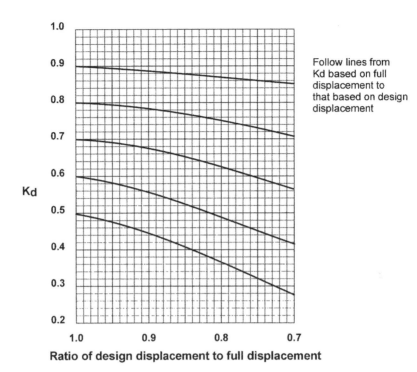

Follow lines from
Kd based on full
displacement to
that based on design
displacement

Ratio of design displacement to full displacement

Fig. 3.2. K_d versus full displacement.

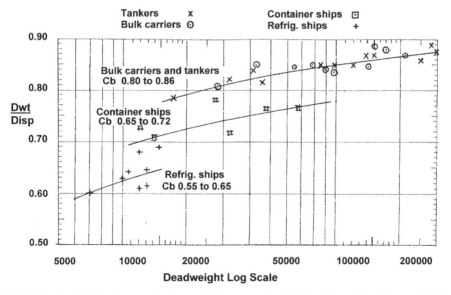

Fig. 3.3. Deadweight/displacement ratio vs displacement. The deadweight used in this graph is the full deadweight *not* the design deadweight.

The line for bulk carriers and tankers seems quite satisfactory with few data points deviating by more than 0.025 from the line. If this deviation is related to a "line" value of say 0.825, the percentage error in a displacement estimated in this way will not exceed about 3%. If this error is distributed pro-rata between the three dimensions of length, beam and draft, the error in each of these at this earliest step in the design process will be less than 1%. In fact it may be even less than this, since it seems likely that divergences from the mean line may be due more to the ships in question having block coefficients differing from that given by the mean line in Fig. 3.12 than because their lightship weights differ markedly from the "line" value.

If at the next step in the design process a block coefficient corresponding to the mean line is used when deriving the main dimensions, there should be the happy result of two deviations at least partially cancelling one another, leaving only a small adjustment to be made to C_b when final design weights become available.

Although the deadweight/displacement ratio of container ships is not of much practical use since, as already noted in §2.9.2, these vessels are volume or stability critical rather than weight based, an approximate line for this type of ship is given on Fig. 3.3 but not surprisingly there is quite a large scatter of data points. The same applies to refrigerated cargo ships which are also volume-based designs.

A deadweight/displacement ratio for either of these ship types and indeed for a tanker should be corrected using Fig. 3.2 before being used in association with the design deadweight.

It may seem strange that the base used for the plot in Fig. 3.3 should be the deadweight itself, but in this case it is used primarily as an indicator of size and because at the time when a designer is still trying to establish K_d it is the only parameter available to him.

A number of other lessons can be learnt from Fig. 3.3. Firstly, it seems clear that the biggest factor in determining the deadweight/displacement ratio is not the ship type as might have been expected but the block coefficient, whilst next in order of importance seems to be ship size, with ship type coming third, suggesting that a regression analysis of this type of data might produce an interesting result.

When drawing Fig. 3.3, only a limited number of data points were used and even so a number of the points diverge significantly from the suggested lines. Without plans of the ships concerned it was impossible to establish whether there are good explanations for these divergences but as the figure is only intended to provide a quick approach to initial dimensions, it is probably accurate enough and a reasonably good example of "the art of drawing sufficient conclusions from insufficient premises" quoted on the title page. It was interesting to note that points relating to double skin tankers and bulk carriers plotted quite close to the line indicating that with good design the weight penalty — if not the first cost penalty — for this type of improved construction is remarkably small.

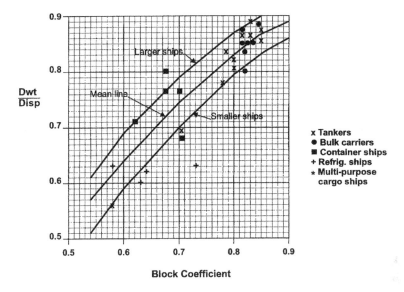

Fig. 3.4. Deadweight/displacement ratio versus block coefficient.

When a single K_d value for a particular ship type is known the trend of the lines on Fig. 3.3 can be used to correct this to a different deadweight.

Although the correlation between deadweight/displacement ratio and block coefficient shown in Fig. 3.4 is interesting, it is unfortunately not of any practical use to a designer as C_b is still an unknown quantity when a designer is trying to decide on an appropriate deadweight/displacement ratio.

3.2 THE VOLUME EQUATIONS

The dimensions of a volume carrier are determined by the equations:

$$V_h = C_{bd} \cdot L \cdot B \cdot D_c \tag{3.6}$$

and either

$$V_h = V_m + V_o \tag{3.7}$$

or

$$V_h = (V_r - V_u) / K_c \tag{3.8}$$

where

D_c = capacity depth in metres
           = $D + C_m + S_m$
D = depth moulded in metres
C_m = mean camber in metres = $2/3C$ for parabolic camber
S_m = mean sheer in metres   = $1/6(S_f + S_a)$ for parabolic sheer
C_{bd} = block coefficient at the moulded depth
V_h = total moulded volume of the ship below the upper deck and
             between perpendiculars in cubic metres
V_r = total cargo capacity required in cubic metres
V_u = cargo capacity above the upper deck in cubic metres
S = deduction for structure within the cargo space expressed as a
             proportion of the moulded volume
V_m = moulded volume equivalent to required cargo capacity below upper
             deck = $(V_r - V_u)/(1 - S)$
K_c = ratio of cargo capacity below the upper deck, to the total moulded
             volume = $(V_r - V_u)/V_h$
V_o = other volume required for accommodation, stores, machinery,
             tanks and other non-usable space within the volume V_h in cubic
             metres (non-usable space depends on the type of cargo carried
             and corresponding type of capacity measurement).

It is interesting to note that the draft T does not appear in these equations, although it is implicit as a second-order term in the difference between the value of C_{bd} and that of C_b at draft T which is established by the form required to suit the Froude number of the ship.

In a way analogous to that given for the weight equation, the volume equation (3.6) can be converted to:

$$L = \left(\frac{V_h (L/B)^2 \cdot (B/D)}{C_{bd}} \right)^{1/3} \tag{3.9}$$

A relationship between C_b at the moulded draft and C_b' at a depth D is given by eq. (3.10).

$$C_b' = C_b + (1 - C_b)\frac{(0.8D - T)}{3T} \tag{3.10}$$

There is some ambiguity in this book in the treatment of block coefficient at other than the load draft. In some chapters C_b' has been taken at the moulded depth D, and it would have been better if it had been possible to stick to this throughout the book both for the sake of uniformity and because it has a better theoretical

Both cargo capacity and the total hull volume are those below the upper deck.
The capacity measurements are those appropriate to the ship types concerned.

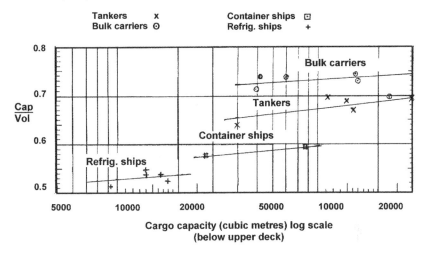

$$\text{Fig. 3.5.} \quad \frac{\text{Cargo capacity}}{\text{Total hull volume}} \text{ versus cargo capacity.}$$

basis. Unfortunately, some of the available data is based on C_b' at $0.8D$ so it has
been necessary to use this in some places.

It can be quite useful to express C_b' as a multiple of C_b which can be done by
converting eq. (3.10) as follows:

$$C_b' = C_b \left[1 + \frac{1}{3} \frac{(1 - C_b)}{C_b} \frac{(1 - T/D)}{T/D} \right]$$

Taking an average value of $T/D = 0.70$, this reduces to

$$C_b' = C_b[1 + 1/21(1 - C_b)/C_b]$$

which gives the values at D and $0.8D$, respectively, shown in Fig 3.6.

3.2.2 Choosing between eqs. (3.7) and (3.8)

The calculation of V_h using eq. (3.7) requires a lot of work as it is necessary to put
reasonably accurate figures on all the many items included in V_o. Its use is necessary
for passenger ships and for ships with a variety of different spaces in addition to their
main cargo capacity. The calculation of V_h in this way is dealt with in Chapter 5.

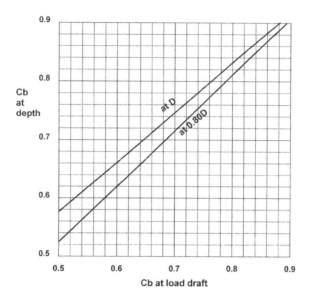

Fig. 3.6. C_b at depth versus C_b at draft.

3.2.3 Equation (3.8) and the capacity/total volume ratio K_c

This method provides an easier approach to arriving at an approximate value of V_h for the more standard types of ship for which data is given in Fig. 3.5. The data used in this plot is given in Table 1.3.

This is mainly based on data from Lloyds register although for container ships it was necessary to go to *Significant Ships* to get the number of containers below deck as the data in *Lloyds Register* only gives the total number of containers. The plot for this type of ship should be used with care as it is based on only two data points and the two ships used are in fact both particularly efficient users of space and it will be difficult to achieve their 0.58–0.59 K_c values in a new design unless this is also very carefully optimised. It may be more prudent to use a K_c value of 0.55 this type of ship.

The capacity for container ships has been based somewhat arbitrarily on an assumed capacity of 38.5 m³ for a 20 × 8 × 8.5 ft container multiplied by the TEU.

Bulk carriers show up as much the most efficient type of ship from a capacity point of view attaining a K_c value of about 0.73.

Tankers, largely because of the need to provide a large clean ballast capacity to meet MARPOL rules, tend to have a K_c value of from 0.66 to 0.69.

Refrigerated cargo ships with relatively high speed, fine lines, high machinery power, space requirements of insulation, etc., tend to have a K_c value of about 0.53. Because no plans of the ships concerned were available the total cargo capacity quoted has, in each case, been assumed to be below the upper deck, although it is possible that in some cases part of the capacity may be provided by a long cargo forecastle, which is not an unusual feature of this type of ship. This could introduce a small error and the warnings given about K_d in the previous section apply equally, if not more so, to K_c.

3.3 DIMENSIONS AND DIMENSIONAL RELATIONSHIPS

3.3.1 General discussion

The fact that there are six dimensional relationships linking the four main ship dimensions of L, B, D and T and that it is necessary to use three of these to solve either the weight or volume equations has already been noted. The relationships are:

$$B = f(L) \quad D = f(L)$$
$$D = f(B) \quad T = f(L)$$
$$T = f(D) \quad T = f(B)$$

Essentially a ship is a container and, as the straight-side container which has the least surface area for a given volume is a cube, it appears that for economy of construction a ship should approach this shape as closely as such other considerations involved in ship design as stability, powering, manoeuvring capability, etc., permit. An approach to a cubic shape requires that draft T (the smallest of the dimensions) should be the maximum permitted by L, B and D; that depth D (the next smallest dimension) should be the maximum permitted by L and B; that breadth B should be the maximum permitted by L and finally that the block coefficient C_b should be as full as possible. The statements "permitted by" and "full as possible" must of course be interpreted as meaning without incurring a significant operational penalty.

In the next few sections the values of each of these ratios suggested in the author's 1962 and 1975 papers are examined together with those which appear to apply today. Such an historical treatment may seem out of place in a technical book but is included because it seems likely that the changes during this period will provide some guidance to the continuing changes there are bound to be in the future.

It is interesting, although not surprising, to note that there has been continuing development in the ratio L/B where the main control is economic, but very little change in the ratios B/D and T/D which represent physical constraints.

3.3.2 Breadth/length ratio $B = f(L)$

In 1962 it was suggested that the relationship between L and B was of the form $B = L/M + K$, with different values of M and K quoted for passenger liners; cargo ships and tankers. The corresponding *L/B* ratios varied from between 6.2 for smaller ships of up to about 400 ft (120 m) to 7.6 for the biggest ships then being built.

By 1975, when Fig. 3.7 was originally presented, ships of more than about 130 m in length were almost invariably being built with an *L/B* ratio of 6.5; ships of up to 30 m in length, such as fishing boats, usually had an *L/B* ratio of 4; whilst vessels whose length lay in the range between 30 and 130 m followed a linear interpolation pattern between *L/B* values of 4 and 6.5.

One of the contributors to the discussion of the 1975 paper drew attention to the then recent development of very large tankers with a limited draft dictated by the depth of the ocean itself in certain areas. These tankers had very low *L/B* values of the order of 5.1 to compensate for the limited draft, and a line indicating this has been added.

Another comment made in 1975 was that there seemed no reason for the *L/B* ratio being different for ship types, unless this was because different ship types tended to concentrate in groups of much the same size and speed.

The frigate line which has been added to the revised version of the *L/B* plot given in Fig. 3.7 shows that these ships, which have much higher speeds in relation to their size than merchant ships giving them Froude numbers in the range 0.40 to 0.50, have *L/B* ratios of about 8.5. Larger warships such as aircraft carriers, having a lower Froude number of about 0.33 and the need for breadth for both arrangement and stability reasons, tend to have an *L/B* ratio of about 7.0 coming closer to the merchant ship 6.5 ratio.

As came out very clearly in the famous "short fat ship" controversy it is not a low *L/B* ratio that is undesirable *per se* but the fact that a short length and therefore a high Froude number for a given displacement are an unavoidable corollaries of having a large beam. For a given speed the high Froude number of the "short fat ship" necessitated such a large increase in power compared with that required for a "long slim ship" that the economies in structural cost quite correctly claimed for the former were more than offset, even in building cost, by the extra cost of the machinery required — whilst an almost doubled life-time fuel bill clinched the matter.

The 1991 values of *L/B* for a number of merchant ship types are shown in Fig. 3.8. In spite of the view expressed in 1975, these have been separated into ship types to show present day practices. For almost all tankers the value appears to have settled down to a figure of 5.5.

Although, as Table 1.1 shows, the Froude numbers of tankers and bulk carriers are closely similar, the *L/B* ratios of most bulk carriers have remained at the more conservative value of 6.25, a figure which also seems to apply to the majority of container ships and refrigerated cargo ships although these have much higher Froude numbers.

The different method of plotting used in Figs. 3.7 and 3.8 is worth a brief comment. The direct plot of two prime variables against one another has much to

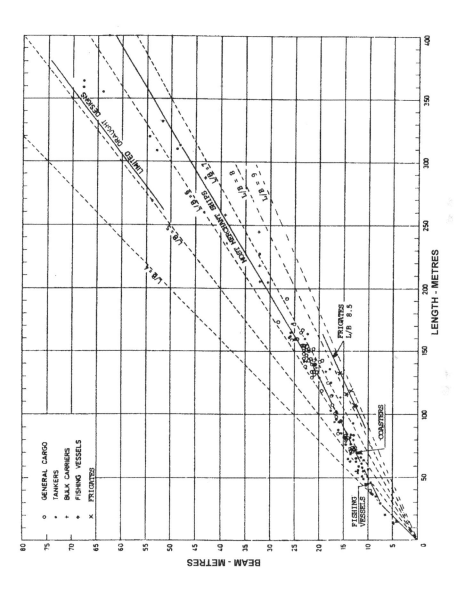

Fig. 3.7. Length–breadth relationship.

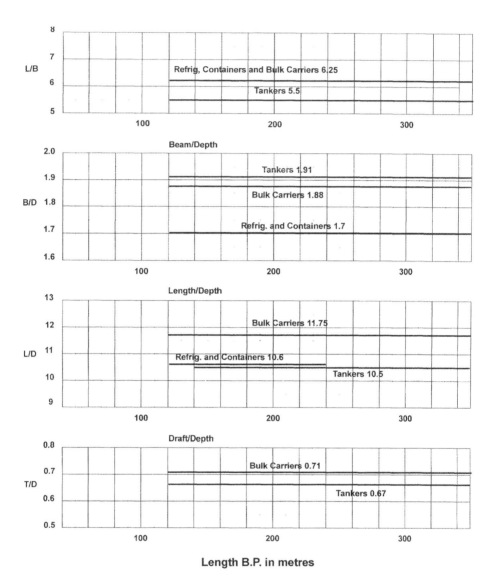

Fig. 3.8. Modern dimension relationships.
(Data from *Lloyds Register* and *Significant Ships*.)

commend it and helps to avoid false conclusions. However, the plot of a ratio such as *L/B* enables a more precise value to be determined, if this is justified by the data. The fact that it makes it possible to bring in another variable as the abscissa can also be helpful.

3.3.3 Depth/breadth relationship D = f(B)

This relationship is closely related to stability since KG is a function of depth and KM is largely a function of beam.

In the 1962 paper the relationship was presented as:

$$D = \frac{B - K}{1.4}$$

with two different values of K stated as representing moderate and good stability, respectively.

The B/D values varied from 1.5 for a large ship with "moderate" stability to 1.8 for a small ship with "good" stability.

By 1975 better standards of stability were demanded by International rules and the distinction between moderate and good stability had become academic and the B/D values had increased.

Figure 3.9, which was originally presented in the 1975 paper, reverts to lines of constant B/D and shows a plot of depth against beam for a number of ship types. It was found that there were two distinct groups. The first group consisted of deadweight carriers comprising coasters, tankers and bulk carriers had an B/D ratio of about 1.9. The second group consisted of volume carriers comprising fishing vessels and cargo ships whose depth was limited by stability considerations and which had a B/D ratio of about 1.65.

The 1991 plot included in Fig 3.8 largely confirms these groupings with tankers and bulk carriers again averaging at a B/D of 1.9.

The second group brought container ships and refrigerated cargo ships together at the slightly increased B/D value of 1.7. The higher B/D value (1.7 vs 1.65) for volume carriers in 1991 may be a consequence of the need to limit the depth of these ships because of the stability implications of making provision for the carriage of containers on deck.

Factors which in general may require an increased B/D value include: higher standards of stability for whatever reason these may be needed; the carriage of deck cargo; reductions in machinery weight raising the lightship KG; and the finer lines needed for high speeds giving reduced KM for a given beam.

Factors which may permit a reduction in B/D include the provision of a large ballast capacity in the double bottom; absence of deck cargo; relatively light superstructure and cargo handling gear; absence of sheer and/or camber; and lines designed to provide a particularly high KM value (see Chapter 8).

Although L and B values are freely available for warships, figures for D are rarely quoted. Because of the need to allow for the considerable variety in the amount of superstructure and other topweight on these ships, there tends to be quite a wide scatter in the B/D ratios used, but a new line for frigates has nevertheless been added to Fig. 3.9 showing a mean B/D ratio of 1.55.

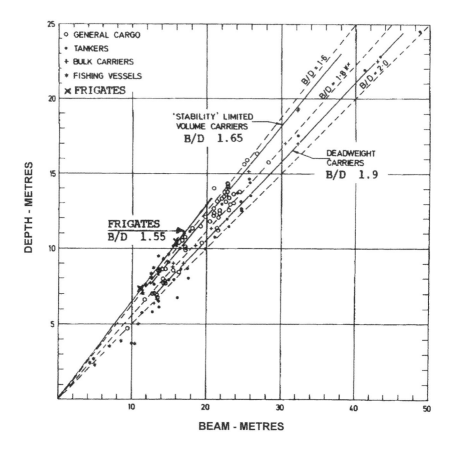

Fig. 3.9. Breadth–depth relationship.

3.3.4 Draft/depth relationship T = f(D)

This relationship, which for merchant ships embodies the freeboard rules, has changed with alterations in these and related rules — notably the 1966 Freeboard Convention and the 1973 IMO MARPOL rules for tankers, which are dealt with in some detail in Chapter 11.

There are no freeboard rules for warships but the design needed to achieve a high degree of seaworthiness in these ships imposes a similar constraint.

In the 1962 paper, the relationship for merchant ships was expressed as

$$T = 2/3 \, [D - (h)] + K$$

$K = 1.2$ m (4 ft) for a closed shelter deck cargo ship or a tanker; $K = 1.9$ m (6.3 ft) for an open shelter deck ship; h = tween deck height of open shelter deck ship.

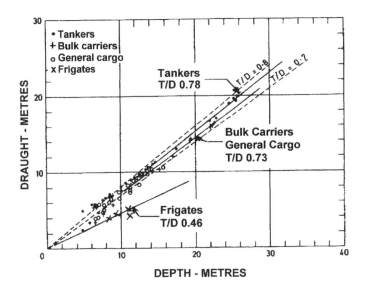

Fig. 3.10. Draft–depth relationship.

This formula is now historic with the unlamented demise of the open shelter deck ship with its undesirable features. For a depth of 15 m the ratio T/D given by this formula for a closed shelter deck ship would be 0.746 which is quite close to the value which applies to a cargo ship today.

In the 1975 plot, reproduced as Fig. 3.10, two lines for $T/D = 0.7$ and 0.8 were shown together with a mass of data spots which mainly lay somewhere between these lines, with tankers with a class "A" freeboard tending towards a higher value of about 0.78, whilst cargo ships with a class "B" freeboard were nearer the lower line and a T/D value of 0.73.

The scale of this plot is not suitable for reading off an accurate value emphasising the advantage of the ratio presentation in Fig. 3.8. This shows bulk carriers and refrigerated ships averaging at a T/D ratio of 0.71. The reason for this being a little less than the 1975 figure for cargo ships may be partly due to the fact that the ships are larger and partly because many of the 1975 ships had sheer, whereas most ships in 1991 do not. The T/D value for tankers now appears to average at 0.67, contrasting with earlier practice when an "A" class freeboard was general and confirming the statement that tankers are now "volume design" ships.

The T/D value of container ships varies quite a lot but the average value is about 0.62.

With designers now having access to computer programmes which makes the calculation of an accurate rule freeboard easy and quick, the use of a T/D ratio value is confined to preliminary estimates.

Although the drafts of warships are kept carefully concealed by the world's navies (or are quoted as drafts to the propeller tip, which are of little value for design purposes), a line for frigates has been added to Fig. 3.10 and shows a mean *T/D* value of 0.46. The high freeboard that this low ratio indicates shows the concern for seaworthiness that is so necessary a feature of the design of these ships.

3.3.5 Depth/length relationship D = f(L)

In the discussion about the *B/D* ratio it was noted that deadweight carriers have a higher value of this ratio than capacity carriers.

In deadweight carriers, stability is generally in excess (sometimes greatly) of rule requirements and depth and breadth are therefore independent variables. For these ships, control of the value of *D* is exercised more by the ratio *L/D* which is significant in relation to the structural strength of the ship and in particular to the deflection of the hull girder under the bending moment imposed by waves and cargo distribution. The largest *L/D* ratios were formerly used on tankers whose "A" type freeboard needed a comparatively small depth for the required draft and whose favourable structural arrangements with longitudinal framing on bottom, deck, ship sides and longitudinal bulkheads together with the fact that this type of ship has minimum hatch openings meant that the steel-weight penalty for an unfavourable *L/D* value was minimised.

When higher tensile steel is used to save weight, it is generally desirable to use a smaller *L/D* value in order to limit the deflection of the hull girder.

L/D values as presented in the 1975 paper are shown in Fig. 3.11, whilst the values in use in 1991 are shown in Fig. 3.8. A comparison of these figures shows little change in bulk carriers, with the *L/D* ratio averaging at about 11.8 in both cases. Tankers, however, show a striking change from a value of about 12.5 in 1975 to one of 10.5 in 1991, a change brought about by the need to provide separate clean ballast capacity.

The line for frigates which has been added shows a mean value for these ships of *L/D* = 13.3. This comparatively high value would appear to be the consequence of the need for length which has already been the subject of comment together with the need to limit depth which is the corollary of the limited beam. The fact that warships do not have large hatch openings reduces the adverse structural effect of a high *L/D* ratio.

3.3.6 Draft/length relationship T = f(L)

This is essentially a secondary relationship resulting from either of the following combinations of relationships:

$$T = f(D) \quad \text{or} \quad T = f(D)$$
$$\text{and } D = f(L) \quad \text{and} \quad D = f(B) \text{ and } B = f(L)$$

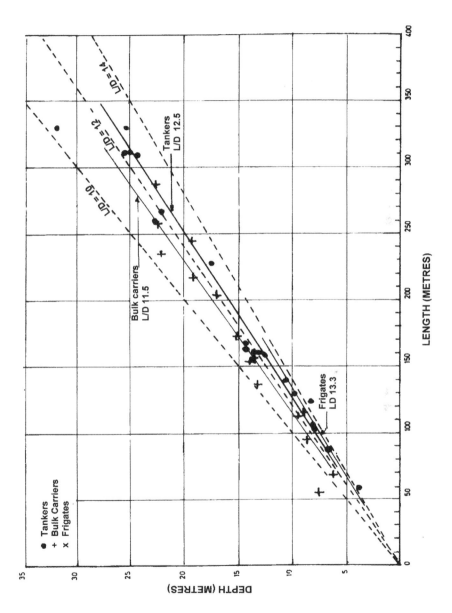

Fig. 3.11. Length–depth relationship.

Table 3.1

Optimisation of main dimensions

	Capital cost		Operational cost
	Hull	Machinery	
Increase L	Most expensive way to increase displacement; increases cost	Reduces power and cost	Reduces fuel consumption and cost
Increase B	Increases cost (but less proportionately than L). Facilitates increase in D by improving stability	Increases power and cost	Increases
Increase D and T	Cheapest dimensions to increase; reduces cost	Reduces power and cost	Reduces
Increase block coefficient	Cheapest way to increase displacement and deadweight	Increases power. Above a certain relationship of F_n to C_b can cause rapid increase in power	Increases

3.3.7 Draft/breadth relationship $T = f(B)$

This is again a secondary relationship, resulting in this case from either of the following combinations of relationships:

$$T = f(D) \qquad \text{or} \qquad T = f(D)$$
$$\text{and } D = f(B) \qquad \text{and} \qquad D = f(L) \text{ and } B = f(L)$$

3.3.8 Optimisation of main dimensions

Table 3.1 summarises the principal effects on building and operational costs respectively of alterations in the main dimensions.

3.4 BLOCK COEFFICIENT

3.4.1 General

The last factor required to complete the equation linking dimensions and displacement is the block coefficient. A first principles approach to the determination of the optimum block coefficient for a ship would involve a trade-off calculation in which the increment in building cost resulting from the increased dimensions required for a fine block coefficient is compared with the saving in operational cost obtained as a result of the reduction in power which fining the lines achieves. This is a major exercise but fortunately it is rarely necessary to adopt such an approach, the more general procedure being the use of an empirical relationship between block coefficient and the Froude number (F_n), which represents the state of the art.

3.4.2 The Alexander formula

One of the oldest such relationships is the Alexander formula. In the 1962 paper this was expressed as:

$$C_b = K - 0.5 \, V_k / \sqrt{L_f} \tag{3.11}$$

where L_f = length in feet. It was suggested that the value of K should vary between 1.03 for high speed ships to 1.12 for slow ships.

3.4.3 The Katsoulis formula

By 1975 it was clear that changes in L/B ratio together with the big increase in the size of many ships demanded a new approach. A proposal made by Katsoulis was studied with great interest as it seemed to involve all the right factors. Katsoulis suggested that C_b as well as being a function of F_n should also be a function of L/B and B/T, since both of these affect the resistance of the ship and the flow of water to the propeller and hence the QPC. He suggested an exponential formula of the form:

$$C_b = K \cdot f \cdot L^a \cdot B^b \cdot T^c \cdot V^d \tag{3.12}$$

and went on to show that this could be transformed into:

$$C_b = k \cdot f \cdot [V / \sqrt{(L)}]^{(d)} \cdot [L/B]^{(-b-c)} \cdot [B/T]^{(-c)} \cdot L^{a+b+c+d/2} \tag{3.13}$$

Katsoulis deduced values of the constants from regression analysis, but unfortunately his values did not appear to give satisfactory agreement with the block coefficients of a wide variety of ship types for which good data was available.

3.4.4 The Watson and Gilfillan C_b / F_n relationship

A plot of block coefficient against F_n was therefore made using all available data and it was found that with very few exceptions all the plotted values lay within a band of ±0.025 from a mean line with the majority of the points within much closer limits.

Whether or not this line represents an optimum depends on whether the large number of different naval architects whose designs provided the data made wise judgments — but it does seem likely to be quite close to an optimum which is not a highly tuned one.

It was disappointing to be unable to detect any significant effect of either L/B or B/T, although it must be noted that different types of vessels, each of which generally has its own particular range of L/B and B/T tend to be concentrated at different parts of the F_n range. This remark also applies to twin screw propulsion, which is generally confined to high speed container ships, passenger ships, ferries and warships.

The types of ships used in the plots are indicated showing the areas in which each predominates. Ship type may have some significance in relation to selection of C_b because of the different practices in relation to service margin which seem to apply to different types of ships.

A slightly modified version of the 1975 plot is presented in an enlarged format in Fig. 3.12.

The use of the Watson–Gilfillan.line in a computer programme was made much more convenient by Dr Townsin who devised the following formula which agrees almost exactly with the mean line:

$$C_b = 0.70 + 1/8 \tan^{-1} \frac{(23 - 100F_n)}{4} \text{ radians} \tag{3.14}$$

Adding recent (1991) data to this graph seems to confirm its continuing validity and prompts the question whether all naval architects are now using the 1975 paper!

Although the block coefficient given by the mean line in Fig. 3.12 should result in fairly near optimum powering, there can be other factors which may make a different block coefficient desirable, and it is as well to remember that powering is more closely related to the prismatic coefficient than it is to the block coefficient — and implicit in any selection of block coefficient must be an intention to associate it with an appropriate midship section coefficient.

Both the design of lines and powering are dealt with in later chapters, but before leaving this discussion of C_b it may be worth mentioning that most warships have much finer values of C_b than might appear necessary for their F_n values purely on a powering criterion. The main reason for this is the requirement that these comparatively small ships have to maintain as much of their service speed as possible in whatever weather they may encounter which together with the modest draft of most warships means that much of the fore body can come out of the water as the ship pitches with a consequent danger of severe slamming.

As the best way to reduce the incidence of slamming is to give the ship a high rise of floor, these ships tend to have a small midship section coefficient which for a given prismatic coefficient gives a reduced block coefficient. The increase in draft obtained in this way can have other advantages such as improving course stability.

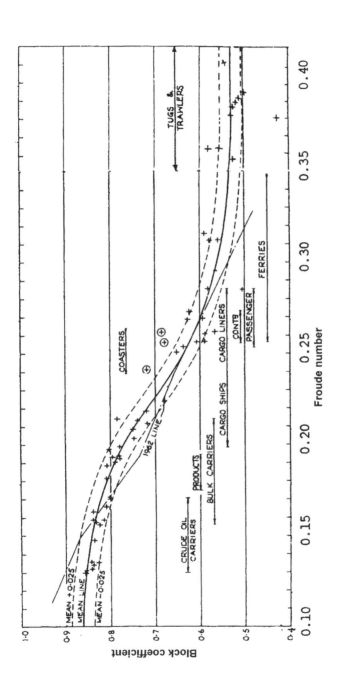

Fig. 3.12. Block coefficient–Froude number relationship.

Mean line formula is given in eq. (3.14).

3.5 APPENDAGE DISPLACEMENT (1 + S)

3.5.1 A first approximation

In order to obtain the full displacement at the desired draft, it is necessary to add a correction to the moulded displacement to allow for shell and appendages. The obverse of this is that when the full displacement which will provide the required deadweight is known, these corrections must be deducted to arrive at the moulded displacement and block coefficient used to determine the ships dimensions.

Whilst appendages are a comparatively small factor in the displacement calculation, it can be important where the deadweight is small and margins are tight to have a good approximation for these items, at least in the later stages when the design is being refined. If this can be done easily, there seems every reason to use the same approximations in the preliminary design stage.

For a single screw ship with an all-welded shell, the simplest approximation is 0.5% of the moulded displacement.

If draft is limited, designers should remember to allow for the keel thickness when setting the draft moulded.

If a more exact estimate of appendage displacement is required, the various appendages should be considered individually.

3.5.2 Individual items

(i) Shell displacement $= \dfrac{t}{380} \cdot (\Delta L)^{1/2}$ (3.15)

where t = mean shell thickness in mm.

(ii) Stern displacement $= [(T/H)^x - 1] \cdot \dfrac{\Delta}{1000}$ (3.16)

where

x = from 2.5 for "fine" sterns to 3.5 for "full" sterns, and
H = height of counter.

(iii) Twin screw bossing displacement $= K_b(d)^3$ (3.17)

where

d = propeller diameter in metres
K_b = 0.2 for stub bossings, open shafts and "A" brackets, and
           from 0.7 for fine bossings
           to 1.4 for full bossings.

(iv) Rudder displacement = $0.13 (A_r)^{3/2}$ (3.18)

where A_r = rudder area in m^2.

(v) Propeller displacement = $0.01 (d)^3$ (3.19)

Other items which may affect the displacement are sonar domes, and the lost buoyancy of bow and stern thrust tunnels, recesses for stabiliser fin stowage, recesses for dredge pipe trunnion slides, moon pools, etc. It may be noted that lost buoyancy effects which at a later stage and in "as fitted" documentation are usually dealt with as lost buoyancy and deducted from the hydrostatics are, in general, better treated as added weights in design work.

Chapter 4

Weight-Based Designs

4.1 INTRODUCTION

The first part of this chapter deals with the estimation and calculation of the lightweight and deadweight which make up the displacement.

The components of the lightweight in merchant ship practice consist of the structural weight, the outfit weight, the machinery weight and the margin. Warship designers use a larger number of weight groups, as shown in Fig. 4.14, but for convenience, warships are also considered here under the three merchant ship groups.

A number of approximate estimating methods are given for each of the weight groups and these are followed in each case by a suggested format for the detailed calculations which should follow as the design is developed.

The later sections of the chapter deal with approximate methods of estimating the lightship VCG and LCG and the maintenance of control over the weight as the design develops.

The term "weight" has been used as this is common parlance although scientifically the term "mass" would, of course, be correct.

4.1.1 Demarcation of the weight groups

The demarcation between the three weight groups of structure, outfit and machinery is not as obvious as it might appear at first sight as there are several items which could logically be placed in more than one group. It is therefore very desirable to have a demarcation that is standard at least within a design office.

In general, the structural group includes all steel or other structural material worked by the shipyard plus such items as deposited weld metal or rivet heads.

The following items of steelwork which are more usually bought from a subcontractor are generally included in the outfit weight and for consistency this should remain so even if for a particular ship they have been fabricated by the shipyard itself:
- – sternframe, rudder, rudderstock, shaft brackets and similar structures whether these are castings or fabrications;
- – steel hatch covers for cargo hatches (covers for access hatches are, however, usually in the structural weight);
- – bollards and fairleads.

Within the outfit weight, two other items which can cause demarcation difficulties are plumberwork and electrical work —systems which are partly in and partly out of the engine room. The demarcation here usually divides the systems at the engine room bulkheads or engine casings. Everything outside the engine room is taken as hull outfit and everything within as machinery weight, with the same demarcation generally applied when writing the specifications of these systems.

There are a number of other items part of which is often fitted within the engine room which for simplicity are usually dealt with wholly as hull items: refrigerating machinery whether for cargo, stores or HVAC; sewage systems; watertight doors; casing insulation and paint.

4.2 STRUCTURAL WEIGHT APPROXIMATIONS

4.2.1 Lloyd's equipment number method

In the author's 1962 paper, the use of Lloyd's equipment numeral was advocated as a basis for estimating steel-weight in preference to the numerals $L \times B \times D$ or $L(B + D)$ which were in common use at that time.

The reasons given for this were that the equipment number introduced allowances of approximately the correct order for changes in draft and in the extent of erections, and avoided making the choice of the deck to which D was measured as critical as it is with both the other numerals.

The Lloyd's equipment numeral (E) of 1962 is no longer in use for the determination of ship's anchors and cables, hawsers and warps; it was replaced in 1965 by a new numeral, which is now common to all classification societies having been agreed to be a more rational measure of the wind, wave and current forces which act on a vessel at anchor. The new numeral, however good it is for its primary purpose, is not a suitable parameter against which to plot steel-weights, but the reasons given for the use of the old numeral still stand. For those not familiar with the old E number, the formula for this is as follows:

$$E = L(B + T) + 0.85 \, L(D - T) + 0.85\{(l_1 \cdot h_1) + 0.75\{(l_2 \cdot h_2) \tag{4.1}$$

where
  l_1 and h_1 = length and height of full width erections, and
  l_2 and h_2 = length and height of houses.

If a numeral of this sort was being devised for the specific purpose of steel-weight estimation, the constants would undoubtedly have been different, but the availability of a great deal of data collected over many years in the E form has become, for the author, a major influence in retaining its use.

4.2.2 Invoiced or net weight

The question of whether to plot invoiced or net weights is worthy of some debate. The net weight is that initially arrived at by detailed calculations based on ship's plans and it is the weight that is required for the deadweight calculation. The invoiced weight, on the other hand, is the weight recorded in a shipyard's steel order books and the one used for cost estimates. In earlier days the invoiced weight was the one that was known more accurately and data was generally presented in this way.

More recently, with shipyards ordering much of their steel in standard plates for stock, the reliability of invoiced records have declined, whilst with prefabricated units now often being weighed, net steel-weight records have improved in accuracy and seem the better choice nowadays.

4.2.3 The effect of the block coefficient on steel-weight

Since the parameter E attaches no significance to the fullness of the ship, a factor which clearly has an appreciable effect on the steel-weight, all steel-weights are corrected to a standard block coefficient before plotting. By the same token all steel-weights read from the graph must be corrected from the standard to the desired block coefficient.

The standard block was set at $C_b' = 0.70$ measured at 0.8D.

Corrections to the steel-weight for variations in C_b from the standard 0.70 value can be made using the following approximate relationship:

$$W_s = W_{si}[1 + 0.05(C_b' - 0.70)] \tag{4.2}$$

where
  W_s = steel-weight for actual C_b' at 0.8D, and
  W_{si} = steel-weight at $C_b' = 0.70$ as plotted/lifted from graph.

In this case C_b' has been taken at 0.8D, because the available data is on this basis.

C_b' at 0.8D can be calculated from the known value of C_b at the load draft using the formula given as eq. (3.10) or read from Fig. 3.6.

4.2.4 Plotting the steel-weight against E

Because of the wide range of ship sizes that this book tries to cover, the only satisfactory plot of steel-weight that can be achieved without resort to an outsize piece of paper requires the use of a log–log scale, as shown in Fig. 4.1. It will be seen that this achieves reasonable accuracy for small ships whilst allowing the steel weights of the largest vessels to be included in the plot. The accuracy with which steel-weights can be read off from this small graph is however limited by the scale and the difficulty in showing the large number of closely spaced lines needed to differentiate between different types of ship. As today's computer users prefer

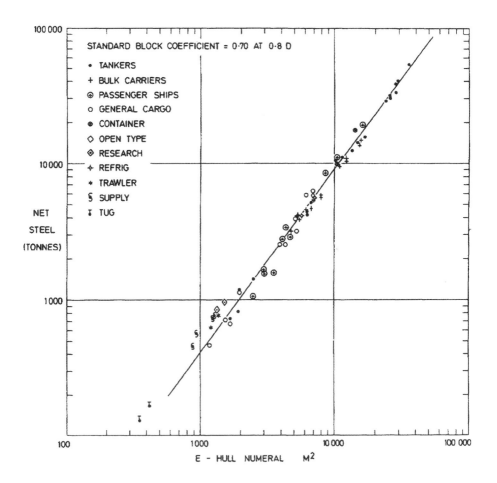

Fig. 4.1. Net steel weight vs Lloyds equipment number E. The line shown is a mean through most of the spots irrespective of ship type and has the formula $W = 0.33\ E^{1.36}$.

Table 4.1

Type	K		Range of E	No. of ships in sample
	Mean value	Range		
Tankers	0.032	± 0.003	1500–40000	15
Chemical tankers	0.036	± 0.001	1900–2500	2
Bulk carriers	0.031	± 0.002	3000–15000	13
Container ships	0.036	± 0.003	6000–13000	3
Refrigerated cargo	0.034	± 0.002	4000–6000	6
Coasters	0.030	± 0.002	1000–2000	6
Offshore supply	0.045	± 0.005	800–1300	5
Tugs	0.044	± 0.002	350–450	2
Research ships	0.045	± 0.002	1300–1500	2
Ro-Ro ferries	0.031	± 0.006	2000–5000	7
Passenger ships	0.038	± 0.001	5000–15000	4
Frigates and corvettes	0.023	not known		

formulae to graphs in any case it was a pleasure to find that all types of ship could be represented by a series of lines with the same slope, or in log–log terms with the same index, making the following formula applicable to all ship types:

$$W_{si} = K \cdot E^{1.36} \qquad (4.3)$$

The values of K are given in Table 4.1. Some words of caution to users of this table. Firstly, for some classes of ship the samples on which it was based were rather limited and there may be ships whose weights are appreciably further from the mean value than the table suggests. Secondly, the data on which the table is based are now somewhat dated. For most ships this probably means it will overestimate the steel-weight, but this should not be too confidently assumed. Thirdly, it is intended to provide an all mild steel structural weight, whereas for some ships the use of higher tensile steel may provide a better design solution. In other ships some parts, usually of the superstructure, may be constructed of aluminium and/or fibre reinforced plastic (FRP), with significant weight savings.

When calculating K, weights of high tensile steel, aluminium and FRP used in the basis ships were converted to equivalent weights of mild steel. On the merchant ship design sheet there is a space for the opposite process to be carried out if these materials are to be used in a new design.

A rough basis for conversion to these alternative materials, on the assumption that these materials are being used to full advantage (which is not always possible) is:

1 tonne of high tensile steel will replace about 1.13 tonnes of mild steel

This conversion is based on high tensile steel with a yield stress of 315 N/mm^2

(Lloyds AH 32) as compared with mild steel of 245 N/mm^2 and a plating thickness ratio based on the square root of the ratio of the yield stresses.

The optimum use of higher tensile steel is for the plating of the strength deck, followed by the bottom shell. It can also be used for the side shell but here the weight saving is less.

1 tonne of aluminium will replace about 2.9 tonnes of mild steel

This is based on a volumetric substitution of aluminium for steel which approximates to the usual practice for those parts of a ship for which aluminium can be considered.

Depending on the method of construction and the materials used, fibre reinforced plastics (FRP) can also be substituted on the basis of 1 tonne FRP to about 2.9 tonnes of mild steel.

A steel-weight from this table is intended for use in the initial stages of design and should be replaced by a more detailed estimate as soon as the design is properly detailed with a general arrangement plan, a body plan and a midship section or similar structural plan.

Ways of minimising structural weight and cost are discussed in Chapter 10.

In the author's 1976 paper it was noted that there had been a reduction in the steel-weights corresponding to a particular value of E of between 15% and 20% from the figures that had applied in 1962.

The reasons for the reduction between 1962 and 1976 were summarised as follows:

(i)   The changes in the ratios L/B, B/D, T/D which had occurred meant that for the same E number, more recent ships had a reduced length, but a larger beam and depth than earlier ships.

(ii)   There was less internal structure, fewer decks and bulkheads in the more recent ships.

(iii) Superstructure had been reduced with the reduction in crew numbers and simplified with the elimination of the overhanging decks which used to be a common feature.

(iv)   The reduction (almost elimination) of "owners' extras" which were once a common feature.

(v)   Rationalisation of Classification Society rules and reductions in permitted scantlings, some of which had followed the introduction of better calculation methods whilst others may, less desirably, have been a consequence of competition between the Societies.

(vi)   Changes in demarcation of work, which had removed from the steel-weight some of the items mentioned in §4.1.2 which used to be made in shipyards but were now almost invariably bought in.

Lack of data prevents the calculation of a corresponding figure for the changes which have undoubtedly occurred since 1976. It seems likely there was a further reduction possibly of about 5% for mild steel construction in the early years of this period — with another reduction if higher tensile steel was used to a major extent as was increasingly the case for large ships. In the last five years, concern over recent ship casualties has reduced the trend towards an increased use of higher tensile steel and seen the recognition of the need for some structural redundancy with a consequent increase in the structural weight of ships now under construction towards, but keeping a little below, the 1976 figures.

4.2.5 Scrap and invoiced steel-weight

Whilst the net steel-weight is required for calculating the light weight, it is still necessary to consider the scrap allowance to arrive at the invoiced weight needed for cost estimation.

In 1962, a figure of 12% of the invoiced steel was suggested as an average scrap percentage. By 1975 scrap percentages had been considerably reduced particularly for larger ships and it was suggested that the amount of shape in the ship was the major relevant factor, with full ships having a lower scrap percentage than fine lined ships.

In general the following factors were seen as likely to influence the scrap percentage:

 - shipyard steel ordering methods: the use of standard plates; the necessity of ordering sections for stock to ensure supply when needed;
 - shipyard constructional methods: the allowance of overlaps on prefabricated units to be cut at the ship to ensure a good fit; the use of numerical or optical lofting methods involving nesting procedures; extra lengths on sections to suit the operation of cold frame benders;
 - the effect of increases in the cost of steel enforcing economy in its use; the skill of draughtsmen in utilising material, particularly by the use of nesting when ordering plates; the accuracies of the calculations and the weighing methods employed to assess net and invoiced weights, respectively.

Figure 4.2 shows a plot of scrap percentage against block coefficient at 4/5 depth with some suggestions for fine tuning in relation to size and some other factors.

When making up the lightship weight from a calculated net weight an addition of 1% should be made to allow for deposited weld metal and rolling margin on the steel.

4.2.6 Other approximate formulae for steel-weight

In a 1974 R.I.N.A. paper, K.W. Fisher summarised a number of alternative steel-weight estimating methods. Most of the formulae quoted appear to have been

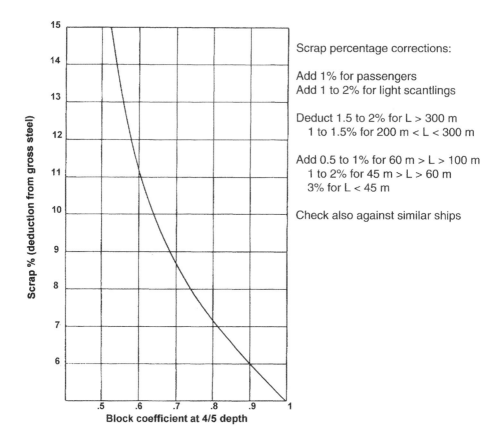

Fig. 4.2. Scrap percentage versus block coefficient.

derived by regression analysis techniques and the indices allotted to the dimensions of L, B, D and C_b vary widely, and unfortunately in most cases the resultant figures appear to have little physical significance.

The base numeral suggested in §4.2.1, although better than most others in general use, lacks theoretical justification and it seems worth investigating whether a better data base can be devised.

In trying to do this, naval architects have become divided between methods based on volume and methods based on beam analogy. The truth appears to lie somewhere between, with part of the weight being volume-dependent and part modulus-dependent — a concept recognised by Eames and Drummond in their 1976 R.I.N.A. paper "Concept exploration" and by Sato in a 1967 S.N.A.M.E. paper "Effect of principal dimensions on weight and cost".

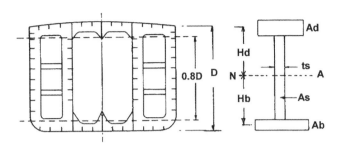

Fig. 4.3. Midship section compared with an I beam.

4.2.7 A rational base for approximate structural weight estimation

Following the modulus-dependent thesis, the midship section of a ship may be compared with the cross section of an I beam as shown in Fig 4.3.

The hull steel-weight per metre $= R(A_d + A_b + A_s + A_t)$ (4.4)

where

A_d = area of deck plating + deck longitudinals + other longitudinal material above $0.9D$ from base

A_b = area of bottom plating + bottom longitudinals + other longitudinal material below $0.1D$ from the base

A_s = area of shell plating and longitudinals + area of longitudinal bulkhead plating and longitudinal within the middle $0.8D$

A_t = area of transverse material per metre of ship's length

R = weight per metre per unit area.

The deck modulus of the hull girder is approximately

$$Z = \frac{A_d(H_d)^2 + A_b(H_b)^2 + 1/12 A_s(0.80D)^2}{H_b}$$ (4.5)

If $k_d \cdot D$ is substituted for H_d and $k_b \cdot D$ for H_b, then

$$A_d + A_b(k_b/k_d)^2 + A_s(1/4.33\ k_d)^2 = Z/k_d \cdot D$$

If we approximate and take $k_d = k_b = 0.45$, then

$$A_d + A_b + A_s/3.8 = K(Z/D)$$

or

$$A_d + A_b = K(Z/D) - A_s/3.8$$

substituting this in eq. (4.4) gives

steel-weight per metre = $R[K(Z/D) + 0.74\,A_s + A_t]$

or

hull steel-weight = $R \cdot I_f \times L \cdot [K(Z/D) + 0.74\,A_s + A_t]$

where I_f = integration factor which is a function of C_b.

A formula for the modulus Z from Lloyds rules which applies to ships with a still water bending moment not exceeding 70% of the wave bending moment is:

$$Z = C_1\,L^2\,B(C_b + 0.7) \text{ cm}^3$$

In this formula C_1 is not a constant but varies from 7.84 to 10.75 as L changes from 90 m to 300 m, at which point it becomes substantially constant. For present purposes it is treated as constant, although its variation with length may explain the slightly higher index of L which Sato suggests.

$A_s = 0.80\,t_s\,D$ and $t_s = f(L)$ from which $A_s = n_2(L \times D)$ and $A_t = n_3(B \times D)$

Hence

steel-weight of hull = $R \cdot I_f \cdot L\{n_1 \cdot C_b(L)^2 \cdot B/D + n_2 \cdot (L \times D) + n_3 \cdot (B \times D)\}$

To obtain the total steel-weight it is necessary to add three more items for each of which a rational expression is suggested:
(i)   weight of bulkheads = $n_4 \times C_b \times L \times B \times D$
(ii)   weight of platform decks = $n_5 \times C_b \times L^2 \times B$
(iii) weight of superstructure, masts and deck fittings = $n_6 \times (V)$ or $n_7 \times B^2 \times L$

where V = volume of superstructure and n_1 to n_7 are constants.
An expression for hull structure weight may be deduced as:
  (i) modulus related
+ (ii) side shell and longitudinal bulkheads
+ (iii) transverse frames, beams and bulkheads
+ (iv) platform decks and flats
+ (v) superstructure and deck fittings

$$W_s = n_1 \cdot (L)^3 \cdot B/D \cdot (C_b)^X + n_2 \cdot (L)^2 \cdot D \cdot (C_b)^Y + n_3 \cdot L \cdot B \cdot D \cdot (C_b)^Y$$
$$+ n_4 \cdot (L)^2 \cdot B \cdot (C_b)^Z + n_5 \cdot (V) \text{ or } n_5\,B^2\,L \qquad (4.6)$$

It should be noted that constants n_1 to n_5 in eq. (4.6) are not identical with those in the earlier expressions as some of the items have been combined.

In this expression the indices of C_b in the various terms have been left as alphabetical symbols. It would appear from inspection that X might have a value close to unity as it has components both from the integration factor and from Lloyd's modulus formula; both Y and Z are clearly fractional indices.

An extrapolation on log–log paper of information on integration factors similar to that presented in Fig. 4.8 indicated that overall the steel-weight is proportional to the square root of the block coefficient.

If it is accepted that for any one type of ship the dimensions L, B, and D are related, the formula can be simplified to:

$$W_s = C_b^{1/2} \cdot L \cdot B \cdot [K_1 \cdot L(L/D) + K_2 \cdot D] \tag{4.7}$$

This has one modulus-related term and one volume-related term and is very similar to Sato's expression:

$$W_s = C_b^{1/3} [w_1(L)^{3.3} \cdot B/D + w_2 \cdot (L)^2 \cdot (B + D)^2] \tag{4.8}$$

Determining values of K_1 and K_2 is left to readers, but the author believes that this method applied to a range of good data would lead to more accurate steel-weight estimation.

4.3 DETAILED STRUCTURAL WEIGHT CALCULATIONS —ALL SHIP TYPES

The principal difficulty in making detailed calculations arises from the fact that this more accurate weight is almost invariably required at a time when very few plans have been drawn. Plans which must be available for a detailed calculation include a reasonably complete general arrangement plan, a body plan and a midship section with the scantlings. Improved accuracy could be obtained if structural sections away from midships, a shell expansion plan, steel deck plans, bulkhead plans are also available, but this is unlikely to be the case, although scrap drawings of these can be made specifically to assist in the calculation.

4.3.1 Use of the midship section weight per unit of length for merchant ship weight calculations

Without the plans just mentioned, calculation of the weights of shell, ship side framing and the double bottom is almost impossible and the best alternative would seem to be the use of an integration factor in association with a calculated weight per metre of the midship section.

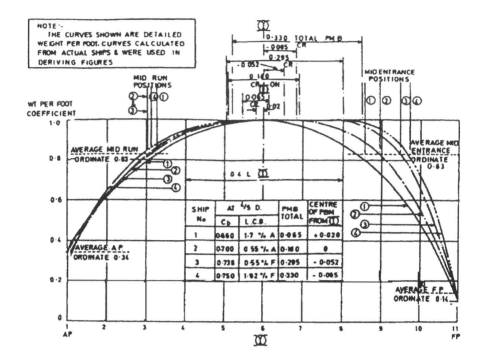

Fig. 4.4. Steel-weight distribution for a number of ships.

The author did a great deal of work on this subject many years ago which was published in 1958 as BSRA Report No. 266 as a method of deriving a hull weight distribution "Coffin" diagram which took into account the ship's form in a way that standard coffin diagrams fail to do. Use of this method provides a way of increasing the accuracy with which the LCG of the hull steel-weight can be calculated and therefore of the accuracy of strength calculations based on weight distribution.

A plot of the weight per metre of the steel-weight of the hull for a number of ships was found to give diagrams very similar to the sectional areas of the ships in question, indicating that block coefficient and centre of buoyancy were the prime factors involved (see Fig. 4.4). In order to eliminate the effect of different draft/depth ratios applicable to different types of ship, the block coefficient and LCB position used relate to a fixed proportion (0.8) of the depth to the uppermost continuous deck, rather than to the draft.

An analysis of the plot in Fig. 4.4 showed that quite simple formulae could be derived for synthesising a hull weight distribution with a correct total weight and correct longitudinal centre of gravity.

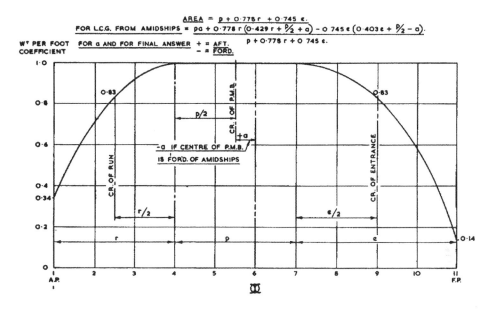

Fig. 4.5. Steel-weight distribution for ships with parallel middle body.

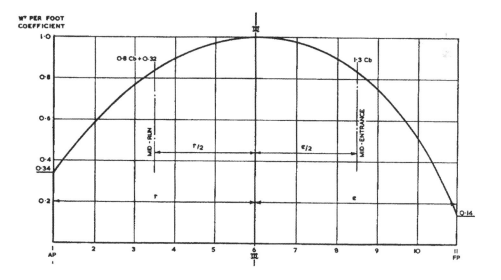

Fig. 4.6. Steel-weight distribution for ships with no parallel middle body.

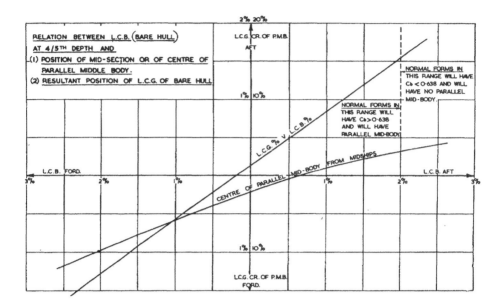

Fig. 4.7. Longitudinal centre of gravity and centre of parallel middle body versus longitudinal centre of buoyancy.

A method of constructing a weight distribution curve for a ship with parallel middle body is given in Fig. 4.5, and for those without parallel middle body in Fig. 4.6. In the former case, use is made of Fig. 4.8, which shows the extent of parallel middle body associated with block coefficient values and of Fig. 4.7, which shows where the centre of the middle body will be for any required LCB position. For the latter case use is made of Fig. 4.9 which gives mid-entrance and mid-run factors based on a matrix of C_b and LCB values.

Figures 4.10 and 4.11 show how well this method of producing a "coffin" diagram agrees with detailed calculations of the weight distribution of two vessels of very different types, block coefficients and LCB positions.

Figure 4.8 shows, in addition to the extent of parallel middle body, the integration factor, which corresponds of course to the area of the weight distribution diagrams.

This can be used to calculate the weight of the hull as follows:

$W_s = I \times L \times$ weight/metre of midship section

and $I = 0.715\ C_b + 0.305$.

If this method is used to estimate hull weight, the resultant weight must be corrected for any significant changes in the structure forward or aft from that included in the midship section weight per metre.

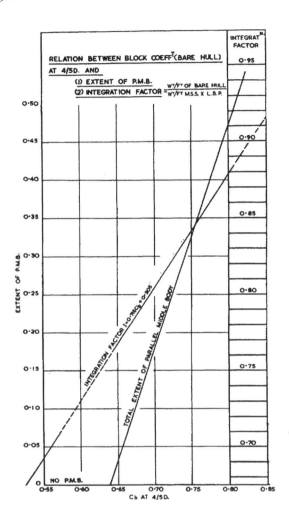

Fig. 4.8. Integration factor and extent of parallel middle body versus block coefficient at 0.8D.

The "weight per metre" can with advantage be calculated by summing steel volumes, conveniently in the form of $m^2 \times$ mms of thickness and converting to weight only after summation. The weight per metre of transverse material such as frames should be obtained by dividing the actual weight by the frame spacing in metres.

It is convenient to group the weight per metre into five items comprising:

(i)   bottom shell and longitudinals up to $0.1D$
(ii) double bottom structure
(iii) side shell and framing between 0.1 and $0.9D$, longitudinal bulkheads and stiffening between the same limits

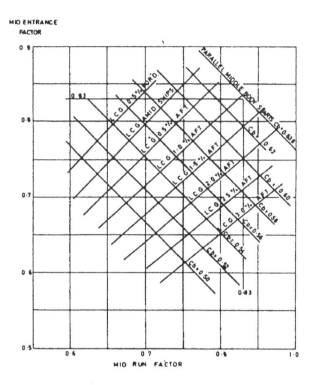

Fig. 4.9. Mid-entrance and mid-run factors vs block coefficient and longitudinal centre of buoyancy.

(iv)　upper deck plating and support structure shell above $0.9D$

(v)　continuous lower decks and support structure.

These weights appear to plot quite well against the parameters given below and records kept in this way can provide a useful cross check on detailed calculations or a method of making a quick approximation.

Weight (ii) against $B \times$ double bottom height

Weight (iii) against D

Weight (i) + (iv) against the midship section modulus

Weight (v) against $B \times$ number of decks.

The VCG of the hull structure can be calculated in a similar way by first calculating the VCG of the midship section structure and then proportioning this by the ratio of the VCB of the ship at the mean depth, allowing for sheer, to the height of the centroid of area of the midship section.

$$\text{VCG (hull)} = \text{VCG (midship section)} \times \frac{\text{VCB at depth } D}{D}$$

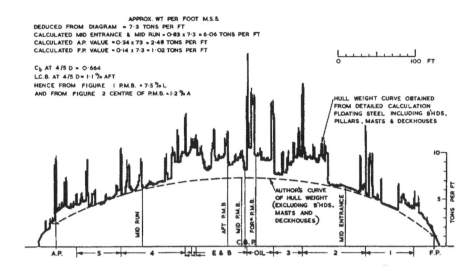

Fig. 4.10. Steel weight distribution by proposed method compared with actual for a shelter deck cargo liner.

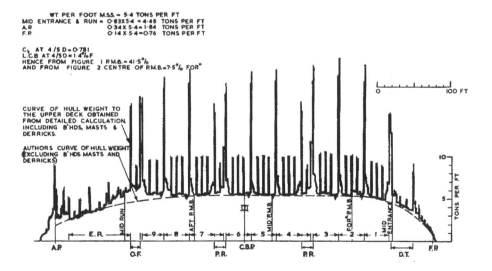

Fig. 4.11. Steel weight distribution by proposed method compared with actual for a tanker.

(Analysis for Figs. 4.10 and 4.11 carried out by Professor A.M. Robb for BSRA.)

4.3.2 Items not in the weight per unit length

Items within the main hull not included in the weight per metre such as platform decks, bulkheads and casings together with all items such as superstructure, masts, etc. outside the main hull must be added to give the total net structural weight.

Rapid and accurate calculation of the weight of all the other structural weight items demands the maintenance of careful records of weights from ships for which the designer has complete plans. These will usually be best recorded as weights per square metre, but sometimes weights per lineal metre are appropriate and occasionally it is best to record the total weight of an item. If at all possible, these weights should be plotted against a suitable base, with the length B.P. with which many scantlings scale often being the best parameter, so that values appropriate to the dimensions of a new design can be selected.

If ingenuity is used in selecting suitable numerals for these plots, they can be remarkably accurate and very useful.

For example, the numeral $N = B \times D^2 \times C_b$ (at $0.8D$) was found to give a remarkably good plot of the average weight of a watertight bulkhead for a particular ship type, with different lines being identified for ships with different numbers of decks supporting the bulkhead and influencing stiffener spans and therefore scantlings and weights. The use of this numeral avoids the necessity of drawing each bulkhead so that its area can be calculated.

In this calculation, as in all detailed calculations, probably the most important thing is to make sure that no items are completely omitted — an error in estimating can be bad enough but a complete omission can be disastrous and to avoid this it is important that a standard checklist, such as that given below, is followed.

1.   Corrections for local structure
     1.1 Shell
     1.2 Decks
     1.3 Double bottom
2.   Side to side erections: forecastle, bridge, poop
     2.1 Front, sides, end
     2.2 Decks
3.   Houses
     3.1 Front,sides and end
     3.2 Decks
     3.3 Machinery casings
     3.4 Other internal steel casings
     3.5 Doors, access hatches, etc.
4.   Fittings on or above the upper deck
     4.1 Bulwarks
     4.2 Hatch coamings

4.3 Bollards and fairleads, chain pipes
4.4 Seats for windlasses, capstans, winches, cranes, steering gear
4.5 Masts and derrick posts

5. Shell fittings
   5.1 Bilge keel
   5.2 Bulbous bow
   5.3 Hawsepipes, thruster tunnels

6. Structure below upper deck
   6.1 W.T. bulkheads
   6.2 Tank bulkheads
   6.3 Casings below upper deck
   6.4 Decks and flats not in weight per metre
   6.5 Pillars and girders
   6.6 Sparring, ceiling

7. Machinery space structure
   7.1 Main engine seats
   7.2 Auxiliary seats, engineers tanks, sewage tanks

The net weight calculated in this way must be corrected for weld metal and rolling margin before it is used when making up the lightship, whilst for cost purposes a suitable scrap percentage must be added.

4.3.3 Detailed structural weight calculations — warships

The midship section weight per metre method outlined above can be used for warships in the preliminary stages of design but in later stages it is usual to draw a number of structural sections to investigate the structure required at different stations along the length of the ship. When such sections are available their use provides the possibility of a more accurate structural weight estimate.

Warship structural weights are generally estimated and recorded in the categories shown in Group 1 on Fig. 4.14, although warship designers use more detailed three-digit weight groups for their calculations.

4.4 OUTFIT WEIGHT CALCULATIONS

4.4.1 Approximate methods — all ship types

The traditional method of estimating the outfit weight for a new merchant ship was by proportioning the outfit weight of a similar ship on the basis of the relative "square numbers", i.e., $L \times B$, and then making corrections for any known differences in the specifications of the "basis" and "new" ships.

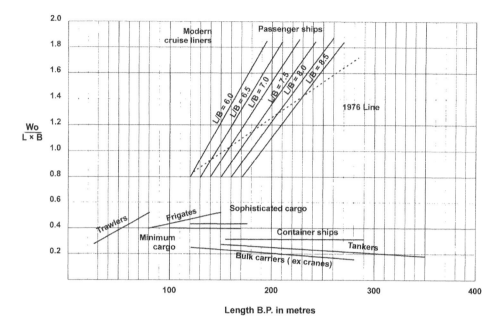

Fig. 4.12. Outfit weight/($L \times B$) versus length and ship type.

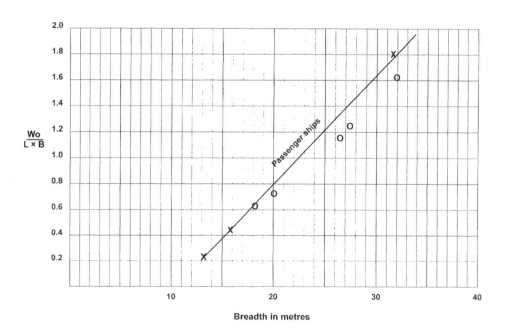

Fig. 4.13. Outfit weight/($L \times B$) versus breadth for passenger ships. o — as plotted in 1976 paper;
× — 1994 additions.

Provided a good "basis" ship is available and the corrections for known differences are made with care the method is the best available short of detailed calculations (see later), which are time consuming and difficult to make with worthwhile accuracy at the early design stage.

A warning about the accuracy of square number proportioning may be taken from Fig. 4.12, which is a modified version of the figure presented in the 1976 paper. This shows that even for a particular type of ship the ratio outfit weight/ square number is not always constant, although near constant values do seem to apply to general cargo ships and container ships.

On the other hand, values slowly diminishing with length seem to apply to tankers and bulk carriers, possibly because some items such as the accommodation on these ships vary only slightly with ship size.

In the 1976 paper the comment was made that the value of the ratio for passenger ships increased quite rapidly with length, probably reflecting the increase in the number of decks which tended to go with increasing length and breadth emphasising the fact that for these ships volume is a better parameter than area.

When data for modern cruise liners was added to this graph it was found to be much higher than the 1976 line suggesting that the latter was now out of date. The reason for this difference appears to be the extra decks which ships of a given length now have, which naturally increases the outfit weight per square metre of $L \times B$. To maintain satisfactory stability, the breadth of these modern ships has been increased relative to their length reducing the L/B ratio to near 6 as opposed to about 8+ for the ships which formed the basis of the 1976 plot. It may be worth noting that the lower speeds of the cruise liners permitted this change without there being an unacceptable penalty on powering.

This reasoning suggested that the new data might be reconciled with the 1976 spots if both were plotted on a base of beam rather than length and the result is Fig. 4.13.

The passenger line of the original 1976 plot has now been replaced in Fig. 4.12 by a series of lines of different L/B values with the 1976 spots falling happily into place. There is a lesson to be learnt here. In 1976 the available data seemed to justify the line given at that time but the base used can now be seen to be inherently wrong. It is vitally important that all approximate formula have a rational scientific basis if they are to continue to be relevant with significant design changes.

Reverting to Fig. 4.12, it may be noted that a line for frigates and corvettes has been added.

The use of this type of graph for warships is complicated by the fact that, as discussed in §§4.1.1 and 4.4.6 and again in relation to specification writing in Chapter 17 and to cost estimating in Chapter 18, outfit is not a concept used by warship designers. But as a number of warship designers are used to merchant ship practice and because outfit can be a helpful concept at the initial design phase a line

1 HULL	2 PROPULSION	3 ELECTRICAL	4 C³I
10 Hull structure	20 Nuclear propulsion	30 Power generation	40 Navigation systems
11 Superstructure	21 Non-nuclear propulsion	31 Distribution equipment	41 Internal communication
12 Structural bulkheads	22 Propulsion units	32 Distribution cabling	42 Ship & mech. control
13 Structural decks	23 Condensers & air ejectors	33 Lighting systems	43 Weapon control
14 Doors, hatches & scuttles	24 Shafting & propulsors		44 Ship protective systems
15 Seats, masts & supports	25 Exhaust & air supply system		45 External communication
16 Control surfaces	26 Steam systems		
17 Structural castings	27 Water cooling systems		
18 Buoyancy & ballast	28 Fuel service systems		
19 Fastenings	29 Lub. oil systems		

5 AUXILIARY	6 OUTFITTING	7 ARMAMENT	8 VARIABLE LOAD
50 Air condition. & ventilation	60 Hull systems	70 Surface to air	80 Officers, crew & effects
51 Fuel systems	61 Boats & life saving equipment	71 Surface to surface	81 Ammunition
52 Sea & fresh water systems	62 Minor hull bulkheads	72 Surface to subsurface	82 Aircraft
53 Air & gas systems	63 Storeroom furnishing	73 Sub launched anti surf/sub	83 Military vehicles
54 Hydraulic systems	64 Living space furnishing	74 Sub launched anti air	84 Victualling stores
55 Aircraft systems	65 Office furnishing	75 Air launched arm.	85 Naval stores & spare gear
56 Waste Disposal systems	66 Galley/workshop equipment	76 Mine warfare equipment	86 Weapon stores
57 Aux. steam systems	67 Superstructure partitions	77 Small arms & rockets	87 Operating fluids
58 Lub. oil systems	68 Portable fire fighting equip.		88 Stowed liquids
	69 Load handling & RAS equip.		89 Cargo

Fig. 4.14. Warship weight and cost groups.

for these ships has been drawn on the basis of the weight included in warship cost sections 3, 4, 6 and 7 (see Fig. 4.14) which seems the nearest equivalent to merchant ship outfit.

This frigate line and the line for trawlers both indicate an outfit ratio increasing with length showing the same trend as that noted as applying to passenger ships and presumably for the same reason.

4.4.2 More detailed outfit weight estimation — merchant ships

When the outfit weight is a significant proportion of the lightship weight, the importance of accuracy in the outfit weight is emphasised and it is best to make a more detailed weight estimate as early as possible in the design process.

As an intermediate step between calculating the whole outfit weight by the use of the square number and a fully detailed calculation, the outfit weight can be divided into a number of groups each of which can be proportioned on different bases appropriate to its content. A possible subdivision might use four groups related respectively to:

(i)   structure,
(ii)   cargo capacity,
(iii) accommodation area or complement,
(iv) deck machinery.

This concept is discussed in more detail in the next section.

4.4.3 Detailed outfit weight calculations — merchant ships

The desirability of dividing the outfit into a limited number of groups as a way to improve approximate weight estimation suggested in the previous section would in principle also be helpful in specification writing and in cost estimating although unfortunately the ideal grouping for each of these purposes differs.

For specification writing the best grouping would be one that brought together everything that is made by one shipyard department or one subcontractor.

For cost estimating it is desirable that the grouping should bring together items whose cost per unit weight is similar whilst for both weight and cost it is essential that there should be a parameter which provides a good measure for the group as a whole and is reasonably easy to calculate.

In the past, shipyard trades were prominent amongst the outfit categories with such items as smithwork, carpenterwork, sheet metal work, joinerwork, plumberwork and electrical work. The first three of these have largely disappeared with changes in ship design, but the other three remain.

For a modern rationale of outfit a division into the following four groups seems sensible, although there are in each of the groupings some items which might equally well be included in another group. The numbering is arranged so as to leave group 1 for structure to conform to the cost estimate sheet in Chapter 18.

Group 2. Structure related
- structural castings or fabrications (sternframe, rudder, etc.)
- small castings or fabrications (bollards, fairleads)
- steel hatch covers
- W.T doors

This group is related primarily to overall ship size with length B.P., displacement or steel-weight being suitable parameters.

Group 3. Cargo space related
- cargo insulation and refrigeration machinery
- cargo ventilation
- firefighting
- paint
- cargo fittings, sparring, ceiling eyeplates, etc.
- 3(a) plumberwork

This group is related to ship type but with this qualification cargo capacity may be a suitable parameter.

Plumberwork was added to this group with some misgivings as its weight and cost depend also on the accommodation, and it may be better to deal with it as a separate item.

Group 4. Accommodation related
- joinerwork
- upholstery
- deck coverings
- sidelights and windows
- galley gear
- lifts
- HVAC
- LSA(lifeboats, davits, etc.)
- nautical instruments
- stores and sundries
- 4(a) electrical work

A suitable parameter for this group is the complement of crew and passengers, or the deck area or volume of the accommodation spaces. Electrical work is tentatively attached to this group but it may be better to treat this as a separate item as suggested for plumberwork, as its weight and cost may be considerably affected by the amount of deck machinery.

Group 5. Deck machinery
- steering gear
- bow and stern thrusters
- stabilisers
- anchoring and mooring machinery
- anchors, cables and mooring ropes
- cargo winches, derricks and rigging
- cranes

For both weight and cost this group should be estimated by a summation of the individual items, but as a cross check there should be reasonable correlation with the same group for a similar type of ship.

Whether the suggested groupings are used (or some other variant), practising naval architects should make a point of equipping themselves with a notebook or a computer file in which to record weights. Such records should include both weights of individual items and rates per square metre for such items as deck coverings, joiner bulkheading, etc.

Individual weights should be accompanied by sufficient information to enable them to be used intelligently in estimation. Great care should be taken to keep this record up-to-date, as new materials and products are constantly coming into use. Chapter 18 recommends that costs should be kept in the same record.

In the 1976 paper, the author noted changes in outfit weight of approximately similar ships built from those built in 1962 and similar changes have continued in more recent years. The reasons for these changes seem worth noting as a guide to future trends:

Factors leading to increases in outfit weight noted in 1976:
- higher standards of crew accommodation — all ships
- fitting of air-conditioning and sewage systems — most ships
- fitting of more sophisticated cargo gear — cargo ships
- fitting of stabilisers, bow thrusters — passenger ships
- patent steel hatch covers now in outfit — cargo ships, container ships, bulk carriers

Noted since 1976:
- fitting of self-tensioning mooring winches — most ships
- new IMO rules, MARPOL, etc.

Factors leading to reductions of weight noted in 1976:
- reductions in weights of most deck machinery for the same duty — all ships
- reductions in weight of deck coverings — cargo ships
- elimination of wood decking, ceiling and sparring

Noted since 1976:
-   reductions in crew numbers and corresponding reductions in accommodation area — most ships

4.4.4 Detailed outfit weight calculations — warships.

It has already been noted that outfit is a merchant ship concept and that its use for warships must be subject to some definitions.

On warships both weight and cost recording and estimating systems make no demarcation between the hull and the machinery spaces, with everything being divided into eight groups irrespective of location. This is undoubtedly a very sensible procedure for warships where machinery, accommodation and weapon systems are very much intermingled.

These groups are shown in Fig. 4.14, which has been abstracted from David Andrew's 1993 R.I.N.A. paper "Preliminary warship design" and are used in the warship preliminary design sheet (Fig. 4.20).

For the reasons advanced in §4.4.1, the warship weight groups have been divided in this book into an approximation to the three merchant ship categories. This involves taking Group 1 as structural, Groups 2 and 3 as machinery and Groups 4, 5, 6 and 7 as outfit, with Group 8 being the warship equivalent of the merchant ship deadweight.

It must be admitted that the equivalence is by no means accurate as Group 3 certainly includes items outside the machinery spaces whilst Groups 4 and 5 include machinery items as will be clear from the subdivision of these groups given in Fig. 4.14. This subdivision stops at a limited number of main headings but warship designers use a much more detailed standardised format.

The reasons for the changes in outfit weights of approximately similar ships which have occurred in recent years are worth noting.

Factors leading to increases in weight:
-   the increased weapon carrying ability now demanded; more bangs for fewer bucks!
-   emphasis on zoning and survivability;

Factors leading to reductions of weight:
-   significant reductions in crew numbers;
-   reductions in accommodation joinerwork (this change has been made primarily to reduce the fire hazard in these ships);
-   reductions in the weight of weapon control systems and data highways using modern computer techniques;
-   the much greater attention now paid to detail in the design of many relatively unimportant items of outfit, which were unreasonably heavy in the past for ships where weight was of such importance.

4.5 MACHINERY WEIGHT

4.5.1 Machinery type

The first step towards assessing the machinery weight is, of course, the calculation of the required power and methods of power estimation appropriate to the design stage are given in Chapters 6 and 7.

The second step involves taking a decision on the type of machinery best suited to the service conditions of the ship under consideration. This subject is dealt with in Chapter 9, but it may be helpful to give a simplified statement here:
- the almost universal choice for the machinery of most medium to large cargo ships is a slow speed diesel engine;
- medium speed geared diesels are the general choice for smaller cargo ships, ferries, tugs and supply boats;
- large cruise liners are frequently fitted with diesel electric installations as are many specialist vessels such fishery research and oceanographic vessels.

Gas turbines and/or high speed diesels are the choice for warships where the need for a high power/weight ratio is all important . An unusual feature of warship machinery is the fact that it usually has to provide both a high speed sprint capability and a reasonable endurance at a slow to medium speed. The machinery provided for these two roles may be arranged so that the two component parts always operate separately (the "or" configuration) or combine together (the "and" configuration) for the high speed role.

Obviously the aggregate power for both configurations must be used as the basis for estimating the machinery weight.

As the weights per unit of power vary considerably between the extremes of slow speed diesels and gas turbines, a decision on machinery type is a necessary preliminary to the assessment of the machinery weight.

As with outfit weights, accurate machinery weights are best obtained by synthesis from a number of group weights and a suggested system for this is given later in this section.

4.5.2 Approximate machinery weight estimation

The simplest possible way of estimating the machinery weight is by the use of a graph of total machinery weight plotted against total main engine power (MCR), with a line for each of the four different main machinery types described in the last section and preferably with the data spots on it identified with the ship types to which they refer. This last suggestion stems from a recognition that the different auxiliaries required by different ship types can exercise a considerable influence on the total weight.

4.5.2 Two or three group methods

A slightly more sophisticated treatment would divide the machinery weight into two components: propulsion machinery and remainder. In the 1976 paper where this two-group method was suggested, the propulsion machinery was limited to the dry weight of the main engine which can be obtained from manufacturers' catalogues with everything else being taken with the remainder.

Largely because of the availability of data in this format, this demarcation is followed again in this book, although the author can now see advantages in the alternative three-group demarcation. In this demarcation the propulsion group is enlarged to include main engine lubricating oil and cooling water, any gearing, the shafting, bearings and glands and propeller(s) as well as the dry main engine weight. The second group would then consist of generators, boilers and heat exchangers, all pumps, valves and piping, compressors and other auxiliaries. The third group would consist of items such as ladders, gratings, uptakes and vents, the funnel, sundry tanks, etc. As these items are generally a good deal cheaper per tonne than machinery items, keeping these separate helps cost estimating. As the total weight of these items will generally be more dependent on the propulsion machinery type and power and the general size of the engine room than on the auxiliaries fitted segregating them into a separate group helps to improve weight estimation.

The original two-group treatment has the advantage that the weight of the main engine can usually be obtained from a catalogue and this significant portion of the weight can therefore be presumed to be correct limiting any error in the machinery weight estimation to that occurring in the estimation of the remainder, the treatment of which as a single entity has the merit of simplicity.

The two-component demarcation is, however, unsuitable for recording the weights of diesel–electric machinery installations in which an aggregate of generators provides both the propulsion power and the electricity supply for other purposes. Weights for diesel–electric machinery seem best kept as a single unit and plotted against the total power which can be generated with all engines on full load.

4.5.4 Propulsion machinery weight

If catalogues giving dry machinery weights are not readily available approximate values for slow and medium speed diesels can be obtained from Fig. 4.15 which is a modified version of a plot from the 1976 paper. The base parameter used in this plot is the maximum torque rating of the engines as represented by MCR/RPM and in 1976 it was commented that most of the current engines conformed remarkably closely to a mean line represented by the formula:

$$W_d = 12 \left(\frac{MCR}{RPM} \right)^{0.84} \tag{4.9}$$

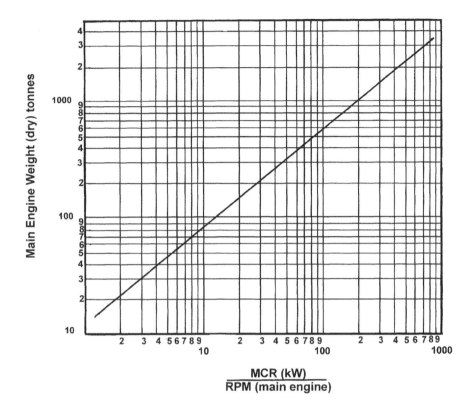

Fig. 4.15. Main engine weight – slow and medium speed diesels (dry).

where MCR = maximum continuous power in kilowatts and in this case RPM is engine RPM and *not* propeller RPM.

The weight given by this equation is about 5% higher than that represented by a line through the data spots to allow for the fact that the graph really ought to be a stepped line corresponding to cylinder numbers with approximately 10% weight steps for the addition of each cylinder.

The constant in the formula now quoted has been modified to allow for the power being in kilowatts and for a slight change in the line to accord with 1992 data but the index remains unchanged.

Apart from its use when catalogues are not immediately handy it may be useful when it is thought wise to use a "mean" weight figure in advance of taking a decision on the make of diesel which will be fitted.

An alternative approach to dry machinery weights is provided by the use of average weights in tonnes per kilowatt, values for each of the main types of engine being as follows:

Slow speed diesels:	0.035–0.045, most usual value 0.037 tonnes/kW or 22 to 28 kW/tonne
Medium speed diesels:	0.010–0.020, most usual value 0.013 tonnes/kW or 50–100 kW/tonne; vee engines tend to be lighter and in-line engines tend to be heavier
High speed diesels:	0.003–0.004 tonnes/kW or 250–330 kW/tonne
Gas turbines:	0.001 tonnes/kW or 1000 kW/tonne

For some reason the reciprocal kW/tonne figures seem to be easier to remember.

4.5.5 Weight of the remainder

In the 1976 paper two alternative parameters were considered as possible bases for plotting the remainder of the machinery weight. These were:
(1)   The Maximum Continuous rating of the main engine(s); and once again
(2)   The engine torque as represented by the quotient MCR/RPM.

The argument for the use of the first of these parameters lies in the fact that the shafting and propellers and many of the auxiliaries, exhaust gas boilers, uptakes are related to MCR of the propulsion machinery.

The argument for the second parameter lies in the fact that the use of torque as a base reduces the parameter of a medium speed engine and still more that of a high speed engine when compared with that of a slow speed engine of the same power in a way that may correspond approximately to the reduced weight of auxiliaries that can be expected in such installations and the smaller size of engine room required with correspondingly reduced weight of piping, floorplates, ladders and gratings, vent trunks, etc.

The best "fit" with the data available was, however, obtained when MCR was used as the base. Figure 4.16 is a revision of the figure presented in the 1976 paper with the MCR altered to kilowatts and with account taken both of the altered demarcation now suggested and of some additional data.

Expressed as a formula:

$$W_r = K \cdot \text{MCR}^{0.70} \tag{4.10}$$

where MCR is in kW.

The constants noted below have been updated to allow for the power now being in kilowatts and weights to 1992 practice.

K = 0.69 for bulk carriers and general cargo ships
= 0.72 for tankers
= 0.83 for passenger ships
= 0.19 for frigates and corvettes

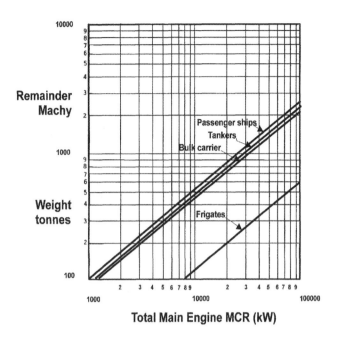

Total Main Engine MCR (kW)

Fig. 4.16. Weight of "remainder" of machinery weight versuss main engine MCR (kW).

4.5.6 Weight of diesel–electric installations — merchant ships

The plot of the total machinery weights of a number of diesel–electric installations given as Fig. 4.17 showed that these could be represented fairly well by the equation:

$$W_{mt} = 0.72 \, (\text{MCR})^{0.78} \tag{4.11}$$

where W_{mt} = total machinery weight and MCR = aggregate MCR in kilowatts of all generator machinery.

4.5.7 Approximate machinery weight — warships

Although the ideas presented in the previous paragraphs of this section were originally developed for merchant ships, the same procedure can be used for warships with the weight of the propulsion machinery and the remainder being estimated separately.

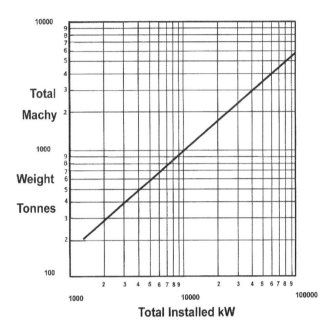

Fig. 4.17. Total weight of diesel electric machinery.

If the there is to be more than one type of propulsion machinery —e.g., gas turbines and diesels — then the weight of these components should be separately assessed.

Catalogue data should be used if available, but if not the weight per kilowatt for high speed diesels and gas turbines given in §4.5.4 can be used.

The weight of the remainder can be estimated using eq. (4.10) with the appropriate K value. It will be noted that the K value for warships is much smaller than that for any merchant ship type, although a warship has many more machinery items and more complicated systems. This indicates the great care that is taken to minimise weight on these fast fine lined ships.

If the machinery is an "and" configuration the power used for estimating the weight of the remainder should be the total installed power machinery, but for an "or" configuration the power used should be that for high speed operation.

4.5.8 Detailed machinery weight calculations — merchant ships

For detailed machinery weight calculations it is essential to have a standard list of items so that calculations are always made in the same order and the danger is reduced of something being omitted.

A suggested list for merchant ships is divided into three groups which are numbered 6,7 and 8 to follow the sequence of the outfit groups. The items in the groups are generally arranged on a functional basis, a secondary motive being a desire to keep items with approximately the same cost per tonne together to facilitate the use of these groups in approximate cost estimates as described in Chapter 18.

Group 6. Propulsion machinery
- 6.1 main engine(s)
- 6.2 main engine lubricating oil and water
- 6.3 main engine control systems
- 6.4 gearing
- 6.5 shafting and bearings, etc.
- 6.6 propeller(s)

Group 7. Auxiliary machinery
- 7.1 generators
- 7.2 compressors
- 7.3 boilers
- 7.4 heat exchangers
- 7.5 purifiers
- 7.6 pumps
- 7.7 pipework
- 7.8 lubricating oil and water in auxiliary machinery and systems
- 7.9 cranes, workshop plant, spare gear

Group 8. Structure related
- 8.1 floorplates,ladders and gratings
- 8.2 engineers tanks
- 8.3 uptakes
- 8.4 vents
- 8.5 funnel

4.5.9 Detailed machinery weight calculations — warships

As already noted, warship designers do not recognise a split in outfit and machinery and instead use the weight groups shown in Fig. 4.14.

Following a similar procedure to that suggested under outfit, it is convenient to take the sum of Group 2, propulsion, and Group 3, electrical, as being approximately equivalent to the merchant ship machinery grouping.

4.6 MARGIN, DEADWEIGHT AND DISPLACEMENT

4.6.1 General

The final item required to make up the lightship is the margin. The purpose of having a margin is to ensure the attainment of the specified deadweight even if there has been an underestimate of the lightweight or an overestimate of the load displacement. The size of the margin should reflect both the likelihood of this happening and the severity of the penalties which may be exacted for non-compliance. When the design is well detailed and clearly specified and the light-weight has been calculated by detailed methods, the margin should in principle be reduced. When the ship type is novel or the design and/or the specification are lacking in precision, larger margins are appropriate.

4.6.2 Margin — merchant ships

Where the lightweight forms a high proportion of the load displacement and the deadweight a correspondingly low proportion, the percentage loss of deadweight which would result from an error in the lightweight estimation can be very serious, and a prudent designer will therefore wish to provide a higher than normal margin. However, it is in just this sort of ship that there is likely to be the greatest pressure to limit the displacement to minimise the power required.

In the 1976 paper, the figure recommended for the margin for merchant ships was 2% of the lightweight. Subject to the qualifications made above this still seems as good advice as can be given.

An alternative to a single percentage weight margin would be the aggregation of a margin based on different percentages of the various weight items depending on the accuracy with which each of these weights is known.

4.6.3 Margin — warships

As well as a margin of the type described above, whose purpose is essentially to take care of errors in the weight estimation, two other types of margin are applied to warship weight estimates to British Ministry of Defence (MOD) rules. These are:

1. A board margin — this is an allowance for additional weight due to changes in the design which may be made by Naval staff during the course of construction; and
2. A growth margin — this is an allowance for the increase in weight due to additions and alterations which may be made during the life of the vessel and to the "natural" weight growth due to the accretion of paint, etc.

A possible figure for the board margin might be 2%, whilst the growth margin might be based on 1/2% per annum of the intended life, in both cases based on the lightweight.

4.6.4 Margins on centres of gravity

Although the estimation of the lightship VCG and LCG will not be dealt with until a later section, it is convenient to deal here with the margins that it is wise to have on these figures.

A margin on the VCG is sometimes established by adding the weight margin at a high centre of gravity but this does not really take account of the main reason for having a margin on the VCG which is to offset a possible underestimate of the centres of gravity of some of the weights that make up the lightship rather than to deal with a consequence of underestimating the lightweight itself.

Whilst the use of an excessively large margin may cause design problems and possibly increase the cost of the ship in question (if for instance it results in the beam being increased unnecessarily) it is vitally important that the margin is big enough to counteract errors in the VCG calculations which generally seem to result in an underestimate — or is it only when this is the case that they are noticed? The margin should scale to some extent but less than directly with ship size and should be bigger in the early stages of design when the calculations are approximate and decrease as these become more accurate. To meet the first of these criteria, it is suggested that the margin should be a based on the square root of the moulded depth D starting say at $0.1(D)^{0.5}$ and decreasing to $0.06\,(D)^{0.5}$ as confidence grows in the calculations.

It is unusual to apply a margin to the calculated LCG position but, on the basis that a small increase in an estimated trim by the stern will usually be acceptable, whereas a change from an estimated level keel to a trim by the head will not be, it may be wise to base preliminary trim calculations on an LCG a little further forward (say $0.5\%\ L$) than the calculated figure.

4.6.5 Deadweight and displacement — merchant ships

If a total deadweight is stipulated the required full displacement is the sum of this and the lightweight.

From many points of view, it is better for an owner to specify the required cargo deadweight and put the onus on the designer to allow for all the non-cargo deadweight items needed to perform the specified service.

The non-cargo deadweight items consist of the fuel for both main engines and generators, fresh water for all purposes, engineers sundry tanks, stores of all sorts, crew and effects, passengers and baggage, water in swimming pools, etc.

The items which commonly make up the total deadweight are:
- cargo deadweight
- passengers and baggage
- crew and effects
- stores of all sorts

- fuel for main engines
- diesel oil for generators
- sundry engineers tanks
- fresh and feed water
- water in swimming pools
- total deadweight

The deadweight used to determine the load displacement must be the maximum one that may occur at any point in a service voyage. This will generally occur at the point of departure from the main fuelling port, but special features of a particular service in which the quantity of cargo carried differs at different stages of the voyage may need to be considered. For example, the maximum deadweight of a fishing vessel will occur at sea when the catch is complete although at this time the fuel, water and stores can be assumed to be at a reduced level just sufficient to meet the requirements of the return voyage to port. The same sort of philosophy applies to tankers bringing cargoes from offshore oil platforms.

4.6.6 Variable loads — warships

Variable loads, which are shown as weight Group 8 in Fig 4.14, are the warship equivalent of the merchant ship deadweight.

4.7 STANDARD CALCULATION SHEETS FOR INITIAL DESIGN

All design calculations are most conveniently carried out on a standard calculation sheet although nowadays this may be a prompt on a computer screen instead of a printed form. One great virtue of a standard format is its ability to ensure that all significant items are remembered in the hurry in which designs always seem to be prepared.

4.7.1 Standard design sheet for merchant ships

The standard sheet for the design of merchant ships given in the author's 1976 R.I.N.A. paper is presented as Fig. 4.18. This sheet and its predecessors has been used for many designs and its use can be recommended with confidence. It has to admitted, however, that it is now somewhat out-of-date and an alternative is presented as Fig. 4.19.

Opposite: Fig. 4.18. Merchant ship preliminary design sheet (as presented in Watson & Gilfillan 1976 R.I.N.A. paper).

DIMENSIONS	metres	STEEL					OUTFIT WEIGHT	tonnes
Length OA		$L(B + T)$					Structural Castings	
L Length BP		$0.85L(D - T)$					Small Castings	
B Beam		$f = 0.85$	l	h	f		Smithwork	
D Depth to		for Super-					Sheet Iron	
Depth to		structures					Carpenter	
T Draft (Scantling)		$f = 0.75$					Plumberwork	
Draft (Design)		for Deck					Electrical	
WEIGHTS	**tonnes**	Houses					Paint	
Invoiced Steel		STEEL NUMERAL					Joinerwork	
Scrap (%)	−	$0.8D - T$					Upholstery	
Net Steel		$(1 - C_B)/3T$					Decorator	
Electrodes	+	δC_B					Deck Coverings	
Steel for Lightship		C_B @ $T =$					Casing Insulation	
Outfit		C_B' @ $0.8D$					Sidelights	
Machinery		STEEL WEIGHT			tonnes		W.T.F.R. Doors	
Margin		From Graph					Firefighting	
Lightship		$1 + 0.5(C_B' - 0.7)$					Galley Gear	
DEADWEIGHT		Steel at C_B'					Refrig. Machinery	
Displacement		Corrections					Cargo/Stores Insul.	
Appendages	−						Ventilation A/C	
Displacement (Mld)							Steering Gear	
DRAFT							Anchors, Cables	
Block Coefficient		TOTAL STEEL WEIGHT					Mooring Machinery	
POWERING		**RESISTANCE**					Cargo Winches	
Trial Speed		$©_{122}$ Basis					Cargo Gear	
Service Speed		B' : T' for $L = 122$					Rigging	
$V/\sqrt{L}$: F_n		$B'/17$: $T'/8$					Canvas	
$K = C_B + 0.5V/\sqrt{L}$		Mumford Indices					Hatchcovers	
© or C_t		$©_{122}$ Corrected					L.S.A.	
$\Delta^{2/3}$		$\delta© = 4(122 - L)10^{-4}$					Nautical Inst.	
V^3		$© = ©_{122} + \delta©$					Stores & Sundries	
P_E		APPENDAGE RESIST.			%		Special Items	
$(1 + a/100)$		Bossing					TOTAL OUTFIT WEIGHT	
$(1 + x)_F/(1 + x)_{ITTC}$		Thruster					DEADWEIGHT	tonnes
$QPC = \quad - (N\sqrt{L})/10^4$		Stabiliser					Oil Fuel	
η_t		Twin Rudder, etc					Diesel Oil	
P_s		TOTAL APP. RESIST. (a)					Fresh Water	
Margin		$A_c = \dfrac{427.1 \times QPC \times \eta_t}{©(1 + x)(1 + a/100)}$					Engineers Tanks	
P_s (Trial)							Stores	
Service Margin		MACHINERY WEIGHT			tonnes		Crew & Effects	
P_s (Service)		Main Engine					Passengers	
Derating		Gearing					Swimming Pools	
M.C.R.		Boiler & Condenser					Cargo	
Main Engine		Shafting & Propeller					TOTAL DEADWEIGHT	
N R.P.M.		Generators					S.W. Ballast	
Fuel/Day		Auxiliaries					CAPACITY	metres³
Range		Piping, Ladders, Gratings					Gross Volume	
Miles/Day		Funnel Uptakes					Deduction	
Days at Sea		Remainder					Net Volume	
Days in Port		TOTAL MACHY WEIGHT					Cargo Cubic ( )	

Some comments first of all on Fig. 4.18. The powering on this sheet uses Ⓒ either to Froude or to ITTC with a corresponding need to invoke an appropriate $(1 + X)$ value. If the three trial ships method mentioned in §3.1.1 is used, each trial ship can be designed on a page of this type, or alternatively a revised version of the sheet can be drawn up with three or four columns. Whilst the sheet was originally designed for use in preliminary design, it can be used to record the main particulars of a design as these change throughout the design process and can also be used to store "as fitted" data on completed ships in a form particularly useful for design work.

The updated version presented as Fig. 4.19a–d is a computer spreadsheet which greatly increases the speed and improves the accuracy with which all the calculations can be made. There is, of course, automatic addition of each of the columns of weights together with transfer of totals to the lightship summation giving a progressive updating as the design proceeds. All the formulae used are built in to the program — some of these formulae appear on the spreadsheet, but others which would have taken up too much space are summarised below.

Address	Formula
c23	eq. (3.5)
c24	eq. (3.9)
c26	eq. (3.14)
m28	eq. (4.2)
x30	eq. (6.38)

Much of the data required when using the spreadsheet are given in tables or graphs in the book generally under the following references.

Address	Data
c12	Fig. 3.3
c16	Fig. 3.5
c18,19,20	Figs. 3.7, 3.8, 3.9, 3.10
c22	§3.5
f34	§3.5.2
j20	§4.3.2
j25	§4.6
m7	§4.2
m23	Fig. 4.1 and Table 4.1
m37	Figs. 4.12 and 4.13 and §4.4
m40	Figs. 4.15 and 4.16 and §4.5
m41	Fig. 4.17
q4	§4.3
q8	Fig. 4.8
q26	Fig. 4.2
q29	§4.5
t4	§4.4

A	B	C	E	F
	Data, "first" design		Consumptions, deadweight appendage disp	
7	date		consumptions etc.	
8	design no			
9	ship type		range	
10	service speed — Vk		miles per day	
11	deadweight		days at sea	
12	dwt/disp ratio — Kd		days in port	
13	disp — Δ		oil fuel/day at sea	
14	cargo capacity — Vr		oil fuel/day in port	
15	cap. ab upper dk — Vu		diesel/day at sea	
16	cap. ratio = (Vr – Vu)/Vh		diesel/day in port	
17	mld vol —Vh		fresh water/day	
18	L/B			
19	B/D			
20	T/D		oil fuel	
21	r (fb = 1.025)		diesel oil	
22	(1 + S)		fresh water	
23	L based Δ		engineers tanks	
24	or L based Vh		stores	
25	Fn = 0.164 Vk/$\sqrt{L}$		crew and effect	
26	Cb		spassengers etc.	
27	d(Cb) = (1 - Cb)(.8D – T)/3T		swimming pools	
28	Cbd		cargo	
29	iterate b23..b28		water ballast	
30	L			
31	B		total deadweight	
32	T			
33	D			
34			shell	
35			stern	
36			propeller(s)	
37	All formulae shown are built		rudder(s)	
38	into the adjoining column.		bossing/brackets	
39	Other formulae from book are		stabilisers	
40	also built in — see Chapter 4.		thrusters (lost buoy.)	
41	Much of the data required is			
42	given in the book.		total appendage dispt.	

Fig. 4.19(a). Spreadsheet for merchant ship design.

H	I	J	L	M
	"Second" design dimensions, lightship, load dispt		approximate weights structure, outfit, machinery	
7 8	dimensions	metres	structural numeral	
9	length O.A.		$L(B + Ts)$	
10	length B.P. — L		$0.85 L (D - Ts)$	
11	beam — B		f | | | h	
12	depth to dk — D			
13	depth to dk			
14	draft scantling — Ts			
15	draft design — Td			
16				
17 18	weights	tonnes		
19	net struct weight			
20	electrodes			
21			steel numeral — E	
22	total struct weight			
23	outfit weight		const — K	
24	machinery weight		$Wsi = K*E^{1.36}$	
25	margin		for Cbd = 0.70; all M.S.	
26				
27	lightship		Cb at Td	
28	deadweight		d(Cb)	
29			Cbd at 0.8D	
30	full displt			
31	deduct appendages		Ws corrected for Cb	
32			other corrections	
33	mld displt			
34			Ws (all mild steel)	
35	block coefficient			
36			$(L \times B)$	
37			Ko	
38			approx. outfit weight	
39				
40			main engine (MCR =)	
41			remainder	
42			approx. machinery weight	

Fig. 4.19(b). Spreadsheet for merchant ship design (continued). Revises first design prepared in Fig. 4.19(a) with rounded-off dimensions and more accurate weights.

O	P			Q	S	T
	Detailed structure weight				detailed outfit weight	
7	weight/metre MSS				sternframe, rudder etc.	
8	integration factor				bollards, fairleads etc.	
9					steel hatch covers	
10	hull structure (basic)				W.T. doors	
11	corrections (+/−)				structure related	
12	side to side erections				cargo, stores insulation	
13	houses				cargo, stores refrig	
14	fittings above upper deck				cargo ventilation	
15	shell fittings				cargo space fittings	
16	structure below upper deck				paint	
17	machinery space structure				firefighting	
18					cargo related	
19	Ws (all materials)				plumberwork	
20					electrical work	
21	divide to proposed materials				joinerwork bulkheading	
22		MS	HT	Alum	furniture, upholstery	
23	net weight				deck cover, casing ins	
24	roll marg				galley gear	
25	inc R.M.				sidelts, windows, F.R. dr	
26	scrap				lifts	
27	invoiced				HVAC	
28					LSA	
29	main engine(s)				nautical inst	
30	gearing				stores and sundry	
31	propeller(s), shafting				accommodation related	
32	(or) propulsion				steering gear	
33	generators				thrusters	
34	other auxiliaries				anchor and moor machinery	
35	piping in engine room				anchors, cables, warps	
36	fluids in systems				winches, derricks	
37	ladders, gratings				cranes	
38	funnel(s), uptakes, vent				deck machinery related	
39	other sundries					
40	(or) remainder				total outfit weight	
41						
42	machinery weight					

Fig. 4.19(c). Spreadsheet for merchant ship design (continued). Revised weights.

V	W	X	Z	AA
	Resistance		Powering	
7	trial speed Vt		Ct inclusive	
8	service speed Vk		$Pe = 0.0697 \cdot Ct \cdot S \cdot V^3 (kW)$	
9	Froude $Fn = 0.164\ Vk/\sqrt{L}$		prop diameter	
10	$Rn = 2.636 \times 10^6 \cdot Fn \cdot L^{1.5}$		service RPM	
11	L		$QPC = 0.84 - (N\sqrt{L})/10000$	
12	B		$Ps = Pe/QPC\ (kW)$	
13	T		trial margin	
14	Cb		Ps (trial) (kW)	
15	Δ		service margin	
16	C		Ps (service) (kW)	
17	$S = C/\Delta L$		derating	
18	$Ⓢ = 1.0166\ S/\Delta^{2/3}$		MCR (kW)	
19	Lb (basis or model)			
20	Bb		C app as % Ct	
21	Tb		rudder 1.5%	
22	Rnb		twin rudders 2.8%	
23	(1 + K)		shaft bkts and shafts 6%	
24	Ctb		stabiliser fins 2.8%	
25	$Cfd = 0.075/\sqrt{(\log Rn - 2)^2}$		bilge keels 1.4%	
26	Cfd (use Rnb)		powering assumes SW (r = 1.025)	
27	$Ctd = ctb + (1 + K)(Cfd - Cfb)$		fb - fall back value	
28	$B' = B(Lb/L)$		machinery type chosen	
29	$T' = T(Lb/L)$			
30	Mumford corrections B & T			
31	Ct corrected L,B,T		approx capacity check	
32	$\Delta C\ (fb = 0.10 \times 10^{-3})$			
33	C app		depth D	
34	C air		sheer camber	
35	Ct inclusive		D' corr S & C	
36			gross vol $L \cdot B \cdot D' \cdot Cbd$	
37			deductions	
38			additions above upper deck	
39			gross cargo cubic	
40			structure deduction	
41				
42			cargo cubic	

Fig. 4.19(d). Spreadsheet for merchant ship design (continued). Powering and approximate capacity calculations.

x16	Fig. 6.1
x23	Figs. 6.3 and 7.4
x24	Figs. 7.5–9 and Table 7.4
x30	§6.7 and Table 6.1
x33,34,35	§6.9

In general, the sheet remains closely similar to the earlier version, but there are notable differences which are worth some comment.

The column marked "first" design enables the designer to use either the deadweight/displacement ratio or the cargo capacity/hull volume ratio to make an initial assessment of hull dimensions using methods described in Chapter 3. This involves making an initial guess of either the length or the block coefficient and provision is made for a short iterative process until the calculated value of L ceases to differ significantly from the input value.

In the third column the dimensions arrived at in the first column can be rounded off before they are used in a second design iteration which follows the procedure used in the earlier sheet.

The fourth column enables approximate calculations of the structural, outfit and machinery weights to be made.

More detailed calculations of these weights can be made at a later stage using the fifth and sixth columns. The layout of the outfit and machinery weight calculations in these columns has been revised along the lines, and for the reasons outlined in §4.4 approximations are included as well as the detailed calculations.

The powering calculations in the seventh and eighth columns has been modified to permit the use of Cr'78 and the powers are all now in kilowatts. C_{tm} values can be obtained from Chapter 7, where the author's reasons for preferring this presentation is explained. C_f values used can be either ITTC'57 or Grigson values and $(1 + K)$ values can be deduced values or figures derived using the Holtrop and Mennen formula. (The format used for the spreadsheet was conditioned by the constraint imposed to suit publication in this book.)

This spreadsheet has gone through several versions during its development but has not had the lengthy practical use that its predecessors had. Users will probably prefer to make some further changes, but should nevertheless find it a good starting point.

4.7.2 Standard design sheet for warships

A standard calculation sheet for the design of warships is given as Fig. 4.20. This sheet which could readily be converted into a spreadsheet follows much the same lines as the earlier merchant ship sheet, but uses warship weight groups. In one respect it goes further than the merchant ship design sheet as it also includes a cost estimating section.

DIMENSIONS		METRES	
Length OA			
Length WL			
Length BP	L		
Breadth at WL	$B.$		
Breadth at Dk	B_2		
Depth to Dk	$D.$		
Depth to Dk	D_2		
Draft Design	T_1		
Draft Scantling	T_2		
WEIGHTS		TONNES	
1			
2			
3			
4			
5			
6			
7			
Lightship (1-7)			
Design Margin			
Board Margin			
Growth Margin			
Lightship with			
Margins			
8 Variable Load			
Displ't Full			
Appendages			
Displ't Mid			
Draft			
Cb LBp & BWL			
Cb LWL & BWL			
Cp			
Cm			
L/B			
B/D			
T/D			
Appendage Displ.			
Shell			
Stern			
Propellers			
Rudders			
Bossing/Bkts			
Bilge Keels			
Stabilisers			
Sonars			
Total			
Lost Bouyancy			
Total Net			

POWERING	FULL SPEED	CRUISE
Speed Trial		
Speed Service		
Months O/D		
Seastate		
$V/\sqrt{L}$		
Fn		
Basis Lb		
Bb		
Tb		
B_1 for Lb/L		
T_1 for Lb/L		
Beam Ratio		
Beam Index		
Draft Ratio		
Draft Index		
© (Lb) or Cт(Lb)		
Corrected B&T Index		
Corrected for L		
Cт		
© = Cт 39.75 Ⓢ		
$\Delta^{2/3}$		
S		
Ⓢ		
V^3		
$P_E = \dfrac{0.748 \, © \, \Delta^{2/3} \, V^3}{431.7}$ (kW)		
1 + a		
1 + X		
QPC		
Ps (0 Margin) kW		
Trial Margin		
Ps Trial kW		
Service Margin		
Ps Service kW		
Derating		
MCR kW		
Engine Type		
Propeller RPM		
Propeller Dia		
Propeller Type		
Appendage (1 - a)		
Bossing		
Rudders		
Stabiliser		
Sonar		
Total 1 + a		

Fig. 4.20 (*above and opposite*). Warship preliminary design sheet.

WEIGHTS	TONNES		TONNES		TONNES		SPACE	M3
GROUP 1		GROUP 3		GROU° 7		1	Aviation	
HULL		ELECTRICAL		ARMAMENT		2	Armament	
						3	Sonar	
						4	Radar/Dir	
						5	Operation	
						6	Communic'ns	
						7	Navigation	
						8	Machinery	
						9	Electrical	
						10	Offices	
						11	Workshops	
						12	Stores	
		GROUP 4		GROU° 8		13	Accommod'n	
		CONTROL &		VARIABLE		14	HVAC	
		COMM.				15	Passages	
				Range		16	Misc	
				Speed		17	Fuel	
				No Days		18	Fresh Water	
				Ps		19	Salt Water	
				Gms kW/Hr		20	Dock	
				EL Load		21	Vehicles	
						22	Cargo	
						23	Voids	
						24		
		GROUP 5		Crew & Effects		25		
		AUX'Y SYST.		Ammunition				
				Aircraft				
				General Stores				
GROUP 2				Naval Stores				
PROPULSION				Spares		TOTAL		
		GROUP 6		Operating Fluids		L		
		OUTFIT		Fresh Water		B at	Dk*	
				Dieso		D to	Dk*	
				Lub Oil		Cb at	Dk*	
				Sanit'y Tanks		Vol to	Dk *	
				Avcat		Superstructure		
				Water Ballast				
				Margin				
				TOTAL		TOTAL		

COST ESTIMATE	1 Hull	2 Propulsion	3 Electrical	4 Control & Comm'ns	5 Aux'y System	6 Outfit & Furnish'g	7 Armament	Services & Misc	TOTALS
Weight									
M Hours/Tonne									
Man Hours									
Labour Rate £/hr									
Materials/Tonne									
Direct Labour									
Overheads									
Materials									
TOTAL									

COST BASE DATE

Contingencies
Total Cost
Profit
Price

4.8 LIGHTSHIP CENTRES OF GRAVITY

4.8.1 General discussion

There can be little doubt that most errors in stability calculations arise from incorrect estimates of the lightship VCG. Much the same could be said about errors in trim with this being generally attributable to errors in the lightship LCG, although fortunately these errors are rarely as serious as errors in the lightship VCG so often are.

Estimates of the lightship centres of gravity have to be made early in the design process and generally have to be based on incomplete plans and specifications; at this stage few decisions have been taken about machinery and equipment suppliers; the only structural plan available is likely to be an outline midship section; a body plan may be available but there is unlikely to be a lines plan at this stage.

Against this back ground, or lack of it, an estimate must be made of the lightship centres of gravity and in particular of the VCG and a number of ways of doing this will be discussed. The results of the stability calculations made at this early stage in the design are used to confirm or amend the preliminary dimensions and arrangement.

In the next stage, the design will be progressed with decisions being taken on many of the major items of equipment and machinery; a lines plan will be drawn and the midship section updated and possibly supplemented by an outline structural profile and decks. Along with these developments progressively more accurate weights and centres of gravity will become available and the design stability and trim calculations can be updated. Provided this process can be completed before a shipbuilding order is placed it is fairly easy, if the stability is found to be inadequate, to take remedial measures such as increasing the beam and/or reducing the depth by a small amount.

Unfortunately shipbuilding orders are sometimes placed on designs that have not been fully developed, or significant changes affecting the centre of gravity may have been made at a late stage in the tender negotiations. In these cases the time allowed for verifying the stability in detail will often be severely limited by production priorities demanding that steel, machinery and outfit ordering and loft work etc be put in hand as quickly as possible. Establishing a need to make changes in dimensions once these processes are under way is highly unpopular to say the least, and it is not unknown for designers to turn a blind eye to revised estimates which give unpalatable answers hoping that weight reductions can be made as the design develops, which needless to say is a recipe for trouble.

See also §4.6. on the subject of margins.

4.8.2 A first approximation of the lightship VCG and LCG

A first estimate of the VCG is generally made by the use of the ratio VCG/D. This is simple but crude. Provided the data from the basis ship has been established by a reliable inclining experiment, the two ships are of the same type and there are no obvious differences which could be expected to alter the ratio, this crude method can give quite a good answer.

If there are obvious differences, corrections can be made. Items fitted to the basis ship which do not appear on the new ship should be deducted at their basis ship centres before the ratio is used with items which only appear on the new ship being added at their centres after the proportioning to the new ship depth.

Although this method is crude, it does have some advantages. The basis VCG is a proven figure from the inclining experiment of an actual ship whereas a calculated VCG is derived from many weight and centre of gravity assumptions in which there may be both errors and omissions particularly in calculations made early in the design process. So, even when detailed calculations have been made it is still no bad idea to compare the answer from these with that given by the approximate method and try to rationalise the difference if any — and if this proves difficult, further detailed checks may be wise!

4.8.3 A second approximation

Intermediate between the method outlined above and calculations based on detailed weights and centres is the Volume–Density method presented in the Watson–Gilfillan 1976 R.I.N.A. paper. In the simplest version of this method the VCG (and LCG) are calculated using the volumes and volumetric centroids of the various spaces which make up the ship. In a more sophisticated version of the method each volume is multiplied be a density value and each volumetric centroid is corrected by a factor to correlate with the centre of gravity position.

If the densities and the centroid/VCG ratios selected for each compartment are 100% correct, the calculation becomes a completely accurate weight calculation. Whilst this is unlikely to be the case, intelligent selection of these factors based on an analysis of ships for which accurate information is available can enable this method to give quite good results. It may be worth noting that the accuracy of the final centre of gravity depends on the relative values of the densities used rather than on their absolute values.

A standard sheet, again readily convertible to a spreadsheet, which can be used for this type of calculation is presented as Fig. 4.21 and should be used in association with the following notes.

- Lines 1 to 8 give the input of dimensions and various hydrostatic particulars.
- Lines 9 to 16 deal with the calculation of the weight and centres of gravity of the superstructure.

Opposite page: Fig. 4.21. Calculation sheet for lightship centres of gravity using volume density
method.

- Column 1 gives the mean height to the deck on which each erection tier is
  built.
- Column 2 gives the tween deck height.
- Column 3 gives the factor relating the VCG of each layer to its centroid.
- Column 4 gives the VCG of each layer above the deck on which it is built.
- Column 5 gives the volume of each erection.
- Column 6 the density and column 7 the weight.
- Column 8 gives the VCG of each erection above the base.
- Column 9 gives the vertical moment.
- Columns 10, 11 and 12 deal with the corresponding LCG and moments.

The table at the bottom left deals with:
- the mean depth corrected for sheer and camber;
- the total hull volume;
- the hull VCB at the mean depth;
- the hull LCB as a percentage of the ship's length.

Below this are noted the R_h, K_h and F_{gh} values used, or analysed.
- Line 17 gives the hull weight, centres and moments.
- Lines 18 to 24 concludes the calculation.

Some values which may be used in the calculations are:
- R_s for accommodation constructed in steel and fitted out to normal cargo ship
  standards appears to have a fairly constant value of about 0.13 tonnes/m^3;
- K_s for accommodation of this type generally has a value of about 0.6;
- R_h varies not only with ship type, but also with ship size and should be
  assessed with care.

Where the hull below No. 1 deck and the superstructure both contain accommodation as on a passenger ship it may be reasonable to make $R_s = R_h = 1.00$ at
lines 1 and 8 and apply a correction factor to the weight obtained at line 10 to give
the corrected hull weight at line 12.

With R_s, K_s and R_h known, the value of K_h for a suitable basis ship can be
determined by analysis starting at both top and bottom of the table. Using a suitable
value of K_h from a good basis ship, the light ship VCG of a new design can be
calculated.

As noted above, this method can also be used to establish the LCG position,
with that of the hull being assumed to be largely determined by LCB position at
mean depth.

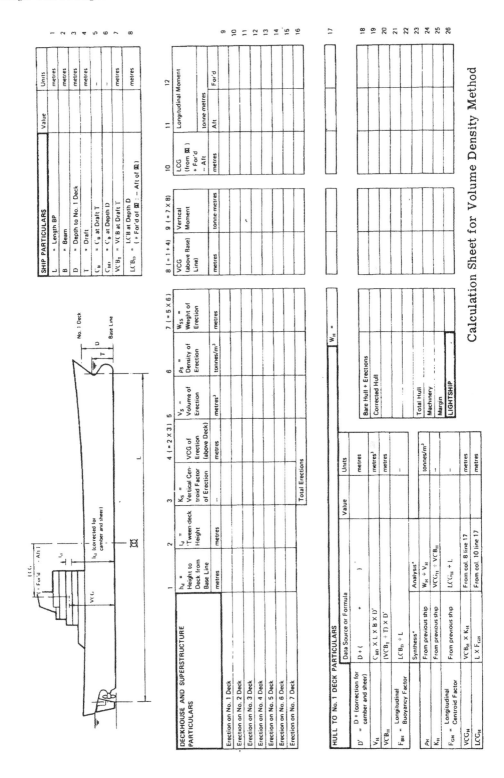

Calculation Sheet for Volume Density Method

4.8.4 Detailed calculations

Finally of course the centres of gravity can be calculated using detailed weights, a procedure which often has to be carried out again and again as the design develops and should certainly be updated when the design nears finality but while it is still possible to make changes to ensure satisfactory stability and trim if the calculations show these to be required. The weight components used in this calculation have been examined under the sections dealing with structural weight, outfit weight, machinery weight and margin. In the calculation each of these weights must be given appropriate centres of gravity, some of which will require calculations whilst others can be lifted from an accurately drawn profile plan, but a good knowledge of ship outfitting is needed to supplement the data shown on the plan if accurate results are to be obtained.

4.9 WEIGHT CONTROL

Although weight control is a remove from preliminary design which has been the theme of this chapter, the chapter has also concentrated on the importance of weight and for this reason it seems appropriate to conclude it with a discussion of weight control.

Weight control is the process whereby the intent of the specification in respect to the deadweight carrying ability of the ship embodied in the design is maintained during the development of the detailed plans, the ordering of outfit and machinery and the building of the ship.

The effectiveness with which weight control can be carried out depends to a major extent on the accuracy of the final weight estimate on which the load displacement and the ship's lines are based.

In the monitoring process thereafter the weight committed by each plan is calculated and, if it is more than that allowed for in the estimate, the question of whether the increase is necessary to meet a specified requirement is raised. Similarly the weight of each bought-in item is obtained before the order for it is confirmed and again if this exceeds the estimate the reason for this is probed.

Changes requested by the owners are evaluated for their weight effects at the same time as their cost and delivery implications are estimated.

Sometimes weight increases, for whatever reason they arisen, have to be accepted but in this case there must be a search for compensatory weight savings.

For some ships it is a specified requirement that every item going on board is weighed, but it has to be admitted that the knowledge thus gained generally comes so late in the construction process that remedial measures are very difficult and expensive. If, however, there is going to be a serious weight increase, then even belated knowledge from weighing is better than the awful truth only coming to light at the inclining experiment when the ship is practically finished and the owner has taken on operational commitments.

Along with the direct loss of carrying ability that weight growth causes, there can be and usually are indirect effects on stability, trim, athwartships moment and even on structural strength which can be every bit as serious or more so. The monitoring of weight should therefore also take note of the effects that weight changes are having on the centres of gravity.

Some remedial measures which can be taken to rectify major contractual failures stemming from weight and/or VCG growth are suggested in Chapter 20.

Chapter 5

Volume, Area and Dimension-Based Designs

5.1 VOLUME-BASED CARGO SHIPS

The design of capacity carriers poses two different types of problem. In the first of these the required capacity is known and the problem is that of finding a solution to the volume equations given in §3.2, whilst in the second the problem lies in establishing the required volume and this is dealt with in §5.2.

Reverting to the first case, the first step is to convert the required cargo capacity from whatever type of measure it is specified in to a corresponding moulded capacity by dividing by the following coefficients:

$V_g/V_m = 0.98$

$V_b/V_m = 0.88$

$V_r/V_m = 0.72$

where
V_g = grain capacity
V_b = bale capacity
V_r = refrigeration capacity
V_m = moulded cargo capacity.

If the cargo spaces include both refrigerated and general cargo spaces, the total moulded capacity is the sum of the moulded capacities of both types of cargo space. Any cargo space which it is intended to provide above the upper deck is then deducted to give the moulded volume of cargo space required below the upper deck.

In the author's 1962 and 1976 papers, graphs were given from which the dimensions of a ship corresponding to a required capacity could be read. The 1962 graph became outdated well before 1976 because of the growth in ship size and the change to metric measurements. The 1976 graph has now also been left behind by a further growth in ship size and indeed this now deters any attempt at a new plot because the scale required to accommodate the largest ships is so small that the information it would supply would be almost useless for smaller vessels.

The procedure now suggested involves the addition to the moulded volume of cargo space of the space required for machinery, oil fuel, fresh water and water ballast tanks, stores, etc. These can either be estimated as individual volumes and summed, or alternatively the total volume below the upper deck can be estimated by the use of a capacity ratio analogous to the deadweight ratio.

The ratio V_m/V_h, where V_h = total volume below the upper deck, can easily be obtained from any suitable basis ship for which the cargo capacity, dimensions and speed are known, using the Froude number to estimate the block coefficient C_b and correcting this to C_{bd} by the method described in §4.2.3. Although there will inevitably be some error in this calculation, it is unlikely to be significant.

With V_h and C_{bd} known, values of L/B, B/D, can be assumed and eq. (3.9) solved for L.

A quick approximation to the length of a bulk carrier can be obtained from the formula:

$$L = 4.5\,(V_h)^{1/3} \tag{5.1}$$

This is derived from eq. (3.9) by assuming values of $L/B = 6.25$, $B/D = 1.88$ and $C_{bd} = 0.80$.

Similar formulae for other ship types can easily be derived from the basic equation using the constants given in Fig. 3.8 for merchant ships and for warships in Tables 1.4 and 1.5.

5.2 ESTIMATING THE REQUIRED VOLUME

The other problem posed by "volume type" ships is the determination of the volume that these must have if they are to fulfil their function. This is a particular problem in the design of passenger ships and was dealt with at some length in the author's 1962 paper. In that paper he felt he had to apologise for giving a list of area figures for all the different spaces found on a passenger ship, most of which were common knowledge and all of which could be easily obtained from a study of ships plans, and said by way of explanation that he had the alternative of either presenting the bare idea of a volume calculation, which might well have been dismissed as impracticable, or of supporting this thesis with data that proved its feasibility and had elected to do this.

The figures presented in 1962 were revised and metricated in 1976 and are repeated below with the warning that it is advisable to check the latest practice by analysis of recent ships. A few figures which have been added giving the areas of cabins, etc. on recent cruise liners emphasise this point.

5.2.1 Areas and volumes of spaces on passenger ships

The numbers in brackets at each heading in this section refer to lines in Fig. 5.1.

(1)–(4) Passenger cabins (excluding bath or toilet) — cruise liners:

Deluxe suites for two persons:	16 m^2
1st class single:	9 m^2
1st class twin:	13 m^2
Tourist twin:	6 m^2
Tourist three:	9 m^2
Tourist four:	12 m^2

(The above figures are as quoted in the author's 1976 R.I.N.A. paper.) An interesting up-date for these figures is given in the 1992 R.I.N.A paper "From Tropicale to Fantasy" by S.M. Payne.

On "Tropicale" introduced in 1981 the cabin areas were:

Deluxe suites including bathroom:	24.7 m^2
Standard cabins including toilet:	14.6 m^2

(twin, some with additional Pullman beds)

On "Holiday" the cabin areas were increased to:

Verandah suites:	42 m^2
Standard cabins:	18 m^2

Overnight accommodation — ferries	
1st class single:	3.6 m^2
1st class twin:	5 m^2
Tourist twin:	4 m^2
Tourist three:	6 m^2
Tourist four:	6.6 m^2
Private bathroom:	3.8 m^2
Private toilet:	2.8 m^2

(5) Passages, foyers, entrances and stairs
About 45% of sum of items 1–4.

			No. of Units	Unit Area (m^3)	Gross Area (m)	Height (m)	Volume (m^3)		Berths
1	Passengers cabins	Type 1							
2	and private toilets	Type 2							
3		Type 3							
4		Type 4							
5	Passages, foyers, entrances, stairs	$\equiv 45\%\ \Sigma(1\ \text{to}\ 4)$							
6	Public lavatories, pantries, lockers								
7	Dining saloons								
8	Lounges, bars								
9	Shops, bureaux, cinema, gymnasium								
					Total of 1 to 9				
10	Captain's and officers'	Type 1							
11	cabins and private toilets	Type 2							
12		Type 3							
13	Offices								
14	Passages, stairs	$\equiv 40\%\ \Sigma(10\ \text{to}\ 13)$							
15	Public lavatories, change rooms								
16	Saloon, lounge								
					Total of 10 to 16				
17	P.Os' and crew's cabins	Type 1							
18		Type 2							
19	Passages, stairs	$\equiv 35\%\ \Sigma(17\ \text{to}\ 18)$							
20	Lavatories, change rooms								
21	Messes, recreation room								
					Total of 17 to 21				
22	Wheelhouse, chartroom, radio room								
23	Hospital								
24	Gallery								
25	Laundry								
					Total of 22 to 25				
26	Fan rooms	$\equiv 2\frac{1}{2}\%\ \Sigma(1\ \text{to}\ 26)$							
27	Lining and flare	$\equiv 3\frac{1}{4}\%\ \Sigma(1\ \text{to}\ 26)$							
					Total of 26 to 27				
28	General cargo (bale)	metres^3		(m^3) $\div 0.88$					
29	Refrigerated cargo	metres3		(m^3) $\div 0.72$					
30	Mails, baggage and passages								
					Total of 28 to 30				

Fig. 5.1. Calculation sheet for design by volume (*continued opposite*).

		No. of Units	Unit Area (m³)	Gross Area (m)	Height (m)	Volume (m³)		Berths
31	Oil fuel tonnes @ SG		(	t ÷	) ÷ 0.98			
32	Diesel oil tonnes @ SG		(	t ÷	) ÷ 0.98			
33	Fresh and feed water tonnes @ 1.000 SG		(	t ÷ 1.000) ÷ 0.98				
34	Water ballast tonnes @ 1.025 SG		(	t ÷ 1.025) ÷ 1.00				
35	Associated cofferdams, pipe tunnels ≡ 15% Σ(31 to 34)							
36	Solid ballast							
					Total of 31 to 36			
37	Refrigerated stores		(	m³) ÷ 0.68				
38	General stores and stores passages		(	m³) ÷ 0.88				
					Total of 37 to 38			
39	Machinery space to crown of engine room							
40	Casings							
41	Shaft tunnels							
					Total of 39 to 41			
42	Sewage plant, stabilisers, thrust units							
43	Steering gear, windlass & capstan machinery							
44	Carpenter's shop, workshopps							
45	Switchboard rooms, refrigeration machinery							
46	Co₂ room, sprinkler plant							
47	Chain locker							
48	Emergency generator							
49	Swimming pool, trunks, etc.							
					Total of 42 to 49			
					TOTAL VOLUME			

Fig. 5.1. Calculation sheet for design by volume (*continuation*).

(6) Public lavatories
To serve public rooms and any passenger sections without private facilities. Space based on facilities provided. Following rates allow for necessary access space:

bath:	3.3 m^2	shower:	1.7 m^2
WC:	1.9 m^2	washbasin:	1.4 m^2
urinal:	1.0 m^2	ironing board:	1.0 m^2
slop locker:	1.5 m^2	deck pantry:	4.5 m^2

(7) Dining saloon
Area should be based on the numbers eating at one sitting. Where large numbers
are involved two sittings are normal. Areas per person should be about:
1st class:	1.5 m^2 for large numbers to 2.3 m^2 for small numbers
Tourist:	1.3 m^2 for large numbers to 1.6 m^2 for small numbers.

Modern cruise liners:
"Tropicale":	1.44 m^2
"Fantasy":	1.66 m^2

(8) Lounges and bars
Base on aggregate seating required. Usually 100% in tourist and in excess of 100%
for 1st Class
Area per seat:
lounges:	2 m^2
libraries:	3 m^2

Modern cruise liners:
"Tropicale":	seats for 72% at 1.42 m^2 per seat plus 170 seats = 12% in discotheque at 1.47 m^2 per seat
"Jubilee":	seats for 65% in lounges plus 9% in discotheque at an average of 1.48 m^2 per seat

(9) Shops, bureau, cinema, gymnasium
Shops, bureau:	15–20 m^2.
Cinema:	20 m^2 for stage + 0.8 m^2 per seat.

(10)–(12) Captain's and officers' cabins (excluding bath or toilet)
Captain and Chief Engineer:	30 m^2 + Bath 4 m^2 or toilet 3 m^2
Chief Officer, 2nd Engineer, Chief Purser:	14 m^2 + toilet 3 m^2
Other officers:	8.5 m^2 often with toilet 3 m^2

(13) Offices
Captain, Engineers, Chief Steward:	each about 7.5 m^2.
Large ships:	add Chef, Provision master, Laundryman.

(14) Passages, stairs
40% of sum of items (10)–(13)

(15) Officers lavatories
Number of fittings usually in excess of DOT rules. Area per fitting as in item (6).

(16) Dining Saloon, lounge.
    Dining saloon: 1.3 m^2 per seat.
    Lounge: 1.7 m^2 per seat
Dining saloon usually seats 100% officers although some may dine with passengers.
Lounge usually seats about 60% officers.

(17)–(18) P.O.s and crew cabins.
    Single berth cabins (usually senior P.O.s): 7 m^2.
    Two berth cabins (Junior P.O.s Deck and Engine Ratings): 6.5 m^2
    Four berth cabins (Stewards): 10.5 m^2

(19) Passages
    35% of item (17).

(20) Crew lavatories, change rooms
Sanitary fittings to DOT rules.
    WCs: 1 per 8
    washhand basins: 1 per 6 (if not in cabins).
Area per fitting as in item (6).

(21) Messes and recreation rooms.
    Messes for P.O.s, Deck
    and Engine ratings: seating for 100%
    Stewards Mess: seating for 40% (other stewards eat in passenger
                              saloon after the passengers)
    Area per seat: 1.1 m^2.
    Recreation room for Deck
    and Engine Ratings: seating for 50% at 1.2 m^2 per seat

(22) Wheelhouse, chartroom, radio room.
    Wheelhouse: 30 m^2
    Chartroom: 15 m^2
    Radio Room: $8 \text{ m}^2 + 2.5 \text{ m}^2$ per Radio Officer

(23) Hospital.
    Number of berths all hospitals: 2 + 1 per 100 of total complement, 35% of
                                         these may be upper berths
    Area per berth one or two tier: 6 m^2.

(24) Galley.
    Area per person served: 0.65 m^2 for small numbers, reducing to
                                       0.55 m^2 for 1000 or more total complement

(25) Laundry, including ironing room, etc.
   50 m^2 + 0.07 m^2 per person of complement

(26) Air conditioning fan rooms
   2.5% of total ventilated volume.

(27) Lining and flare
   3.4% total ventilated volume (1)–(25).

(28)–(30) Cargo spaces
As specified. Convert to moulded volume by dividing by following constants.
   Grain: 0.98
   Bale: 0.88
   Refrig: 0.72.

(31)–(32) Oil fuel, diesel oil
Calculate for the required endurance at specific consumption rates corresponding
to the engines selected. Allow for port consumption and for a margin remaining on
arrival at bunkering port. Allow for fuel used for heating, distillation and hotel
service purposes.

(33) Fresh/feed water
With distillation or osmosis plants now generally fitted, fresh and feed water
storage capacity is arranged to provide storage to suit the emergency which would
result from a breakdown and this obviously depends on the voyage route.

(34) Water ballast
Only tanks with no other use need be considered. Provision must be made for the
tanks required to maintain stability in the burnt-out arrival condition, plus any
tanks needed to provide flexibility of trim to cope with all loading conditions.
Generally, water ballast capacity should be between 2/3 and 3/4 of the sum of the
oil fuel, diesel oil and fresh water consumption.

(35) Cofferdams, pipe tunnels
   15% of volume of (31)–(34).

(36) Solid ballast
If it is intended to fit this, the necessary stowage space should be allowed.

(37) Refrigeration stores
Allow 0.04 m^3 per person per day of voyage and convert to gross volume by
dividing by 0.72.

(38) General stores
Allow $140 \text{ m}^3 + 0.1 \text{ m}^3$ per person per day. An insight into the variety of refrigerated and general stores carried by a high class cruise liner is given in Table 5.5 compiled from a very interesting hand-out given to passengers.

(39)–(41) Machinery space volume including casings, shaft tunnel
The total volume of these spaces can be estimated from the machinery weights calculated as suggested in §4.5 by the use of a density figure derived from a suitable basis ship whose machinery weight and volume is known. Approximate densities are:

Slow speed diesels:	0.16 tonnes/m^3
Medium speed diesels:	0.13 tonnes/m^3
High speed diesels:	0.11 tonnes/m^3 (on ferries)
Gas turbines:	0.10 tonnes/m^3 (on frigates)

(42)–(49) Miscellaneous spaces
The space provided for each of these items should be assessed on the basis of the specification and measurements from plans of ships which appear similar to the one being designed.

5.2.2 Tween deck heights

To convert the areas into volumes it is necessary to allot to each of the areas an appropriate tween deck height.

Cabin areas:	2.45–2.50 m on larger ships.
	2.60 m on the deck in which the main ventilation
	trunks and main electric cables are run.
Main public rooms:	2.90 m
Galley:	2.75 m

5.2.3 Standard calculation sheet

A standard calculation sheet (from the 1976 R.I.N.A. paper) for the design of a passenger ship or with some adaptation for the design of any ship whose volume needs to be calculated as a preliminary to assessing its dimensions is presented as Fig. 5.1.

Although the passenger numbers will be specified, those of the crew may not be and the completion of this sheet needs an estimate to be made of the crew numbers. This subject along with that of the crew requirements of other ship types is addressed in §5.5.

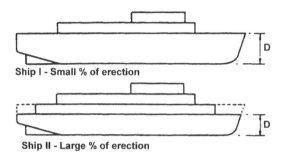

Fig. 5.2. Ships with small and large percentages of erections. With the exception of the dotted parts, Ships I and II are identical, but D and the proportion of the erection volume to total volume are obviously widely different.

5.3 FIXING THE DIMENSIONS OF A PASSENGER SHIP

5.3.1 Main hull and superstructure

In the 1976 paper it was suggested that the next step should be the division of the total volume into a main hull volume and a superstructure volume and that this should be done by assuming that the superstructure volume was a certain percentage of the total volume.

However, as Fig. 5.2 shows, a very small change in the design of a ship can make a very major change in this percentage. To overcome this difficulty it is suggested that the percentage should be derived from a suitable basis ship. If suitable data is not available a reasonable first approximation might assume that the superstructure will provide 25% of the total volume. A ship of this sort will have a relatively high uppermost continuous deck and the minimum of erections. With the volume of the main hull known, the dimensions can be derived in the way outlined in §5.1.

With the main dimensions and the volume of the superstructure known a preliminary profile can be drawn, but in the course of doing this it will usually be desirable to modify the depth D to provide a double bottom, holds and tween decks all of suitable heights. This modification should take the form of reducing the depth and adding the volume thus subtracted from the main hull to the super-structure to maintain the total volume. The ship will consequently change from Type 1 towards Type 11.

5.3.2 Modern passenger ship design

The trend in recent passenger ship design of extending all the superstructure decks from near the bow to the extreme stern, with little of the step back terracing which used to be a feature, suggests a change in design methods, with the use of the following equation:

$$V_t + V_b = C_{btd} \cdot L \cdot B \cdot D_{td} \qquad (5.2)$$

or the corresponding equation for L is

$$L = \left(\frac{(V_t + V_b)(L/B)^2}{C_{btd}(D_{tb}/B)} \right)^{1/3} \qquad (5.3)$$

where
   V_b is the volume taken up by lifeboats if these are to be stowed under
   overhanging superstructure decks, which is now becoming common practice on
   the larger ships of this type (such stowage usually takes two tweendeck
   heights, a breadth for both sides of about 10 metres and a length to suit the
   number of lifeboats to be carried); and
   D_{td} is the depth to the topmost continuous or nearly continuous deck ignoring
   comparatively short houses; and C_{btd} is the corresponding block coefficient,
   which can be estimated, admittedly fairly approximately because of the
   considerable extrapolation involved, by the use of eq. (3.2.4).

It will be noted that this eliminates the need to divide the total volume into hull and a superstructure percentages. If the value of D_{td} is well chosen in relation to the value of B this will go a long way towards ensuring that the chosen dimensions will result in satisfactory stability in the same way that the B/D ratio does for cargo ships.
   For large modern cruise liners and for large passenger ferries the ratio

$D_{td}/B = 1.2 \ (\pm 0.05)$ is remarkably constant.

This ratio is based on data for seven cruise liners: Crown Princess, Crystal Harmony, Fantasy, Horizon, Monarch of the Seas, Asuka and Statendaam, and two passenger ferries: Silja Serenade and Olau Britannia. The D_{td} figures used have been estimated by scaling from small plans or by adding standard tween deck heights to the moulded depth in the absence of more accurate information, but the error is unlikely to be significant.
   Having reached an initial set of dimensions, the depth to the top deck will usually require some adjustment to make it a convenient summation of the required tween deck heights, etc. The beam should then be modified to keep the D_{td}/B ratio within the suggested band and the length of the ship adjusted to give the required volume. This may seem a very approximate method but sensibly used it will generally ensure that the dimensions adopted for a preliminary design need little modification when the results of later more detailed calculations become available.

5.4 OTHER "VOLUME DESIGN" SHIP TYPES

There are quite a number of other ship types which are best designed by calculating the total enclosed volume required to accommodate all the spaces needed for the crew and the various activities which they undertake. These include: fish factory ships, offshore safety ships, livestock carriers, oceanographic and fishery research vessels amongst merchant ships. However, possibly warships of frigate and corvette types are the largest and most important category of volume-based ships, although because these ships are also usually designed for minimum weight some designers tend, erroneously, to regard them as weight-based designs or claim that they are balanced designs in which weight and volume are equally important.

The warship design calculation sheet given as Fig. 4.19 includes a section for the calculation of the total internal volume. It would have been nice to supplement this with a series of guidance notes on the completion of this form to parallel those given for passenger ships, but as all the data that the author has on this subject was derived from plans subject to security classification, readers must be left to formulate their own approximate algorithms.

It can, however, be confidently asserted that this design method can be applied to warship design with considerable advantage.

5.5 CREW NUMBERS

5.5.1 Passenger ship crew numbers

In §5.2 the need to know what constitutes a suitable crew was noted in relation to passenger ship design but of course the same applies to the design of all ship types.

In the 1976 paper it was noted that the passenger/crew ratio for passenger ships had not changed much from that noted in 1962. This seemed surprising when related to the very significant reduction in the crews of cargo ships over the same period, but the explanation lay in the higher standard of hotel services being provided, which offset reductions in deck and engine department manning fairly similar to those on cargo ships.

Ships in 1976 were seen to group into passenger/crew ratios of about 1.7 to 2.2 for ships aiming at the upper end of the cruise trade with ratios of 2.5 to 3.0 applying to ships catering for the more popular end of the trade. In both cases the lower figures applied to the smaller ships and the higher ones to the larger ships.

In 1992 the passenger/crew ratio for "Fantasy" was 2.86.

5.5.2 Cargo ship crew numbers

In 1976 the change in manning since 1962 was seen to have come about as the result of a felicitous conjunction of motive and means: the growing pressure for

Table 5.1

Reasons for changes in manning, 1962–1976

Cost reduction motives	Cost reduction means
Competition from aeroplanes to passenger ships	Improved machinery, requiring less attention and less maintenance
Competition from land routes to container ships	Automation of machinery
Competition between shipping companies as many new nations enter the field	Use of self-lubricating fittings
All of these leading to relatively, if not actually, lower freight rates	Cargo gear requiring less attention
	Patent hatch covers with push-button operation
Better job opportunities ashore leading to the necessity of paying higher wages and providing better conditions for seagoing personnel	Self-tensioning winches, universal fairleads, thrust units
	Modern paint systems, modern plastic accommodation linings
The enormous growth in shipping making the acceptance of reduced manning politically acceptable	Electric galley gear
	The use of work study
	The use of general purpose crews

Table 5.2

Changes in manning 1962–1976 and expectation in 1976 for future

Ship type	1962	1976		Future
	Typical	Typical	Automated	automated
General cargo or bulk carrier	36	30	26	11
Sophisticated cargo liner or container ship	50	36	28	11
Tanker	45	36	26	9

Future figures taken from B.V Report 1976 by Monceaux "A look at the personnel of automated ships".

cost reduction on the one hand and the arrival of a great deal of helpful new technology on the other hand. These factors were presented in a table, which is reproduced as Table 5.1.

The effect of these changes on the manning of some typical ships was given in the table presented as Table 5.2 which has been retained to show the speed of development there has been in this area.

The path the development has followed and may continue to follow is shown in Table 5.3 which is reproduced from the report on a project sponsored by the British Department of Transport in 1986 entitled "Technology and Manning for Safe Ship Operation". It will be seen that the first column of this table shows the number of a conventional crew at 30. By column 4, the crew has reduced to 20 with column 5 taking it to 18, and in fact today most cargo ships have a crew of either 18 or 20,

Table 5.3

Alternative manning for handy-sized products tanker in 1986

Column:	1	2	3	4	5	6	7	8	9	10	11	12	13	14	15
	Conventional (all manual)	GP Ratings	UMS	No C.P.O.	Altered catering	Auto-steering (normal conditions)	No R/O	No Pumpman	C/E to keep watch in emergency manual mode	6 on 6 off watches	Semi-integrated Junior Officers	Operator maintainer	Matrix manning	Polyvalent or dual purpose Senior Officers	Combined Master & Chief Engineer
Master	1	1	1	1	1	1	1	1	1	1	1	1	1	1	1
Mates	3	3	3	3	3	3	3	3	3	2	1	–	1	–	–
Chief Engr.	1	1	1	1	1	1	1	1	1	1	1	1	1	1	–
Engineers	6	6	3	3	3	3	3	3	2	2	1	1	1	–	–
R/O	1	1	1	1	1	1	–	–	–	–	–	–	–	–	–
Pumpman	1	1	1	1	1	1	1	–	–	–	–	–	–	–	–
Bosun	1	–	–	–	–	–	–	–	–	–	–	–	–	–	–
Deck ratings	6	–	–	–	–	–	–	–	–	–	–	–	–	–	–
Eng. ratings	4	–	–	–	–	–	–	–	–	–	–	–	–	–	–
G/Ps	–	7	7	6	6	4	4	4	4	4	4	3	3	3	3
Catering	6	4	4	4	2	2	2	2	2	2	2	2	1	1	1
Dual purpose Junior officers	–	–	–	–	–	–	–	–	–	–	2	2	2	2	2
Dual purpose Senior officers	–	–	–	–	–	–	–	–	–	–	–	–	–	2	2
Total manning	30	24	21	20	18	16	15	14	13	12	12	10	10	10	9

and these numbers seem to apply to all crew nationalities. Progress to still lower numbers is in hand however with the Scandinavian countries generally in the lead.

A Dutch statement quoted in the Department of Transport report sets the following principles:

"When setting minimum safe manning standards, the lowest numbers will only be allowed when a vessel has automatic steering, UMS, mechanical hatches, mechanically operated moorings and systems which will reduce on board maintenance".

As far as the U.K. is concerned, tabular rules prescribing crew numbers are now a thing of the past with safe-manning certificates now only being issued after a

Table 5.4

A survey of crew numbers for various types and sizes of ship

Type of ship	Parameter						
	Crew						
Tanker	Dwt	3160	48966	113131	142000	275782	319600
	Crew	9	20 (30)	32	26	30	31
L.P.G.	m^3	1600	4300	8237	57000	75208	125760
	Crew	25	16	22	25 (30)	32	40
Bulk carrier	Dwt	61687	74000	77500	96725	169178	
	Crew	26	28	25 (31)	33 (38)	29	
Container ship	Cont. nos.	976	1201	1315	1960	3568	4407
	Crew	8	26 (34)	24	(26)	16 (28)	19 (29)
Refrig. ship	m^3	5240	16332	21684			
	Crew	8	16	9* > 6			
Multi-cargo	Dwt	12100	13150	17175			
	Crew	(19)	25	25			
Cruise liner	Passengers	100	584	960	1354	2604	2744
	Crew	65	240	480	642	980	826
Passenger ferry	Passengers	1600	2500				
	Crew	248	264				
Freight ferry	Cont. nos.	120	301	1388			
	Crew	9	7 (10)	18			

Where a number in brackets follows another number, this indicates "accomodation for" and includes repair crew and spare rooms. Suez crew (generally 6 on large ships) are excluded.
*This crew is based on an "integrated ship control system. The present crew of 9 consists of captain and 2 deck officers, 1 cook/steward and 2 g.p. ratings. It is intended that this crew be reduced to 6 in the future.
Data abstracted from *Significant Ships*, 1990 and 1991.

submission to the Department of Transport of all relevant details. Apart from watch-keeping requirements the manning level is often set by the number required for mooring operations.

Up-to-date crew numbers abstracted from *Significant Ships* of 1990 and 1991 are presented in Table 5.4; Table 5.5 presents a fascinating insight into the extraordinary variety of skills which go to make up the crew of Q.E.2.

Table 5.5

Stores list and crew list for Q.E.2 on a transatlantic voyage

Food

Biscuits	2000 lbs	Lemons	5000	Dog biscuits	50 lbs
Cereals	800 lbs	Grapes	2000 lbs	Foie Gras	100 lbs
Tinned fish	1500 tins	Ice cream	5000 qts	Bacon	2500 lbs
Herbs, spices	50 lbs	Beef	2500 lbs	Cheese	3000 lbs
Marmalade	9600 jars	Pork	4000 lbs	Cream	3000 qts
Tea bags	50000	Sausages	2000 lbs	Fish	8000 lbs
Coffee	2000 lbs	Duck	3000 lbs	Crabmeat	1000 lbs
Baby food	600 jars	Potatoes	20000 lbs	Oranges	15000
Caviare	200 lbs	Fresh vegetables	27000 lbs	Melons	1000
Butter	3500 lbs	Flour	3000 lbs	Limes	2000
Ham	1200 lbs	Rice	3000 lbs	Frozen fruit	2600 lbs
Eggs	30000	Tinned fruit	1500 gals	Kosher food	600 lbs
Milk	2500 gals	Jam	300 lbs	Lamb	6500 lbs
Lobsters	1500 lbs	Juices	3000 gals	Veal	6000 lbs
Grapefruit	3000 lbs	Tea	500 lbs	Chicken	5000 lbs
Apples	6000	Sugar	5000 lbs	Turkey	5000 lbs
				Pickles, sauces	2000 bot

Liquor and Tobacco

Champagne	1000 bot	Beer	1200 bot	Vodka	320 bot
Whiskey	1200 bot	Cigars	4000	Liqueurs	360 bot
Rum	240 bot	Tobacco	1000 tins	Port	120 bot
Brandy	240 bot	Assorted wines	1200 bot	Beer (draught)	6000 gals
Sherry	240 bot	Gin	600 bot	Cigarettes	25000 pkts

Cutlery, Glass and Crockery

Glassware	51000 items	Kitchenware	7921	Cutlery	35850 items
Kosher crockery	3640	Crockery	64000 items	Tableware	64531 items

Continued opposite

Table 5.5 (*continuation*)

Queen Elizabeth 2 "At Your Service"
Crew List

Asst. Barkeepers	16	Night Stewards	6	Doctors	2
Baggage Masters	2	Nursery Nurses	2	Deck Ratings	34
Beauticians	2	Philipino Staff	164	Engineer Ratings	53
Bedroom Stewards	26	Printers	4	Executive Chief	1
Bosun	1	Public Room Stewardesses	8	General Manager	1
Captain	1	Security Petty Officers	7	Hotel Officers	44
Chefs de Cuisine	4	Shop Assistants	18	Laundry Staff	17
Chief Barkeepers	2	Staff Bedroom Stewards	3	Librarian	1
Communications Assistants	2	Waiters	218	Masseur Male	2
Cruise Staff	14	Writers	5	Medical Ratings	2
Data Input Clerks	3	Asst. Restaurant Managers	14	Nursing Sisters	3
Deck Officers	8	Bank Staff	4	Orchestra Staff	27
Engineering Officers	26	Bell Boy	1	Photographers	3
Entertainers	5	Bedroom Stewardesses	59	Public Room Stewards	24
Fitness Instructors	4	Bosuns Mate	1	Radio Officers	1
Hairdressers	13	Casino Staff	11	Security Officers	3
Laundry Supervisors	1	Chefs	89	Storekeepers	4
Leading Wine Steward	1	Commis Waiters	17	Staff Bedroom Stewardesses	2
Masseuse Female	2	Crew Administration Asst.	1	Wine Stewards	16
Medical Dispenser	1	Dancers	6		

Grand Total — 1014

5.5.3 Warship crew numbers

The crews of warships are very large by merchant ship standards, partly because of need to man a large number of weapons along with their command and control systems on a 24-hour basis — at least when there is a state of emergency — and partly because of the need to provide for such labour-intensive activities as damage control parties and replacements for casualties. Nevertheless, the need to economise in manpower that has long been recognised as essential in merchant ships is now regarded with the same urgency by both naval staff and their designers. A typical frigate of the decade 1970–1980 generally had a total crew of a little more than 250 persons; in the next decade the crew numbers of ships with very much the same capability had dropped to about 170.

Towards the end of the decade (in 1987) the firm of consultants YARD (now BAeSEMA) for whom the author worked at that time came to the conclusion that it was likely that in the not too distant future there would be a need, for demographic reasons, for very much reduced crews. After considerable study into ways of reducing manpower, they found that a combination of changes in operational procedures and the introduction of new equipment for which the technology already existed or would shortly be available would make it possible to reduce the crew of a frigate to 50 men. They then went on to develop the design discussed in Chapter 16 to prove their case. Whilst this design was a look into the future and needs the full development of equipment then still only at the prototype stage it was, somewhat to the firm's surprise, warmly welcomed by naval staff from several navies, as showing the way ahead.

With these major changes in crew numbers over a short period of time, it would be wrong to try to give any more detailed guidance on warship crew numbers.

5.6 DESIGNS BASED ON DECK AREA

Prominent amongst ships designed on the basis of deck area are train and vehicle ferries. If the space for trains and vehicles is enclosed it can of course be argued that these are volume designs, but this is not strictly true as the volume required for these must be provided on a limited number of decks, generally one or two. These decks must be above the bulkhead deck and have free access to the loading/ unloading doors and/or ramps, making the case for considering them as area designs. Figure 5.3 extracted from a paper by Alan Friis to the Cruise and Ferry 91 Conference shows a typical deck view and midship section of a Danish combined train and vehicle ferry.

In laying out the deck of train ferries it is essential to have good guidance on the minimum acceptable turning radius and on the necessary clearance between the tracks that is associated with this.

For the easier case of road vehicles, the clearances allowed must enable drivers and passengers to use car doors without too much difficulty. This occasional and not particularly fat passenger thinks that these clearances are too frequently skimped and is reminded of the advice he was given as a young man "Go and try it yourself and see what space you need".

5.7 DESIGNS BASED ON LINEAR DIMENSIONS

There are a number of ship types in which the design process proceeds directly from the linear dimensions of the cargo, an item or a number of items of equip- ment, or from constrictions set by canals, ports, etc. and for which the deadweight,

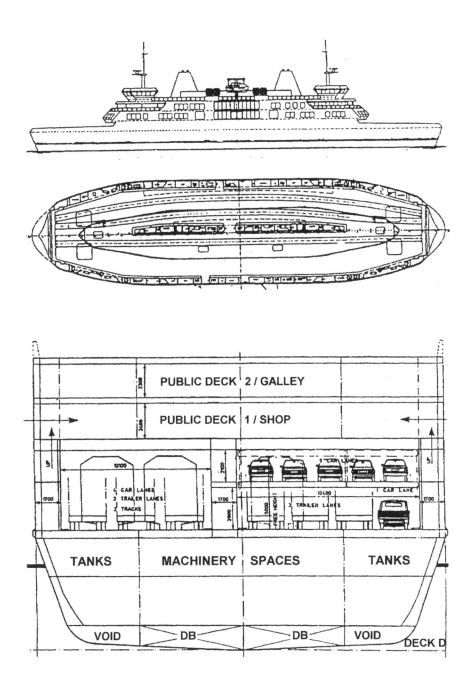

Fig. 5.3. A design based on deck area — a Danish train ferry.

capacity and sometimes the speed are determined by the design instead of being the main factors that determine it.

The design processes for these ships are essentially non-standard and give the naval architect a chance to exercise his ingenuity. Some methods and ideas relating to these types may, however, be of interest.

5.7.1 Container ships

As the design deadweight of most container ships can be carried at a draft less than that obtainable with a type B freeboard, deadweight cannot be used directly to determine the main dimensions.

On the other hand, as container ships carry a substantial proportion of their cargo on deck, it is not possible to base the design on the required cargo volume as this is indeterminate. In these circumstances, stability considerations take over the primary role in the determination of the main dimensions.

For maximum economy in the design of any container ship, the dimensions should be such that containers can be stacked up in tiers to the limit permitted by stability. To maximise numbers, the upper tiers, subject to the owners agreement, should be reserved for relatively lightly loaded (or even empty) containers, whilst heavier containers are directed to the lower levels. It may also be desirable in the interests of maximising container numbers and therefore revenue to design the ship to carry ballast, either permanent or water or both even in the load departure condition — something that would be a heresy in the design of most ship types!

For each number of tiers of containers carried there is an associated breadth of ship which will provide the KM necessary to ensure stability, whether the tiers are enclosed below hatch covers or carried on deck being a second order effect. Ships designed to achieve a particularly high KM for a given breadth obviously have an advantage provided any penalty incurred in powering or seakeeping is acceptable (see Chapter 8, §8.6)

Longitudinal and torsional strength require a proportion of the breadth of the ship to be devoted to structural decks with the balance of the "open" ship providing space for a number of container cells with their cell guides. There is therefore a direct relationship between the number of container tiers and the number of container rows in the breadth.

A first approximation to the length of the ship is then generally determined by what is thought to be an economically desirable value of the length/beam ratio. This is then adjusted in association with the length required for the engine room, peak tanks, cell guides,bulkhead stiffeners etc. so that the cargo spaces are tailored to a multiple of container lengths.

With the number of container tiers "fixing" the number of rows in the breadth and this in turn "fixing" the number of container cells in the length, there is the

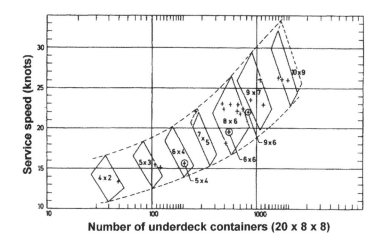

Fig. 5.4. Number of containers carried under deck versus numbers in the midship section and ship speed.

rather surprising possibility that there are a number of ranges of container numbers for which optimum ships can be designed with intervening numbers which require an acceptance of some dimensional proportions that take the ship away from the optimum.

Speed has an effect on the container numbers that can be carried in a ship of certain dimensions partly because of its influence on the block coefficient and partly because of its influence on machinery power and thus on the engine room dimensions.

Figure 5.4 reproduced from the 1976 paper shows the under-deck container numbers which give economic container ships for various speeds. It also shows the tier × row numbers for which the midship section would be arranged. The value of this figure could have been improved, particularly in these days of wholly open container ships, if the abscissa had been total containers rather than under deck containers.

5.7.2 Open container ships

Container ships without hatch covers represent possibly the latest major ship design development and are a nice example of the fact that major improvements stem from lateral thinking rather than from optimisation techniques. One thing this type of ship confirms is the contention that container ships are controlled by stability rather than by volume or weight.

One great advantage of these "open" container ships is the fact that all the containers carried are in cells and no lashing is needed. A second advantage is the fact that no time need be spent opening and closing hatch covers. The absence of hatch covers lowers the centre of gravity of the upper tiers of containers enabling more of these to be carried within a stability limit. There is a cost saving for the hatch covers but this is offset by the cost of the increased depth of the ship and of the additional safety features required.

Chapter 6

Powering I

6.1 AN INTRODUCTION TO POWERING

The subject of powering in all its aspects usually takes up a number of chapters in textbooks on naval architecture and there are also several specialist books which are confined to this one subject. The treatment in this book has therefore been written on the understanding that all naval architects undertaking ship design can be expected to be familiar with the theory of the subject, but that many will nevertheless appreciate some help with its practical application as few books seem to be specifically directed to the primary needs of the designer, which are:

(i) How to estimate with an acceptable accuracy the machinery power which must be fitted to a new ship design to enable it to attain the specified speed.

(ii) How to minimise this power so as to reduce the capital cost and/or improve the fuel economy and therefore the operating efficiency of the ship.

The aim of this chapter and Chapter 7 is to meet these needs as concisely as possible.

The first section of this chapter starts by dealing with resistance in some detail because the treatment of this subject has changed considerably in recent years and few books so far seem to have caught up with this. Because familiarity with the theory is assumed, the next section jumps to providing an *aide memoire* on most of the components of powering, which are then dealt with in more detail in the rest of this chapter and in the subsequent one.

In the course of writing the chapter, however, it became apparent that recent changes in the methods used by test tanks to estimate ship powers had not yet been written up in naval architecture textbooks and this has led to some extension of the original intent of the chapter.

6.2 RESISTANCE AND SHIP MODEL CORRELATION

6.2.1 The classical treatment of resistance

Resistance is the force that the ship overcomes as it moves through the water. The classical treatment of resistance is outlined in this section and is followed in the next section by an outline of the present day treatment of resistance.

In the classical treatment of resistance, this is divided into two components, which are governed by different laws, so that they can be separately extrapolated from model to full-scale ship size:

(i) the skin frictional resistance, which is governed by the Reynolds' number, and

(ii) the residuary resistance, taken mainly to be wavemaking, which is governed by the Froude number.

In the following paragraphs use will be made of the resistance coefficient C. This is related to the wetted surface S, the speed V, and the mass density ρ by the following equation:

$$R = 1/2 \cdot C \cdot \rho \cdot S \cdot V^2 \tag{6.1}$$

The coefficient C is given two types of subscripts. The first of these refers to the subdivision of resistance with t = total; f = frictional; r = residuary; w = wavemaking. The second subscript distinguishes between model resistance = m; and ship resistance = s.

In the classical treatment, the skin frictional resistance coefficient of the model is calculated based on the coefficient of friction applicable to a plank (flat plane) of model length and having the same wetted area as the model. This is then deducted from the total model resistance coefficient to establish the model residuary resistance coefficient.

$$C_r = C_{tm} - C_{fm} \tag{6.2}$$

At a constant Froude number the residuary resistance coefficient remains the same for the ship as it is for the model, so there is no need for a suffix to indicate "model" or "ship" in this case.

The ship frictional resistance coefficient C_{fs} is again calculated using the coefficient of friction applicable to a plank, this time one of the same wetted area and the same length as the ship. This is then added to C_r to arrive at the total resistance coefficient C_{ts} of the ship.

$$C_{ts} = C_{fs} + C_r \tag{6.3}$$

It should be noted that the use of friction coefficients based on a plank for both model and ship implies that the skin friction is independent of the shape of the lines.

The main component of the residuary resistance is the wave-making resistance, but there are also smaller components stemming from eddy-making resistance and from the resistance caused by the movement through the air of that part of the model and of the ship, respectively, which is above the waterline.

For reasons that will be discussed later in this section, a change in the treatment of resistance was made in 1957 when the basis for calculating the friction coefficients was altered from the Froude line which had been used for many years to the 1957 ITTC (International Towing Tank Conference) line.

The 1957 ITTC line is expressed by the formula:

$$C_f = 0.075/(\log R_n - 2)^2 \tag{6.4}$$

R_n = Reynolds' number (see §6.3, ix).

This change decreased the frictional component increasing the residuary component C_r correspondingly. Whilst it reduced the size of the ship–model correlation factor, it did not entirely eliminate the need for this and did not improve the accuracy of power estimation in the way that it had been expected to do and a further change was made by ITTC in 1978 and indeed further changes proposed by Grigson are discussed later.

6.2.2 The present day (ITTC'78) treatment of resistance

The present day treatment recognises that the frictional resistances of both model and ship differ from those of flat plates of the same length and area. The viscous resistance coefficient (C_v), as the frictional resistance of a shaped body is now called, is increased over the frictional resistance coefficient of the corresponding flat plate by a form factor K so that

$$C_{vm} = C_{fm}(1 + K) \tag{6.5}$$

(C_{fm} is still based on eq. (6.4)).

The form factor can either be deduced from model experiments at very slow speeds when C_r is reduced to nearly zero, or in some tanks by the direct measurement of C_r from the energy delivered to the wave systems.

C_r for the model is now calculated from the formula

$$C_r = C_{tm} - (1 + K) C_{fm} \tag{6.6}$$

As K tends to have a value of between 0.25 and 0.35 for most ship forms, a value of C_r calculated from eq. (6.6) is much smaller than one calculated from eq. (6.2)

The total resistance coefficient of a ship is now considered as made up of

$$C_{ts} = (1 + K) C_{fs} + C_r + C_{app} + \Delta C + C_{air} \tag{6.7}$$

where

C_{app} is the resistance of appendages

ΔC is the roughness allowance which is discussed later in §6.2.6

C_{air} is the air resistance coefficient, for which there is the following approximate formula:

$$C_{air} = 0.001 \cdot A_t/S \qquad (6.8)$$

where A_t is the projected cross sectional area of the ship above the waterline.

It should be noted that in association with C_{ts} derived in this way the wetted surface S in eq. (6.1) is the total wetted surface inclusive of the surface area of the bilge keels, if fitted.

$$S(\text{wetted}) = S(\text{naked}) + S(\text{bilge keels})$$

6.2.3 Discussion of the two treatments

The newer treatment is more scientific and is now in fairly general use in tank testing. However, the use of the method for power estimation in advance of tank testing presents some difficulty because there is still a lack of data in the new format compared to the mass that exists in earlier formats.

6.2.4 Hull finish and the importance of skin friction resistance

An understanding of the importance of hull finish requires a knowledge of the proportion of the total ship resistance which is frictional. Ideas on this have changed considerably in recent years with first of all the change from the Froude friction line to the ITTC line and then the introduction of the form factor. In addition some, if not all, of the ship–model correlation factors that have been applied to the total resistance ought almost certainly to have been applied to the frictional resistance only.

In ITTC'78 practice the proportion of the viscous component is

$$\frac{(1+K)C_{fs} + \Delta C}{(1+K)C_{fs} + \Delta C + C_r + C_{air}} \qquad (6.9)$$

This may be compared with the former practice in which the frictional component proportion was

$$\frac{C_{fs}}{C_{fs} + C_r + C_{air}} \qquad (6.10)$$

It is important to note that the C_r values in these two equations differ from one another, indeed the biggest part of the difference between the two formulae occurs when the viscous or frictional components respectively of the model resistance are subtracted from the model total resistance to establish the respective residuary resistance coefficients, with the further change caused by the multiplication of the ship friction coefficient by the form factor being of lesser significance.

An example with some figures may help to make the difference clear. A ship of:

$L = 330$ m, $V = 300,000$ m^3, 15 knot speed, $F_n = 0.136$ (V = volume of displacement)

was tested using a model with the following particulars:

$l = 7$ m, $F_n = 0.136$, speed $= 1.127$ m/s

K from test $= 0.33$
model $R_n = 6.856 \times 10^6$ ship $R_n = 2.14 \times 10^9$
$C_{tm} = 4.309 \times 10^{-3}$
$C_{fm} = 3.203 \times 10^{-3}$
$\Delta C = 0.10 \times 10^{-3}$
$C_{air} = 0.05 \times 10^{-3}$

By the ITTC'57 method:
$C_r = (4.309 - 3.203) \times 10^{-3} = 1.106 \times 10^{-3}$
$C_{fs} = 1.39 \times 10^{-3}$
$C_{ts} = (1.39 + 0.10 + 1.106 + 0.05) \times 10^{-3} = 2.646 \times 10^{-3}$
Proportion of frictional resistance $= (C_{fs} + \Delta C)/C_{ts}$
$= (1.39 + 0.10)/2.646$
$= 56\%$

By the ITTC'78 method:
$C_r = (4.309 - 1.33 \times 3.203) \times 10^{-3} = 0.05 \times 10^{-3}$
$C_{vs} = (1.33 \times 1.39 + 0.10) \times 10^{-3} = 1.949 \times 10^{-3}$
$C_{ts} = (1.949 + 0.05 + 0.05) \times 10^{-3} = 2.049 \times 10^{-3}$
Proportion of viscous resistance $= 1.949/2.049$
$= 95\%$

Apart from the change in the proportion of frictional/viscous resistance, the very large reduction in C_{ts}, from 2.646×10^{-3} to 2.049×10^{-3} or 29% should be noted. This change in value is of course tied to the K value of 0.33 used in this example, a figure which appears to agree with Holtrop and Mennen's formula given in §6.9.

Although most tank authorities appear to have adopted the new method, others are sticking to the use of ITTC'57. Designers can only hope that there will shortly be an end to the succession of changes and variety of methods used by tanks which have caused them so much difficulty in the last two decades.

This hope may, however, be a little premature as a 1993 R.I.N.A. paper by C.W.B. Grigson "An accurate smooth friction line for use in performance prediction" questions the accuracy of the 1957 ITTC line. He is almost certainly right in doing this if this line is to be used as a base for a form factor as the ITTC line was never claimed to be a friction line having originally been introduced as a ship–model correlation line. Grigson's paper shows that at Reynolds' numbers in the model area (4×10^6 to 2×10^7) C_f '57 values are up to 6% higher than what he suggests are the "correct" values, whilst in the ship area which is of the order of (4×10^8 to 2×10^9), C_f '57 values are about 5% below the "correct" values (see Fig. 7.18). This would mean that both C_r and C_{fs} are being underestimated, and as the paper also revises the $(1 + K)$ values upward the overall effect is to increase Pd by about 7% and propeller RPM by about 1.5%.

When relating K new line to K ITTC'78, the physical quantity that remains the same is the viscous drag, so:

$$(1 + K_{new})C_{f.new} = (1 + K_{.78})C_{f.'57}$$

with C_r remaining unchanged.

6.2.5 Ship–model correlation

As well as $(1 + K) C_{fs}$ the total viscous resistance for the ship includes the term ΔC which is intended to allow for the influence on resistance of the roughness of the paint. This is now seen as an addition to the frictional resistance and not, as in the past, a factor $(1 + x)$ applicable to the total resistance.

It is interesting to take a brief look at the history of ship–model correlation which came to prominence with the change from all riveted construction to all welded ship hulls in the early fifties, when it was found necessary to bring in ship–model correlation factors to reflect the differences in smoothness between an all-riveted ship (1.10), a ship with riveted seams and welded butts (1.00) and an all-welded ship (0.95).

By the late fifties/early sixties, most shipyards were building all-welded ships and ship–model correlation seemed to have reached such a satisfactory state that it was possible to start taking a more sophisticated look at the effect of the smoothness of the platework and its paint coatings.

Within a few years, however, trial results from significantly larger ships started to be tabulated, and it was found that many of these vessels had performed much better than had been predicted.

A new factor was accordingly introduced into ship–model correlation — a factor which scaled with ship size. This had a value of 1.00 for a ship length of about 105 m reducing linearly to 0.80 at a length of 275 m.

The necessity for a size-dependent factor clearly indicated a fault in the extrapolation from model to full size with a factor less than unity predicating full-size ships with a finish superior to that of a wax model!

After discussion, tank superintendents agreed that skin friction was being over-estimated by the Froude method, and decided to change to the 1957 ITTC formulation. More recently, appreciation of the effect of the form has changed the emphasis with the frictional element now being seen once again as much the largest part of the resistance of most ships.

6.2.6 Hull finish, ΔC and $(1 + x)$

The ITTC 57 model–ship correlation line was intended to eliminate ship–model correlation making $(1 + x) = 1.00$ for all-welded ships having a shell roughness amplitude of 165 microns which was found to be typical of new construction in the decade 1960–1970.

Prior to the general use of ΔC, various formulae had been suggested for the calculation of $(1 + x)$ for different shell roughness values, with the most authoritative probably being that given in a 1980 R.I.N.A paper "Speed, power and roughness — the economics of outer bottom maintenance" by Townsin et al., although the physics of this have been challenged.

In this paper the value of x is shown to change with a change in roughness (h) in microns mean apparent amplitude (MAA) as follows:

$$\Delta(x) = 0.058 \left[(h_1)^{1/3} - (h_2)^{1/3} \right] \tag{6.11}$$

Applied to a base of $(1 + x) = 1.00$ at $h_2 = 165$, this gives the following values.

MAA	$(1 + x)$
80	0.932
125	0.968
165	1.000
230	1.037
400	1.109

125 microns is quoted in a 1972 NMI Report No. 172 as the best figure achieved on the ships measured in that survey; 230 is quoted as typical of a rather poor performance, and 400 was the worst.

As has already been said, modern practice favours the use of ΔC rather than $(1 + x)$. A formula for ΔC suggested by Townsin and accepted by ITTC, although here once again the physics of this empirical formula have been challenged, is

$$10^3 \, \Delta C = 44 \left[(h / \text{LWL})^{1/3} - 10(R_n)^{-1/3} \right] + 0.125 \tag{6.12}$$

where h is an average measure of the height of the elements of roughness of the entire hull, LWL is the waterline length, both in identical units.

It has been pointed out that ΔC can vary quite considerably even on hulls with the same h, depending on the number and spacing, individual height and shape of the imperfections that give rise to the *h* value. A large number of essentially streamline elements will have less effect than a smaller number of less streamlined solidified droplet-type imperfections.

Values of ΔC for different roughness values for the 330 m vessel used in the calculations in §6.2.4. for a speed of 15 knots are as follows:

MAA	$10^3 \Delta C$
75	0.052
100	0.079
125	0.102
150	0.122

Both $(1 + x)$ and ΔC values make very clear the bonus to be gained from a good paint finish and the penalty imposed by a poor one.

6.2.7 Steelwork roughness

Townsin and his co-authors, encouraged by the success of constant emission toxic coatings in reducing fouling, saw roughness of steelwork and paint finish as a major factor to be attacked in pursuit of fuel economy. Under laboratory conditions, they obtained a roughness of 50 microns with a ship's paint system applied to shot-blasted steel and thought that 80 microns could be attained when applied to a ship in drydock in very strictly controlled conditions.

Grigson, whose paper "The full-scale viscous drag of actual ship surfaces" was published in 1987 in the *Journal of Ship Research*, states that it is not only the height of the average roughness that matters but also the form that the roughnesses take. He also makes the point that with good modern practice we ought not to be speaking of surface roughness but of imperfect smoothness. With shop-blasted and shop-primed smooth steel plate, welded by automatic machines producing smooth rounded weldments and up to four coats of paint applied by airless spraying machines under cover, a very high quality finish is possible. Unfortunately, unless senior management insists on the best practice, airless spray can be incorrectly applied resulting in "dry overspray" which can increase ΔC to as much as 20% of C_{ts}.

6.3 AN AIDE MEMOIRE ON THE COMPONENTS OF POWERING

(i) Effective horsepower (P_e)

This is the power required to tow a ship, overcoming its resistance. It can be expressed either as

$$P_e(\text{kW}) = \frac{0.7355 \cdot \textcircled{C} \cdot \Delta^{2/3} \cdot V_k^3}{426.6} = \frac{\textcircled{C} \cdot \Delta^{2/3} \cdot V_k^3}{580} \qquad (6.13)$$

or as

$$P_e(\text{kW}) = 0.7355 \cdot \frac{1000 \cdot R_t \cdot V_m}{75} \qquad (6.14)$$

in which

$$R_t = C_t \cdot 1/2 \cdot \rho \cdot S \cdot V_m^3$$

Combining these gives

$$P_e(\text{kW}) = C_t \cdot \frac{1}{2} \cdot \rho \cdot S \cdot V_k^3 \times \frac{0.7355 \times 1000 \times (0.5144)^3}{75 \times 9.81}$$

which reduces in salt water to

$$P_e(\text{kW}) = 0.0697 \, C_t \cdot S \cdot V_k^3 \qquad (6.15)$$

In the above equations $\textcircled{C}$ is one of R.E. Froude's "circular" notations which are discussed in §6.3 (xi), and C_t has already been defined in §6.2.

The denominator in eq. (6.13) allows for Δ in tonnes and P_e in metric horse-power before the conversion to kilowatts.

In traditional British units of tons and imperial horsepower the denominator used to be 427.1.

Δ = displacement in tonnes (s.w.)
V_k = speed in international knots
V_m = speed in m/s
S = wetted surface in m^3
r = density   = 1.025 tonnes/m^3 for salt water
                   = 1.000 tonnes/m^3 for fresh water
ρ = mass density   = r/g
1 metric tonne   = 0.984 British tons of 2240 lbs
1 metric horsepower   = 75 kg mass × metres per second
                   = 1000 Newton metres per second
                   = 0.7355 kilowatts
                   = 0.986 British horsepower
1 British horsepower   = 550 lbs mass × feet per second
                   = 0.746 kilowatts

The figures normally used as the equivalent of speeds in knots are:
International   1852 m/h; 30.867 m/min; 0.5144 m/s
Imperial   6080 ft/hr; 101.33 ft/min; 1.69 ft/s

It should be noted that the two knots are not quite identical, and neither are the metric/imperial conversion figures; this can cause problems in complex equations.

Although great care has been taken with the calculation of the various constants quoted in this section the accuracy of many of these figures depend on several conversion figures and exactness cannot be guaranteed, but any error should be less than 1%.

Some figures quoted in this section are intended only to give a "feel" and their use should be limited to approximate calculations.

Because the change from the Froude method to the ITTC'57 method of extrapolation from model to ship results in markedly different P_e and ⓒ values, it has been necessary in recent years when these treatments have been in use in parallel to annotate each of these items as "Froude" or "ITTC" to ensure that the correct ship model correlation factor is used.

From eqs. (6.13) and (6.15) the relationship between ⓒ and C_t can be shown to be

$$ⓒ = 40.46 \, C_t \cdot S/\Delta^{2/3} \tag{6.16}$$

$$= 39.80 \, C_t \cdot ⓢ$$

As eq. (6.15) requires the use of the wetted surface, it may be appropriate to give some approximate formulae for S at this point.

The following formulae, based on metric dimensions, give the value of S in m^2:

Mumford's formula

$$S = 1.7 \, L \cdot T + C_b \, L \cdot B \tag{6.17}$$

It may be noted that Guldhammer and Harvald in the paper discussed in §6.8 suggest increasing the Mumford value by adding a factor of 1.025.

Taylor's formula

$$S = C\sqrt{\Delta L} \tag{6.18}$$

This was originally based on Δ in tons and L in feet but has been metricated in Fig 6.1. For merchant ships of normal proportions $C = 2.55$ can be used as a quick approximation.

Holtrop and Mennen, whose powering method is examined in §6.7, suggest the following formula which, although too complex for use in hand calculations, can be readily incorporated in a computer program.

$$S = L(2T + B) \, (C_m)^{1/2}(0.453 + 0.4425 \, C_b - 0.2862 \, C_m - 0.003467 \, B/T$$

$$+ 0.3696 \, C_{wp}) + 2.38 \, A_{bt}/C_b \tag{6.19}$$

where A_{bt} = transverse sectional area of the bulb at the fore perpendicular.

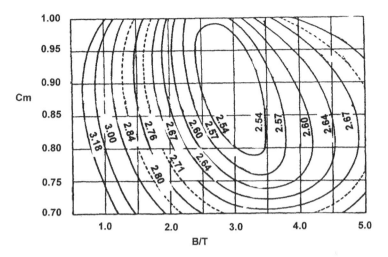

Fig. 6.1. Contours of wetted surface coefficient vs midship area coefficient (C_m) and beam/draft ratio (B/T).

(Taylor metricated) $S = C\sqrt{\Delta L}$. S in metres, L in metres, Δ in tonnes.

The author is unable to offer any advice as to which of these formulae is the most accurate, although he has always had a preference for Taylor's method because this seems to him to have a better scientific basis. Unfortunately, however, its graphical presentation is not computer friendly and most designers will now prefer to use one of the other formulae.

It may be worth mentioning as a minor aside that designers used to make use — for very approximate powering — of the fact that almost all conventional ships except very fast ones have, at their service speed, a Ⓒ Froude value of about 0.70 or a Ⓒ ITTC of about 0.60.

A corresponding statement for today's more usual C_t notation might be:

C_{ts} = about 2.5×10^{-3} (reducing to about 2.4×10^{-3} if $L > 200$ m)

The Ⓒ ITTC value and the first of the C_t values quoted above equate when Ⓢ = 6.03.

The effective horsepower may be "naked", i.e., as given by a tank test conducted with "no" appendages or "inclusive" with the resistance of appendages added, or "ship predicted" with the further addition of a ship–model correlation factor.

Ship predicted inclusive effective horsepower

$$P_e(\text{ship predicted}) = P_e(\text{naked}) \times (1 + x)(1 + A) \qquad (6.20)$$

where

$(1 + x) =$ ship–model correlation factor, if not already included in P_e(naked) (see §6.2)

$A =$ additional resistance of appendages expressed as a fraction of the naked resistance (bilge keels, stabilisers, bow thrust tunnels, twin rudders, shaft brackets, bossings, etc.; see also §7.3).

(ii) Quasi-propulsive efficiency η_d

This is made up of the open water propeller efficiency η_o, the relative rotative efficiency η_r, and the hull efficiency η_h.

$$\eta_d = \eta_o \times \eta_r \times \eta_h \tag{6.21}$$

η_h in turn is made up from the thrust deduction and wake factors as follows:

$$\eta_h = \frac{1-t}{1-w_t} \tag{6.22}$$

where

$$t = \text{thrust deduction factor} = \frac{T-R}{T} \tag{6.23}$$

and

$$w_t = \text{Taylor wake fraction} = \frac{V-V_a}{V} \tag{6.23a}$$

$T =$ thrust and $R =$ resistance
$V =$ ship speed and $V_a =$ speed of advance.

An approximate value of η_d, derived by Emerson is

$$\eta_d = K - \frac{N\sqrt{L}}{10000} \tag{6.24}$$

where N is propeller RPM and K is a constant, originally given as 0.83, but which should probably be increased to 0.84 for single-screw ships with today's more efficient propellers. (See §7.5 for a discussion about this formula.)

(iii) Delivered horsepower P_d

This is the power at the propeller and is given by the formula

$$P_d = \frac{P_e(I)}{\eta_d} \tag{6.25}$$

(iv) Shaft horsepower P_s

This is the delivered horsepower increased to allow for the transmission efficiency η_t.

$$P_s = \frac{P_d}{\eta_t} \tag{6.26}$$

A normal value of η_t for a modern ship with machinery aft is 98.5–99%. For machinery amidships the value may reduce to about 98%.

(v) Brake horsepower P_b

This is the shaft horsepower increased to allow for gearing efficiency, η_g (if gearing is fitted).

$$P_b = \frac{P_s}{\eta_g} \tag{6.27}$$

A fairly usual value for η_g is 96%.

(vi) Service power P_{bs}

So far, the powers calculated apply to a clean ship in "trial" weather conditions. If the power required is for a fouled ship in service conditions a percentage must be added. Possibly the most usual addition is one of about 20%, but see §7.7.

(vii) Maximum continuous power P_{bc}

With diesel engines it is usual to limit the service power to a fraction of the power the engine is capable of developing on a continuous basis in order to improve its life and reduce maintenance. This is known as derating (d_r) and the most usual values of d_r are 85 or 90%.

$$P_{bc} = \frac{P_{bs}}{d_r} \tag{6.28}$$

This appears a useful place to define other factors used in this chapter.

(viii) Admiralty coefficient A_c

$$A_c = \frac{\Delta^{2/3} \cdot V^3}{P_b} \tag{6.29}$$

The Admiralty coefficient is a crude but still useful method of estimating power. Provided its use is confined to cases where there is only a relatively small change in displacement and speed from the "basis" and that other factors like propeller revs are constant, it can give reasonable results.

(ix) Froude and Reynolds' numbers

Froude number $(F_n) = \dfrac{V}{\sqrt{g \cdot L}}$ (6.30)

Volumetric Froude number $F_v = \dfrac{V}{\sqrt{g \cdot \nabla^{1/3}}}$ (6.31)

Reynolds number $(R_n) = \dfrac{V \cdot L}{v}$ (6.32)

where

$\left. \begin{array}{l} L \quad = \text{ship length} \\ \nabla \quad = \text{displacement volume} \end{array} \right\}$ all in compatible units.

V = speed
g = acceleration due to gravity
v = kinematic viscosity $= \mu/r$
μ = viscosity; r = density

In the units normally used:

v = 1.188×10^{-6} m²/S for salt water at 15°C
   = 1.139×10^{-6} m²/S for fresh water at 15°C

$F_n = \dfrac{0.298 V_k}{\sqrt{L_f}} = \dfrac{0.164 V_k}{\sqrt{L_m}}$

where

V_k = speed in knots
L_f = length in feet
L_m = length in metres.

The values of F_n in normal use range from about 0.12 to 0.48. For the BSRA standard 122 m (400 ft) ship these correspond to about 8–32 knots.

$F_v = \dfrac{0.164 V_k}{\Delta^{(1/6)}}$

where Δ = displacement S.W. in tonnes

It can be convenient to express Reynolds' number in terms of the Froude number

$$R_n = F_n \frac{g^{0.5} \times L^{1.5}}{v}$$

As R_n values are not exactly memorable, readers may find a few figures helpful:

	R_n (SW) for 122-m ship	R_n (FW) for 5-m model
For $F_n = 0.12$	4.26×10^8	3.69×10^6
For $F_n = 0.48$	17.05×10^8	14.76×10^6

(x) Water transport efficiency η(wt) (see Fig. 2.1)

$$\eta(wt) = \frac{\text{work done}}{\text{energy used}} = \frac{5.045 \Delta V_k}{P_b(kW)} \tag{6.33}$$

(xi) Froude's circular notation

One of the coefficients Ⓒ which forms part of this notation has already been mentioned, but as a number of the other notations are in use in the presentation of powering data it may be helpful to summarise these and explain their basis.

The "circular" notations which were devised by R.E. Froude in 1888 are a series of non-dimensional coefficients based in general on the dimensions of a cube which has the same volume as the displacement volume of the ship.

The side of this cube $\qquad U = \nabla^{1/3} = 0.9918\,\Delta^{1/3}$

The face of the cube has an area $\quad U^2 = 0.9837\,\Delta^{2/3}$

Where the non-dimensional expressions are converted below to dimensional units, the units used are:

L_m in metres
S_m in square metres
Δ in tonnes
V_k in knots.

Ⓜ the length/displacement ratio relates the ship's length to the length of a side of the cube having the same volume of displacement as the ship.

$$\text{Ⓜ} = L/U = 1.00826\,L_m/\Delta^{1/3}$$

Ⓢ the wetted surface coefficient relates the wetted surface S to the area of one side of the cube.

$$\text{Ⓢ} = S/U^2 = 1.0166\,S_m/\Delta^{2/3}$$

Froude's approximate formula for Ⓢ, which may be compared with Mumford's formula for S (see eq. (6.17)) is:

$$\text{ⓈＳ} = 3.4 + 0.5 \, L\sqrt{\Delta}$$

For the speed/length ratio, Froude introduced two different constants Ⓚ and Ⓛ, whilst a third, Ⓟ, was later added by Baker.

Ⓚ relates the speed of the ship to that of a trochoidal wave having a length of $U/2$

The speed of such a wave is $\sqrt{(g/2\pi)\times 1/2(\nabla)^{1/3}}$

$$\text{Ⓚ} = \frac{V}{\sqrt{(g/4\pi)\times(\nabla)^{1/3}}} = 0.5846 \, V_k /(\Delta)^{1/6}$$

Ⓛ relates the speed to that of a wave of length $L/2$

$$\text{Ⓛ} = \frac{V}{\sqrt{(g/2\pi)\times L/2}} = 0.5822 \, V_k /(L_m)^{1/2}$$

Some useful inter-relationships between Ⓚ, Ⓛ and F_n are given by:

$$\text{Ⓚ} = \text{Ⓛ} \times \text{Ⓜ}^{1/2}$$

$$\text{Ⓛ} = 3.545 \, F_n$$

Ⓟ relates the speed to that of a wave of length $C_p \cdot L$

$$\text{Ⓟ} = \frac{V}{\sqrt{(g/2\pi)\times(C_p \cdot L)}} = 0.412 \, V_k /(C_p \cdot L_m)^{1/2}$$

A more fundamental definition of Ⓒ than that already given is:

$$\text{Ⓒ} = \frac{R_t \times 1000}{\Delta \times \text{Ⓚ}^2}$$

Froude wanted to use R_t /Δ which is the total resistance per unit of displacement weight in identical units, but because the value of this ratio increases quite rapidly at high speeds he divided it by Ⓚ^{2} and multiplied it by 1000 to avoid small numerical values. The rather peculiar denominator D in eq. 6.13. can be derived as follows:

$$D = \frac{\text{Ⓒ} \, \Delta^{2/3} \, V_k^3}{P_e(\text{kW})}$$

Substituting the above formula for Ⓒ and Ⓚ together with the formula for P_e given in eq. 6.14 gives

$$D = \frac{R_t}{\Delta} \frac{1000}{(0.7355 \times 0.5846)^2} \frac{\Delta^{1/3}}{V_k^2} \frac{\Delta^{2/3}V_k^3}{1000R_t} \frac{75}{0.5144V_k}$$

$$D = \frac{75}{(0.7355 \times 0.5846)^2 \times 0.5144} = 580$$

⊙ is used in the Froude treatment of skin friction for the correction from model to ship size or vice versa or between two ship sizes L_2 and L_1.

$$Ⓒ_{L_2} = Ⓒ_{L_1} + \frac{[⊙_{L_2} - ⊙_{L_1}] \times Ⓢ}{Ⓛ^{0.175}}$$

6.4 EFFECTIVE HORSEPOWER CALCULATION METHODS IN GENERAL

Designers have available to them several methods for estimating effective horse-power. The older methods have, however, become outdated in recent years for a number of reasons:

(i) the increased dimensions and in particular the increased length of modern ships;

(ii) the greatly improved smoothness of the hull resulting from the change from riveted to welded construction, from the use of shop-blasted and primed instead of "weathered" plates and from the use of modern paints;

(iii) changes in the design of ship lines which have come from years of tank testing and have greatly improved the performance of modern ships;

(iv) the tank test results which formed the basis of the older methods were recorded prior to the universal adoption of trip wires or studs to eliminate laminar flow;

(v) the fact that the friction line used for the extrapolation from model to ship size has changed with the earliest data still in use being based on the Froude line and later results being calculated using the ITTC 57 line. More recently there has been the further change bringing in the form factor. This means that Ⓒ values ought to be annotated as Froude or ITTC'57, whilst C_t and EHP values have the further possible annotation of ITTC'78 indicating the use of the form factor. These annotations are essential if the correct ship–model correlation factor is to be chosen;

(vi) finally there is the fact that the units in which the data of some of the earlier methods is recorded date back to before the general adoption of SI units.

Altogether, the use of older methods can present a number of pitfalls to those to whom this is unfamiliar territory. Nevertheless, brief descriptions of two of the best known methods, Taylor and Ayre, are included in this chapter as it is thought that some lessons can be learnt from these, and occasionally when a designer has no better data available one of these methods may still be useful.

More modern methods are outlined in succeeding sections, which are arranged in historical order and reflect the technology of their dates of origin.

§6.7 Moor's method uses Ⓒ Froude
§6.8 Guldhammer and Harvald's method uses C_t ITTC'57
§6.9 Holtrop and Mennen's method uses C_t ITTC'78
§7.1 The use of in-house data
§7.2 Moor's data converted to model size for use with $C_t'57$, $C_t'78$ or Grigson friction lines.

Whilst the later methods may be the best in theory there is no doubt that there is more data available in some of the earlier forms and provided the appropriate ship/model correlation factors are used these can still give reasonably accurate results. But it is essential the testing used trip wires.

6.5 TAYLOR'S METHOD

This method was originally presented in a book entitled *Speed and Power of Ships* published in 1910 by Rear Admiral D.W. Taylor of the U.S. Navy. The book, which was revised in 1933 and 1943, was for a long time the best known work on ship powering.

The residuary resistance, which consists of all resistance other than the skin frictional resistance is obtained from graphs of R_r/Δ. Taylor plots values of this, for two values of B/T (2.25 and 3.75) and a range of values of $V/\sqrt{L}$ (from 0.30 to 2.0) against parameters of prismatic coefficient from 0.48 to 0.86 and displacement/length ratio from 20 to 250.

where displacement/length $= \dfrac{\Delta}{[L/100]^3}$ 6.34

Taylor's units are: V in knots, L in feet, and Δ in tons of 2240 lbs. R_r and R_f in lbs. The R_r/Δ values for the actual B/T are then obtained by interpolation.

The skin friction R_f is calculated on the basis of

$$R_f/\Delta = f S V^{1.83} \tag{6.35}$$

Taylor used a value of $f = 0.00904$ as derived by Tideman as applicable to a 500 ft ship of steel construction (riveted construction, clean and well painted) and suggested taking a value of S from the following formula:

$$S = C\sqrt{\Delta L}$$

The coefficient C has an average value of 15.4 in its original units of tons, feet and feet squared which becomes 2.6 for units of tonnes, metres and metres squared.

Taylor's wetted surface formula remains a widely used one and a modified SI unit graph of the coefficient plotted against midship section coefficient and B/T is reproduced as Fig. 6.1.

In their day, Taylor's methods gave quite an accurate estimate of the resistance, although some variables that are now thought to have a significant effect are missing. For example, no account is taken of the position of the longitudinal centre of buoyancy and no distinction is made between single-screw and twin-screw forms, both of which are generally recognised as influencing the resistance. By the same token, the power estimate gives little help towards designing lines to minimise the power required.

A much improved presentation of the Taylor tests is given in "A Reanalysis of the Original Test Data for the Taylor Standard Series" by M. Gertler published in DTMB Report 806 of 1954.

Several tanks to this day present the ratio model result/Taylor prediction as an indicator of quality.

6.6 AYRE'S C2 METHOD

This was presented by Sir Amos Ayre first in 1927 and revised in 1933 and 1948 in papers to the North East Coast Institution of Engineers and Shipbuilders.

Ayre's formula is

$$P_e = \frac{\Delta^{0.64} \cdot V^3}{C2} \tag{6.36}$$

This is very similar to the Admiralty coefficient formula, except that the formula is for EHP and not SHP, and the index of the displacement is 0.64 and not 2/3. Contours of $C2$ are plotted on a base of $V/\sqrt{L}$ for a range of $L/\Delta^{1/3}$ from 10 to 30.

These values are then corrected for block coefficient and position of the longitudinal centre of buoyancy, for both of which optimum values are laid down, so that all corrections are additions.

The corrections for block coefficient are fairly modest if the ship is finer than standard, but severe if it is fuller.

The corrections for the LCB position tend to be more severe if the LCB is forward of the standard position, those applicable if the LCB is abaft the standard position being smaller.

Interestingly, the standard block coefficient for a twin-screw ship is set 0.01 fuller than for a single-screw ship. The data used in preparing the curves of $C2$ is now unfortunately outdated. In its day, the method not only provided an accurate estimate of EHP but also gave some very useful guidance towards the optimisation of the ship's lines.

6.7 MOOR'S METHOD

6.7.1 Single-screw ships

In a paper entitled "The effective horsepower of single-screw ships —average modern attainment" presented to R.I.N.A. in 1959, Moor and Small give Ⓒ values for standard ship dimensions of $400 \times 55 \times 26$ ft ($122 \times 16.76 \times 7.93$ m) for a range of:

(i)   block coefficients — from 0.625 to 0.80 by 0.025 intervals;
(ii) LCB positions — from 2.00% aft to 1.75% forward;
(iii) speeds — from 10 knots to 18 knots, corresponding to $F_n = 0.15$ to $F_n = 0.27$.

Corrections to Ⓒ for other lengths of ship are read from a slightly complicated graph given in the paper, which the author has been able to simplify with only a trivial reduction in accuracy to the formula given below:

$$d\text{Ⓒ} = 4(L_2 - L_1)10^{-4} \qquad\qquad\qquad 6.37$$

where $d\text{Ⓒ}$ is the change in Ⓒ for a change in length from a basis length of L_1 to a new ship length of L_2, both in metres.

The corrections for differences in beam and draft from the standard values are made using indices of a type devised by Mumford, who was at one time Superintendent of the Denny Tank.

Mumford postulated that P_e varies as B^x and T^y. Moor investigated the values of x and y which had been found to apply to 11 different standard series before settling on $x = 0.90$ and a y value varying with Froude number from 0.54 for $F_n = 0.15$ to 0.76 for $F_n = 0.30$.

This can be transformed into

$$\text{\textcircled{C}}_2 = \text{\textcircled{C}}_1 \times [(B_2/B_1)^{0.23} \times (T_2/T_1)^{(1.47F_n - 0.35)}] \tag{6.38}$$

Moor gives limits for the use of these indices of a 10% change in the Beam ratio and a 15% change in the Draft ratio, whilst the F_n range is stated as from 0.15 to 0.27. The author, in his design work, has gaily extrapolated well beyond all three limits without his calculations being found in error to any degree by later tank testing. But caution is of course desirable.

It can be helpful when using Mumford indices to remember that $\text{\textcircled{C}}$ increases with an increase in the beam ratio, but generally reduces with an increase in the draft ratio except at high speeds.

Moor's $\text{\textcircled{C}}$ values are based on model results with turbulence stimulation but the Froude friction line was used and the appropriate ship/model correlation factors must therefore also be used.

The standard ship method was also adopted by B.S.R.A. (British Ship Research Association) for their very considerable programme of tank test research which covered block coefficients from 0.525 to 0.875.

Moor revised his "average" attainment data in a 1969 B.S.R.A. report, which also gives "optimum" values. This time the draft used was 24 ft (7.32 m) and the range of block coefficient was increased to cover from 0.54 to 0.88 and the $V/\sqrt{L}$ values to cover from 0.40 to 1.10. Although the effect of the draft change is not very great, this change needs to be carefully noted if great accuracy is required.

Possibly because of the increasing number of ships with bulbous bows by this time, Moor felt that the standards were better established without reference to the LCB position, although a plot of the LCB positions of the best forms used to derive the optimum curves is given.

6.7.2 Twin-screw ships

In another R.I.N.A. paper "Some aspects of passenger liner design" written in 1962 in conjunction with R.V. Turner and M. Harper, Moor presented what is probably the best of the rather limited data there is available on the resistance and propulsion of twin-screw ships. The data in the paper referred to "average modern attainment", but was revised and extended to give "optimum" data as a B.S.R.A. report in 1968.

For twin-screw ships the standard ship has the same length and beam, but the draft is reduced to 18 ft (5.5 m)

Corrections for beam and draft variations from the standard figures are again made using Mumford indices. These are given for a wider range of beam and draft ratios than for single-screw ships and cover a greater range of $V/\sqrt{L}$ and in this case the index x varies with F_n.

From $F_n =$ 0.21 to 0.29, x is constant at 0.83,
           thereafter it increases steadily to a value at
   $F_n =$ 0.60 of 1.77.

From $F_n =$ 0.21 to 0.25, y is constant at 0.50,
           thereafter it increases steadily to a value at
   $F_n =$ 0.60 of 1.47

A simple formula as given for single-screw ships is not possible, but an abbreviated presentation is given in Table 6.1.

Table 6.1

Mumford indices for twin-screw ships (abbreviated list with rounded figures)

B2/B1	Beam factor					
F_n	0.24	0.28	0.32	0.36	0.40	0.44
0.75	0.95	0.95	0.91	0.86	0.82	0.79
0.80	0.96	0.96	0.92	0.88	0.86	0.83
0.85	0.97	0.97	0.95	0.92	0.90	0.88
0.90	0.98	0.98	0.97	0.95	0.93	0.92
0.95	0.99	0.99	0.98	0.97	0.96	0.96
T2/T1	**Draft factor**					
0.45	1.14	1.02	0.94	0.91	0.81	0.66
0.50	1.12	1.02	0.94	0.92	0.83	0.70
0.55	1.10	1.02	0.95	0.93	0.85	0.73
0.60	1.09	1.01	0.96	0.94	0.87	0.77
0.65	1.08	1.01	0.97	0.95	0.89	0.80
0.70	1.06	1.01	0.87	0.96	0.91	0.83
0.75	1.05	1.01	0.98	0.97	0.93	0.86
0.80	1.04	1.01	0.98	0.97	0.94	0.89
0.85	1.03	1.00	0.99	0.98	0.96	0.92
0.90	1.02	1.00	0.99	0.99	0.97	0.94
0.95	1.01	1.00	1.00	0.99	0.99	0.97

For values of *B2/B1* or *T2/T1* greater than unity use the reciprocal of the parameter and the reciprocal of the factor.

6.8 GULDHAMMER AND HARVALD'S METHOD

A more modern powering method which avoids some of the difficulties mentioned earlier in this section was published by Guldhammer and Harvald in *Ship Resistance — Effect of Form and Principal Dimensions* (1974 Akademisk Forlag, Copenhagen). This defines the total resistance R_t as:

$$R_t = C_t \cdot 1/2 \cdot \rho \cdot S \cdot V^2 \tag{6.39}$$

and

$$C_t = C_r + C_f$$

where
  C_t = total resistance coefficient
  C_r = residual resistance coefficient
  C_f = frictional resistance coefficient
  ρ = mass density = r/g
  V = velocity
  S = wetted surface
(all in SI units).

The C_f value is based on the 1957 ITTC Friction line. A plot of this based on ship length L for a range of speeds from 0.1 m/s to 20 m/s (approx 0.2 to 39 knots) is reproduced in Fig. 6.2.

It is worth noting that the difference between the C_f values for two Reynolds' numbers can be used to correct a C_t value from one ship length to another (see §7.1).

These C_f values do not allow for a form factor, the use of which was not adopted by the ITTC until 1978.

A correction for the increased resistance resulting from any appendages is made by increasing C_f proportionally to the increased wetted surface due to the appendages.

$$C_f' = C_f \times S'/S \tag{6.40}$$

The C_r values are based on vessels with a standard position of LCB, a standard B/T value of 2.5, normal shaped sections and a moderate cruiser stern. They are plotted for a number of values of $L/\Delta * 1/3$ ranging from 4.0 to 8.0 by 0.5 steps. Each graph is on a base of Froude number and is plotted for a range of prismatic coefficients from 0.50 to 0.80 by 0.01 steps.

The graph of C_r is intended to correspond to an LCB position close to optimum. A graph of this standard (optimum) LCB position on a base of F_n is given, together with the correction to be applied if the actual LCB is forward of the standard. Interestingly, there is no correction for the LCB being aft of the standard position.

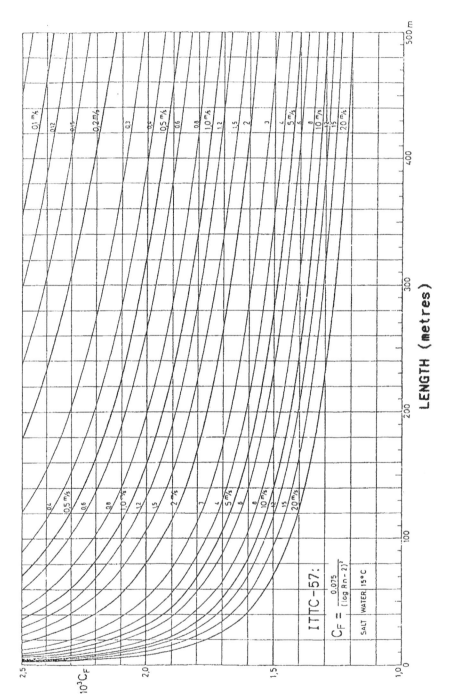

Fig. 6.2. $10^{-3} C_f$ (ITTC'57) versus length in metres and contours of speed in m/s (based on S.W. at 15 C; $r = 1.025$; $v = 1.191 \times 10^{-6}$ m^2 s^{-1}).

Correction for any variation in B/T from the standard value of 2.5 is made as follows:

$$10^3[C_r(B/T) - C_r(2.5)] = 0.16 [B/T - 2.5] \tag{6.41}$$

Corrections for the form of the sections are suggested if these are either extremely U or extremely V-shaped:
  – U-shaped bow sections and V-shaped stern sections reduce C_r;
  – V-shaped bow sections and U shaped stern sections increase C_r.

The standard form has an orthodox non-bulbous bow and corrections are given for bulbous bows of different sizes at a range of Froude numbers. In general a bulbous bow is shown to be advantageous at high Froude numbers relative to block coefficient. The corrections are given for the loaded condition but there is a statement that bulbous bows can give a remarkable decrease in resistance for full forms in the ballast condition.

A ship–model correlation factor for roughness and scale effect based on ship length is applied as a correction factor C_a. The value of this ranges from

$C_a = +0.4 \times 10^{-3}$ for a ship $L = 100$ m

through a zero value for a length of 200 m to

$C_a = -0.3 \times 10^{-3}$ for $L > 300$ m

The similarities which this method has with some aspects of Taylor's method and some aspects of Ayre's method may be noted.

6.9 HOLTROP AND MENNEN'S METHOD

The fact that the C_t method just discussed is already out of date to modern tank test procedures following the introduction of the form factor has been noted. Additionally it is not computer friendly as C_t must be read from a graph.

Holtrop and Mennen's method, which was originally presented in the *Journal of International Shipbuilding Progress*, Vol. 25 (Oct. 1978), revised in Vol. 29 (July 1982) and again in N.S.M.B. Publication 769 (1984) and in a paper presented to SMSSH'88 (October 1988), meets all these criteria with formulae derived by regression analysis from the considerable data bank of the Netherlands Ship Model Basin being provided for every variable. Many naval architects use the method, generally in the form presented in 1984 and find it gives acceptable results although it has to said that a number of the formula seem very complicated and the physics behind them are not at all clear, (a not infrequent corollary of regression analysis).

The total resistance coefficient C_t of a ship is subdivided into:

$$C_t = C_f(1 + K) + C_w + C_{app} + C_b + C_{tr} + C_a \qquad (6.42)$$

where

C_f	= frictional resistance coefficient to 1957 ITTC
$(1 + K)$	= form factor
C_w	= wave-making resistance coefficient
C_{app}	= appendage resistance coefficient
C_b	= coefficient of the additional pressure resistance of a bulbous bow near the surface
C_{tr}	= coefficient of the additional pressure resistance of an immersed transom stern
C_a	= coefficient of model–ship correlation resistance.

This formula is very similar to that given as eq. (6.7) for the 1978 ITTC treatment of resistance but there are a number of differences:

C_r in the ITTC'78 formula has been subdivided into C_w, C_b and C_{tr} whilst C_{air} has been omitted and ΔC has been changed to C_a.

The method provides regression analysis formulae for each of these resistance components and goes on to provide further regression formulae for estimating the propulsion factors of effective wake fraction, thrust deduction fraction and relative rotative efficiency.

Further formulae for the prediction of the propeller open water efficiency enables the calculation of the shaft horsepower to be completed in an expert system type computer calculation.

In the 1984 paper the authors state that they had focused attention on improving the power prediction of high block coefficient ships with low L/B ratios at one end of the spectrum and of fine-lined slender naval ships at the other, so the method has a wide application.

Due to a policy decision by Marin, not all the formula are given for the last reference but those in general use are given in the 1984 paper and there seems no point in repeating these here although as a sample of the great pains these authors have taken to bring in every variable which may affect one of their factors and of the complexity which results, a slightly modified version of the formula for $(1 + K)$ from the 1984 paper is given below.

The modification made to the formula consists of reducing the coefficients and indices from five or six decimal places to three, which seems more appropriate to the probable accuracy of a formula produced by regression analysis. This simplification has the advantage of enabling the formula to be fitted into a page more easily and whilst it may have introduced some error this seems unlikely to be significant.

It was rather disconcerting to find that the formula for $(1 + K)$ in the third reference was significantly changed from that given in the second with new variables being introduced, and even more to find an even greater change in the last reference involving a new factor Y which modifies $(1 + K)$ to $(1 + Y \cdot K)$ with Y varying with Froude number.

As this new factor is associated with formulae for C_w which are not given, most users continue to use the formulae given in the third reference which seems to have given reasonably satisfactory results to date.

The simplified formula for $(1 + K)$ mentioned above is as follows:

$1 + K =$

$$0.93 + 0.487 \cdot (C)_{14} \cdot (B/L)^{1.068} \cdot (T/L)^{0.461} \cdot (L/L_r)^{0.122} \cdot (L^3/\nabla)^{0.365} \cdot (1-C_p)^{-0.604} \quad (6.43)$$

factor no.   1   2   3   4   5   6 (for reference)

In this formula L is waterline length
   C_p is the prismatic coefficient on this length
The length of run L_r in factor 4 is defined as:

$$L_r = L[1 - C_p + 0.06 \, C_p \, \text{lcb}/(4 \, C_p - 1)] \quad (6.44)$$

where lcb is the longitudinal centre of buoyancy forward (+) or aft (−) of 0.5 L as a % of L.

Factor 1 is defined as $C_{14} = 1 + 0.011 \, C_{stern}$
C_{stern} = −25 to −20 barge-shaped forms
       = −10 after body with V sections
       = 0 normal shape of after body
       = +10 after body with U sections and Hogner stern

Figure 6.3 abstracted from the 1988 paper may help interpretation of these values.

Another reason for giving the formula for $(1 + K)$ is because there seems to be little other data on this factor and designers may wish to use it along with other powering data (but see later).

Most users of this formula will tend to use it embedded in a computer program and will thus gain little, if any, knowledge of the relative importance of the various factors or, indeed, of what $(1 + K)$ value to expect for a particular type of ship.

Values have therefore been calculated in Table 6.2 for a number of ship types. The first three ships conform to the standard dimensions of $122 \times 16.76 \times 7.32$ m with two extreme block coefficients and a middle value. The other particulars required being set at values reasonably appropriate to each of the block coefficients.

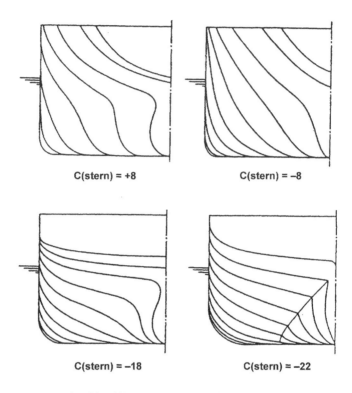

Fig. 6.3. Different stern types and C_{stern} values.

The need for C_p values corresponding to the C_b values chosen was met by plotting Fig. 8.9. This shows values of C_m for ships with no rise of floor, with a low rise of floor and a high rise of floor. The first two representing merchant practice, the latter that of frigates and corvettes.

C_b	0.55	0.70	0.85
C_m	0.97	0.975	0.98
C_p	0.567	0.718	0.867
∇	8232	10477	12722
LCB	−2.5% (aft)	amidships	+2.5% (ford)
Sections	normal	normal	normal

All three ships	$L/B = 7.28$	$T/L = 0.06$

For these three ships with identical main dimensions the factors affecting $(1 + K)$ can be seen to be factors 4, 5 and 6 — the position of the LCB, the volume of displacement and the prismatic coefficient. The fact that the effect of fullness reduces 5 but increases 6 suggests that it might be possible for these two factors to be combined.

Table 6.2

$(1 + K)$ Form factor based on Holtrop and Memmen's formula

A	B	C	D	E	F	G	H	I	J	K	L	M	N	O	P
	Variable studied	C_b	C_m	C_p	LCB (%)	L/B	T/L	L_r/L	C_{14}	$(B/L)^{1.07}$	$(T/L)^{0.46}$	$(L/L_r)^{0.12}$	$(L^3/\nabla)^{0.36}$	$(1-C_p)^{-0.60}$*	$(1+K)$
7	C_b and LCB	0.550	0.970	0.567	−2.5	7.28	0.06	0.432	1.000	0.120	0.273	1.108	7.168	1.658	1.142
8	standard	0.700	0.975	0.710	0	7.28	0.06	0.282	1.000	0.120	0.273	1.167	6.397	2.148	1.188
9	proportions	0.850	0.980	0.867	2.5	7.28	0.06	0.136	1.000	0.120	0.273	1.276	5.965	3.387	1.344
10															
11	Cb and LCB	0.550	0.970	0.567	−2.5	5.5	0.06	0.432	1.000	0.162	0.273	1.108	6.308	1.658	1.181
12	low L/B	0.700	0.975	0.718	0	5.5	0.06	0.282	1.000	0.162	0.273	1.167	5.783	2.148	1.244
13	ratio	0.850	0.980	0.867	2.5	5.5	0.06	0.136	1.000	0.162	0.273	1.276	5.393	3.387	1.435
14															
15	C_b and LCB	0.550	0.970	0.567	−2.5	8.5	0.06	0.432	1.000	0.102,	0.273	1.108	7.378	1.658	1.115
16	high L/B	0.700	0.975	0.718	0	8.5	0.06	0.282	1.000	0.102	0.273	1.167	6.764	2.148	1.161
17	ratio	0.850	0.980	0.867	2.5	8.5	0.06	0.136	1.000	0.102	0.273	1.276	6.308	3.387	1.301
18															
19	Different	0.500	0.840	0.595	−2.5	8.5	0.04	0.404	0.900	0.102	0.227	1.117	8.835	1.727	1.103
20	sterns	0.500	0.840	0.595	−2.5	8.5	0.04	0.404	1.000	0.102	0.227	1.117	8.835	1.727	1.123
21	T/L and	0.500	0.840	0.595	−2.5	8.5	0.06	0.404	1.000	0.102	0.273	1.117	7.635	1.727	1.131
22	C_m values	0.500	0.965	0.518	−2.5	8.5	0.06	0.481	1.000	0.102	0.273	1.093	7.635	1.554	1.107

* Indices rounded from those shown in eq. 6.43.

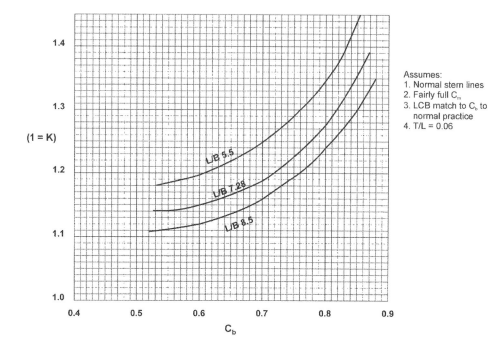

Fig. 6.4. (1 + K) Based on Holtrop and Mennen's formula plotted against block coefficient as primary variable.

These (1 + K) values are plotted in Fig. 6.4 against block coefficient, which seems the most important parameter as changes in most of the other factors such as V, C_p, LCB tend to be associated with a change in block coefficient. As Fig. 6.4, shows (1 + K) increases with the block coefficient.

To explore some of the other factors, further values were then calculated for ships with L/B ratios of 5.5 and 8.5, respectively.

An L/B ratio of 5.5 was chosen as quite usual practice for a modern tanker and only a little lower than that used for a modern cruise liner. An L/B ratio of 8.5 on the other hand represents a value which might apply to a frigate.

In these cases the obvious change at constant block coefficient is that made by factor 2 but there is also in each case a change in factor 5.

As might be expected the (1 + K) factor increases when L/B is reduced to 5.5 and reduces when L/B is increased to 8.5.

Interpreting the rules for ⓒ$_{14}$ presents some difficulty but for most single-screw ships a zero value of (C)(stern) seems appropriate, whilst for twin-screw ships the –10 is probably correct; compare this with Ayre's view that the basic lines of twin-screw ships are better than those of single-screw ships. The effect of this factor can be seen by comparing lines 19 and 20.

Except for shallow draft ships the variation in T/L is limited and factor 3 has little effect as can be seen by comparing lines 20 and 21.

Factor 4 is of course dependent on the LCB position and to a lesser extent on the C_p, both of which may on occasion vary from the norms assumed in the table.

A variation in C_p from the norm, such as applies to frigates with a high rise of floor and a low C_m will also change factor 6 increasing $(1 + K)$, an effect shown in lines 21 and 22 of Table 6.2.

Some further thought suggested that it might be useful to have a basic value of $(1 + K)$ which applies to a ship which is "middle of the range" in all respects and develop a series of factors that correct its $(1 + K)$ value for changes in the various factors from the assumed basic values.

It was hoped that this would give both a better feel for the value of $(1 + K)$ likely to apply to a particular type of ship form and of the relative effects exercised by the various form factors.

For the middle of the range ship it seemed appropriate to use the $122 \times 16.76 \times 7.32$ m ship with $C_b = 0.70$, $C_m = 0.975$, $C_p = 0.718$, normal stern sections, LCB amidships, for which $(1 + K)$ has already been calculated. This gives a standard $(1 + K)$ value of 1.185 and the following guidelines for changes from the standard.

Stern lines:	basic is "normal"
	for U sections add 0.02
	for V sections subtract 0.02
L/B ratio:	basic is 7.3
	normal range is 5.5 to 8.5
	per unit of $(L/B - 7.3)$ subtract 0.02
	per unit of $(7.3 - L/B)$ add 0.03
T/L ratio:	basic is 0.06
	normal range for cargo ships is 0.058 to 0.064
	other ship types may go down to 0.03 (hardly worth consideration)
LCB position:	basic amidships
	per 1% aft of amidships subtract 0.02
	per 1% forward of amidships add 0.04
Block coefficient:	basic 0.70
	per unit of $10(C_b - 0.70)$ add 0.06
	per unit of $10(0.70 - C_b)$ subtract 0.03

Further comments on $(1 + K)$ are given in the next chapter.

Chapter 7

Powering II

7.1 POWER ESTIMATING USING IN-HOUSE DATA

7.1.1 The alternative methods

Most naval architects will have available to them a number of tank test results and may wish to base power estimates for new designs on these in preference to relying on methods employing unknown data. The use of in-house data is of course particularly desirable for a specialist ship if data relating to a similar basis ship is available.

The main difficulty in using in-house data lies in the fact that such data is likely to be in a variety of formats depending on the date of the tests and the tank used. Great care needs to be taken to ensure that the data is used correctly. In principle there are two ways of working:

1. by using the method used in the tank test report, or
2. by converting the data to an up-to-date method.

The first of these methods is probably the easiest but its use means that the estimate for the new design can only be as accurate as the method in use at the time of the tank test and no advantage will be gained from the improvements since that date.

There may also be a need to become familiar with conversion factors for units that are now rarely used — tons, British horsepower, Imperial knots, etc.

Even the oldest of the methods, which used the Froude friction line, can still give good results provided an appropriate $(1 + X)$ correlation value is used. And there is of course a vast amount of data in this format.

If the original tank test results are available, re-analysis of these to the ITTC'78 method is quite easy, but if the results are only available as ship estimates either for actual ship dimensions or for BSRA standard ship dimensions the conversion involves a number of steps.

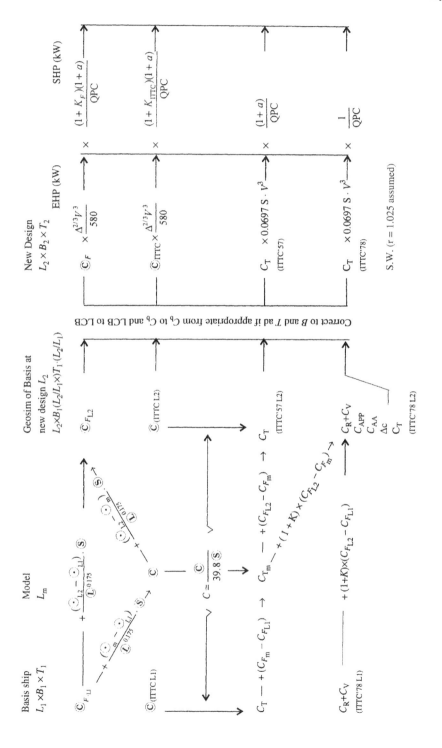

Fig. 7.1. The inter-relationship between different powering methods.

7.1.2 Re-analysis by an up-to-date method

The first step is to recreate the model results by reversing the process used for the ship estimate and then subject these to the ITTC'78 treatment, all generally as illustrated in Fig. 7.1.

Having drawn Fig. 7.1, the author decided to check how well the two E.H.P. formulae agreed with one another. As the process used also illustrates a quick method to approximate ship dimensions it seemed worth including as a digression.

A bulk carrier with a deadweight of 24000 tonnes and a service speed of 15 knots was used as a sample.

From Fig. 3.3 dwt/disp = 0.8 making disp = 30000 tonnes
From Fig. 3.8 $L/B = 6.25$; $B/D = 1.88$; $T/D = 0.71$
From Fig. 3.12 $C_b = 0.75$ assuming $F_n = 0.2$ approx $f = 1.023$; $(1 = S) = 1.05$

Using eq. (3.5)

$$L = \left[\frac{30000 \times 6.25^2 \times 1.88}{1.025 \times 1.5 \times 0.75 \times 0.71} \right]^{1/3}$$

$L = 156.65$ m; $B = 25.06$ m; $D = 13.33$; $T = 9.47$

From Fig. 6.1 $C = 2.55$; $S = 2.55 \, (30000 \times 156.65)^{1/2}$
$$S = 5528 \text{ m}^2$$
From §6.3(x) $\text{(S)} = 1.0166 \times 5528/(30000)^{2/3}$
$$= 5.82$$

Assume $C_{ts} = 2.5 \times 10^{-3}$
Equation (6.16) gives $\text{(C)} = 39.8 \times 2.5 \times 10^{-3} \times 5.82 = 0.579$

$P_e = \dfrac{\text{(C)} \cdot \Delta^{2/3} \cdot V^3}{580} = 3254 \text{ kW}$ ⎫
⎬ remarkable agreement
$P_e = 0.0697 \, C_{ts} \cdot S \cdot V^3 = 3251 \text{ kW}$ ⎭

Returning to the re-analysis:
A computer spreadsheet provides a convenient way of handling what is quite an involved calculation and one that a design office is likely to have to do repeatedly. Figure 7.2 gives the headings of a Lotus 123 spread sheet used by the author. This is reasonably comprehensive and can accommodate a variety of different input formats and give alternative outputs.

The formulae involved in the various steps are as follows:

(M) — Circular M

B	C	D	E	F	G	H	I	J	K	L	M	N	O	P	Q	R	S	T	U	V	W	X	Y
	Basis ship (1)	New design (2)	F_n	Ⓚ	Ⓛ	R_n (L1)	Basis $\bar{C}_F$ (L1)	Froude frict L1 to m	Model R_n Ç m (m)	(m)	C_t (m)	C_f	C_r ITTC·57	R_n (L2)	C_t (L2)	C_t (L2) ITTC·57	1+K	C_t (m) × (1+K)	C_t ITTC·7 8	C_t(L2) ×(1+K) ITTC·7 8	C_t(L2) geosim corr. ITTC·7 8	Mumford corr. B and T	C_t(L2) corr. B and T
L																							
B										or													
T																							
Δ							Basis																
C_b							©ITTC C_t ITTC C_r C_t																
C_m							(L1) (L1) (L1) ITTC·57																
C_p																							
Ⓜ																							
Ⓢ																							
L/B																							
Mumford B																							
T/L																							
Mumford T																							
Basis Model																							
f																							
⊙																							

Fig. 7.2. A spreadsheet for conversion from ©Froude or ©ITTC to C_t ITTC'57 or C_t ITTC'78 and correction to required dimensions.

$$\textcircled{M} = \frac{L}{\nabla^{1/3}} = \frac{L}{[C_b \cdot L \cdot L \cdot B / L \cdot L \cdot T / L]^{1/3}} = \frac{[L/B]^{1/3}}{[C_b \cdot T / L]} \tag{7.1}$$

$\textcircled{S}$ — Circular S

Froude's formula was one of two most commonly used by British tanks to arrive at $\textcircled{S}$ when moving from the measured model resistance to $\textcircled{C}$; the other made use of Mumford's formula for S given in eq. (6.17). When reversing direction from $\textcircled{C}_{Froude}$ back to C_{tm} it is important to use the same formula

Froude's $\textcircled{S} = 3.4 + 0.5\ L/\nabla^{1/3}$ \hfill (7.2)

Another formula for $\textcircled{S}$ can be devised as follows:

$$\textcircled{S} = \frac{1.0166S}{\Delta^{2/3}} = \frac{1.0166c(\Delta L)^{1/2}}{\Delta^{2/3}}$$

$$= 1.0124 \cdot c\ \textcircled{M}^{1/2} \tag{7.3}$$

where c is the constant in Taylor's wetted surface formula metricated (see Fig. 6.1).

This can be transformed to:

$$\textcircled{S} = 1.0124 \cdot c \cdot \left[\frac{L/B}{C_b\,T/L} \right]^{1/6} \tag{7.4}$$

For most ships with C_m between 0.95 and 0.99, $c = 2.55$.

For the BSRA standard ship with dimensions $122 \times 16.76 \times 7.32$ m, $L/B = 7.28$ and $T/L = 0.06$, this transforms to:

$$\textcircled{S} = \frac{5.74}{(C_b)^{1/6}} \tag{7.5}$$

A comparison of the $\textcircled{S}$ values obtained using these three different formulae is given in Table 7.1.

$\textcircled{L}$ and $\textcircled{K}$ — Circular L and K

$\textcircled{L} = F_n \times 3.545;\ \textcircled{K} = \textcircled{L} \times \textcircled{M}^{1/2}$

It is most important that $\textcircled{C}$ data is designated as either Froude or ITTC and is related to a particular ship length or to a model.

$$R_n = F_n \frac{g^{0.5}\,L^{1.5}}{v}$$

Table 7.1

A comparison of Ⓢ values using different formulae based on standard ship dimensions
$L = 122, B = 16.76, T = 7.32, L \cdot B \cdot T = 14967$
Froude Ⓢ $= 3.4 + 0.5 \, L/V^{i/3}$
Mumford Ⓢ $= (1.7 \, L \cdot T + C_b \cdot L \cdot B)/V^{2/3}$
Modified Taylor for standard dimensions Ⓢ $= 5.74/(C_b)^{1/6}$

C_b	V	Ⓢ Froude	Ⓢ Mumford	Ⓢ Taylor
0.55	8232	6.421	6.481	6.342
0.60	8980	6.335	6.352	6.250
0.65	9729	6.258	6.246	6.167
0.70	10477	6.188	6.158	6.092
0.75	11225	6.124	6.086	6.022
0.80	11974	6.066	6.025	5.958
0.85	12722	6.013	5.973	5.898

Maximum difference about 2%.

$$v = 1.188 \times 10^{-6} \text{ for S.W.}$$
$$= 1.139 \times 10^{-6} \text{ for F.W.}$$

If model length is not known, a length of 5 m can be assumed without this being likely to introduce any significant error.

If the input Ⓒ is Froude then this must be taken back to model size (or to the length of a new ship design) by the use of the Froude friction line. The correction from a length $L1$ to a length $L2$ (of the model or new ship) is:

$$Ⓒ_{F.L2} = Ⓒ_{F.L1} + [●_{L2} - ●_{L1}] \, Ⓢ \, Ⓛ^{-0.175}$$

where

$$● = \frac{12.767 \, f}{L_{m}^{0.0875}} \quad \text{and} \quad f = 0.00871 + \frac{0.01616}{2.68 + L_m}$$

these two formulae being metricated versions (L_m in metres) of the Froude friction line formulae.

If the Ⓒ value is ITTC it can be converted to model length (or the length of a new ship design) by the formula:

$$Ⓒ_{L2} = Ⓒ_{L1} + (C_{f_{L1}} - C_{f_{L1}}) \times 39.8 \, Ⓢ \tag{7.6}$$

Once at model length the conversion from Ⓒ to C_t is made using the formula:

$$C_t = Ⓒ/39.8 \, Ⓢ \tag{7.7}$$

Although rarely required, a direct conversion from ⒸF to ⒸITTC or *vice versa* can be made by reversing direction at the model scale or by the use of Fig 7.3,

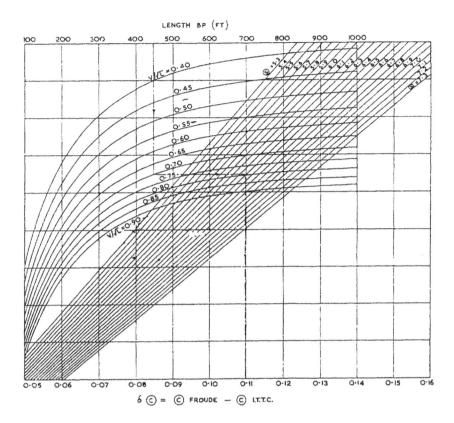

Fig. 7.3. Difference between merchant ship Ⓒ values based on Froude and ITTC skin friction corrections.

abstracted from Lackenby and Parker's R.I.N.A. paper "The BSRA Methodical Series — An Overall Presentation".

In ITTC'57 the values of C_{fm} for the model and C_{fs} for the ship under design (C_{fs}) are calculated using the formula:

$$C_f = \frac{0.075}{(\log R_n - 2)^2} \tag{7.8}$$

At this point the difference between ITTC'57 and ITTC'78 must be taken into account.

$C_r('57) = C_{tm} - C_{fm}$ and $C_{ts}('57) = C_{fs} + C_r('57)$

$C_r('78) = C_{tm} - C_{fm}(1 + K)$ and $C_{ts}('78) = C_{fs}(1 + K) + C_r('78)$

7.1.3 Calculating (1 + K)

There are two possible approaches to obtaining a $(1 + K)$ value to use in a $C_{ts}('78)$ calculation. The first of these is, of course, the Holtrop and Mennen formula given in §6.9 — or any update of this which becomes available.

A second method is to calculate a $(1 + K)$ value using the tank test data. The simplest approximate way is to assume that $C_r = 0$ at the lowest available Froude number (provided this is less than 0.18) and therefore at this point $(1 + K) = C_{tm}/C_{fm}$. A value established in this way will be higher than that given by the more accurate Prohaska method described below as there is likely to be a small residual C_r.

The Prohaska method assumes that C_r is a function of F_n^4.

$$C_{tm} = C_{fm}(1 + K) + k_2 F_n^4 \tag{7.9}$$

The value of K can be obtained at $F_n = 0$, by plotting

$$\frac{C_{tm}}{C_{fm}} - 1 = K + k_2 \times \frac{F_n^4}{C_{fm}} \tag{7.10}$$

as ordinate against F_n^4/C_{fm} as abscissa and finding the intercept at $F_n = 0$.

An approximate formula which avoids plotting uses two data points and the following formula:

$$(1 + K) = \frac{C_{tm1} - C_{tm2}(F_{n1}/F_{n2})^4}{C_{fm1} - C_{fm2}(F_{n1}/F_{n2})^4} \tag{7.11}$$

The data points should be at low F_n values (< 0.18), where Prohaska's line should be straight. Whilst the method is theoretically correct, the absence of the smoothing which plotting provides can introduce error and to minimise this it is wise to make two calculations with two adjacent sets of data. The results will show if the points are out of line and taking an average of two values should increase the accuracy.

A spread sheet for calculating $(1 + K)$ in this way is shown in Table 7.2. The data used in this table relate to the warship powering data presented in Fig. 7.10.

A comparison of the figures calculated in Table 7.2 with the Holtrop and Mennen figures shown in Fig. 6.4 shows very good agreement. It had been intended to make similar calculations for the single screw data plotted in Figs. 7.6 to 7.8 and the twin screw merchant ship data in Fig. 7.9, but on examination this data was found to be unsatisfactory at the low Froude numbers required for this calculation (see also §7.2.2).

Although trying to establish $(1 + K)$ values from some data may be unsuccessful, designers with suitable tank test results, and especially where these relate to specialist ship types, are recommended to try to establish their own $(1 + K)$ values.

Table 7.2

Twin-screw warship.

Calculation of $(1 + K)$ based on $(1 + K) = \dfrac{C_{tm1} - C_{tm2}(F_{n1} / F_{n2})^4}{C_{fm1} - C_{fm2}(F_{n1} / F_{n2})^4}$ (C_{fm} values based on ITTC'57)

C_b	F_{n1}	C_{tm1}	C_{fm1}	F_{n2}	C_{tm2}	C_{fm2}	$\left[\dfrac{F_{n1}}{F_{n2}}\right]$	$[\,]^4$	$1 + K$
0.391	0.144	0.003794	0.003497	0.192	0.003764	0.003316	0.750	0.316	1.063
0.420	0.146	0.003821	0.003489	0.195	0.003792	0.003308	0.749	0.314	1.073
0.451	0.148	0.003849	0.003481	0.197	0.003820	0.003301	0.751	0.319	1.083
0.486	0.150	0.003878	0.003473	0.199	0.003850	0.003294	0.754	0.323	1.094
0.525	0.152	0.003908	0.003465	0.202	0.003881	0.003286	0.752	0.321	1.105
0.567	0.153	0.003940	0.003457	0.205	0.003913	0.003278	0.746	0.310	1.117
0.614	0.156	0.003973	0.003448	0.207	0.003946	0.003271	0.754	0.323	1.128

Having established $(1 + K)$ at a low Froude number, the same value is assumed to apply at higher Froude numbers.

It is worth noting that a C_{ts} value calculated to ITTC'78 will always be less than a C_{ts} based on ITTC'57 with the difference increasing both at the higher values of K that apply to fuller block coefficients and with the reduced values of C_{fs} applicable to longer ships. Both of these are clearly shown in eq. (7.12) (see also §7.2.2).

$$C_{ts'78} = C_{ts'57} - K(C_{fm} - C_{fs}) \qquad (7.12)$$

7.1.4 The Grigson friction line

Another alternative to both ITTC'57 and ITTC'78 involves the use of the Grigson friction line which has already had a brief mention in §6.2.4. This line was presented in 1993 in an R.I.N.A. paper "An accurate smooth friction line for use in perform-ance prediction". Grigson plotted a mass of experimental data on friction coeffi-cients and found that the ITTC'57 line did not provide an accurate representation of this data. To improve the accuracy required a change from the simple formula used for ITTC'57 and the use of two separate formulae for respectively the Reynold's number ranges of tank test models and ships. Both of the formulae proposed are somewhat complex but can be readily built into a spreadsheet where they will do their job without any problems. In considering the merits of Grigson's line it is perhaps worth remembering that ITTC'57 was never claimed to be an accurate friction line but was introduced to improve ship/model correlation. The Grigson formulae are:

For the model range of $1.5 \times 10^6 < R_n < 2 \times 10^7$

$$C_f = [0.93 + 0.1377(\log R_n - 6.3)^2 - 0.06334(\log R_n - 6.3)^4]$$
$$\times (0.075/(\log R_n - 2)^2 \qquad (7.13)$$

For the ship range of $10^8 < R_n < 4 \times 10^9$

$$C_f = [1.032 + 0.02816(\log R_n - 8) - 0.006273(\log R_n - 8)^2]$$
$$\times (0.075/(\log R_n - 2)^2 \qquad (7.14)$$

The last factor in each of these equations is of course the ITTC'57 formula whilst the first is a suitable modifier.

The two lines are shown in Fig. 7.4, and an abbreviated tabular comparison of the two C_f values is given in Table 7.3, from which it will be seen that at low Reynold's numbers corresponding to models the Grigson value is generally less than the ITTC value (from 6% to about equal), whilst at ship size Reynolds numbers it is 5 to 6% more. The former results in $(1 + K)$ Grigson being greater as C_r is the same in both cases. As the Grigson C_f is larger at ship size and is multiplied by a larger $(1 + K)$, a Grigson C_{ts} will be greater than an ITTC'78 C_{ts}, which is less than an ITTC'57 C_{ts} — bringing Grigson and ITTC'57 values fairly near to one another, subject to the $(1 + K)$ value used.

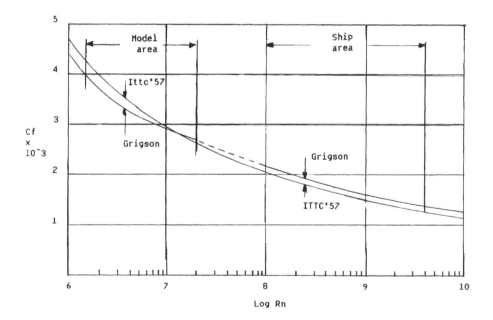

Fig. 7.4. A comparison of Grigson'93 and ITTC'57 friction coefficient C_f values.

Table 7.3

An outline comparison between the C_f values given by ITTC'57 and Grigson TRINA'93

Range	Rn	ITTC'57	Grigson TRINA'93	Ratio 93/57
Model size	4.0×10^6	0.0035143	0.0033347	0.94169
	2.0×10^7	0.0025590	0.0026877	1.0070
Ship size	4.0×10^8	0.0017207	0.0018008	1.0466
	2.0×10^9	0.0014070	0.0014883	1.0578

Clearly compatible $(1 + K)$ values are required if the Grigson friction line is to be used. Until now, this has necessitated designers applying the Prohaska method to the data they are using — and the problems this can bring have already been indicated. Fortunately, however, Grigson has now followed up his earlier work with an investigation into matching $(1 + K)$ values.

Following an analysis based on 78 data points covering a range of C_b from 0.47 to 0.89, Grigson obtained a straight line with an acceptable deviation on three alternative parameters. The one with the smallest deviation (an RMS of 0.033) is shown in Fig. 7.5. This uses a parameter P as the abscissa, where

$$P = (C_b)^{1/3} \cdot S/L^2 \tag{7.15}$$

where S = wetted surface area and L is the waterline length.

It is interesting to substitute Taylor's wetted surface in this which gives

$$P = 1.012 \cdot C \cdot C_b^{5/6} \cdot \frac{(T/L)^{1/2}}{(L/B)^{1/2}} \tag{7.16}$$

Taylor's C in the formula being itself a function of C_m and B/T.

The simplicity of this parameter contrasts with the very involved Holtrop and Mennen formula and yet seems to bring the available data into line whilst invoking what would seem to be the most important criteria.

With the line going through the origin the formula for K could hardly be simpler.

$$K = 1.4 \, (C_b)^{1/3} \cdot S/L^2 \tag{7.17}$$

Grigson, ever a perfectionist, is seeking additional data to confirm this excellent result before publishing it under a title such as "A fresh look at the determination of hull resistance from models", but has, in advance of this, most kindly allowed his $(1 + K)$ value to be given its first publicity in this book. ·

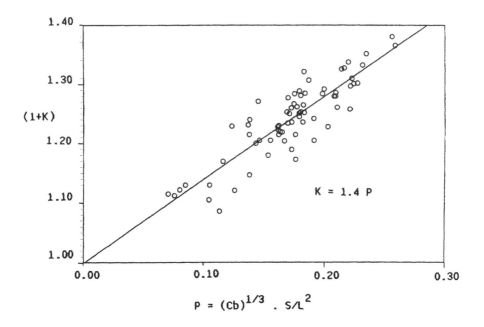

Fig. 7.5. $(1 + K)$ values for use with Grigson friction line.

7.1.5 Corrections for differences in hull geometry

All the C_{ts} values so far obtained relate to a geosim of the basis ship and further corrections must be made for any departures from this geometry.

Although the use of the basis ship data must presuppose that the C_b value and the LCB position of the new design differ to only a minor extent from the values of the basis ship, corrections should be made for the effect of these features if they differ.

A correction for a difference in C_b can be made by using the ratio of the C_t values given for the two block coefficient values on one of the graphs given in this section.

There is less data available to assist in making a correction for a change in LCB position but Guldhammer and Harvald's data and some of Moor's data can be used — again in the form of a ratio.

Corrections for a change in L/B and/or the L/T ratio can be made using the Mumford indices given in §6.7.

7.2 POWERING DATA

7.2.1 A "C_t" Method based on Moor's and similar data

In giving the title "Practical Ship Design" to this book, the author intended not only to present the theory of ship design, but to provide sufficient data to enable a reader to prepare at least an outline design for many types of ship without the need to refer to other data.

The discussion of powering methods has so far stopped short of providing such data, partly because full data on each of the methods described is readily available in the reference documents quoted and partly because an abstract suitable for this book could cover only a limited area of Froude Number, block coefficient etc. It is now time to give some data.

7.2.2 C_{tm} for single screw ships

For single screw ships the author originally intended to draw on his 1981 Parsons memorial paper "Designing ships for fuel economy". In writing that paper he felt it important to establish a "base case" of a thoroughly efficient powering perform-ance for a range of ships against which comparisons could be made, as too many claims of substantial improvements in performance — of increased speed and/or reduced fuel consumption — could be shown on analysis to be based on poor performance by the ship or machinery used as the reference point. For the perform-ance criteria he turned to Moor's work which has already been mentioned and to the B.S.R.A. Methodical series, updating these to transform the lines to Froude number from the historic $V/\sqrt{L}$ basis (V in knots and L in feet), which had become anachronistic in an SI unit age, and changing the ordinate from Ⓒ Froude to C_{ts} ITTC'57, retaining the standard ship dimensions of $122 \times 16.76 \times 7.32$ m.

By the time this book came to be written, C_t ITTC'57 had itself become out of date with the introduction of form factor and C_t ITTC'78. To transform the Moor and BSRA data into this new format was quite easy using a spread sheet and Table 7.4 shows the transformation of Moor's single screw average values.

Because each block coefficient has a different $(1 + K)$ and this value is needed when correcting to the dimensions of a new design, plotting C_t ITTC'78 is impractical. Before discussing the plot which the author ultimately decided to use, it is worth looking at the comparison of C_t'57 and C_t'78 given in Table 7.4. These differ quite considerably, with C_t'57 the greater by from about 5% to 20%. The difference is greatest on full ships at slow speeds reducing for finer ships and faster speeds. The comparison is not strictly correct as ΔC, C_{air} and C_{app} have to be added to ITTC'78 when calculating the total resistance coefficient and this brings the two values a little closer. The fact that there are some negative C_r'78 figures in Table 7.4 appears to stem from an anomaly in the original data in which resistance

Table 7.4. Moor's average single screw transformed from ©F to C_{tm}, C_{ts} '57 and C_{ts} '58.
Ship: $L = 122$ m, $B = 16.76$ m, $T = 7.32$ m, $L \cdot B \cdot T = 14957$. Model: $L = 5$ m, $L/B = 7.32$, $T/L = 0.06$

C_b	V	$\textcircled{S}$	$V_k\sqrt{L_f}$	F_n	L	S.W.R_n(s) (122 m)	F.W. R_n(m) (5 m)	$\textcircled{C}$F(s)	Frict corr $\textcircled{C}$m m–s	C_{tm}
0.55	8232	6.42	0.60	0.179	0.634	635221697	5497096	0.660	0.319	0.979	0.003829
0.55	8232	6.42	0.70	0.209	0.739	741091980	6413278	0.669	0.310	0.979	0.003831
0.55	8323	6.42	0.80	0.238	0.845	846962262	7329461	0.685	0.303	0.988	0.003865
0.55	8232	6.42	0.90	0.268	0.951	952832545	8245644	0.719	0.297	1.016	0.003974
0.55	8232	6.42	1.00	0.298	1.056	1058702828	9161826	0.802	0.291	1.093	0.004278
0.55	8232	6.42	1.10	0.328	1.162	1164573111	10078009	0.882	0.287	1.169	0.004572
0.58	8681	6.37	0.90	0.268	0.951	952832545	8245644	0.747	0.294	1.041	0.004108
0.58	8681	6.37	1.00	0.298	1.056	1058702828	9161826	0.910	0.289	1.199	0.004730
0.58	8681	6.37	1.10	0.328	1.162	1164573111	10078009	0.985	0.284	1.269	0.005007
0.60	8980	6.34	0.60	0.179	0.634	635221697	5497096	0.661	0.314	0.975	0.003868
0.60	8980	6.34	0.70	0.209	0.739	741091980	6413278	0.676	0.306	0.982	0.003894
0.60	8980	6.34	0.80	0.238	0.845	846962262	7329461	0.695	0.299	0.994	0.003941
0.60	8980	6.34	0.90	0.268	0.951	952832545	8245644	0.778	0.293	1.071	0.004246
0.60	8980	6.34	1.00	0.298	1.056	1058702828	9161826	1.020	0.287	1.307	0.005185
0.60	8980	6.34	1.10	0.328	1.162	1164573111	10078009	1.094	0.283	1.377	0.005460
0.65	9729	6.26	0.50	0.149	0.528	529351414	4580913	0.653	0.321	0.974	0.003908
0.65	9729	6.26	0.60	0.179	0.634	635221697	5497096	0.666	0.310	0.976	0.003920
0.65	9729	6.26	0.70	0.209	0.739	741091980	6413278	0.686	0.302	0.988	0.003967
0.65	9729	6.26	0.80	0.238	0.845	846962262	7329461	0.724	0.295	1.109	0.004092
0.65	9729	6.26	0.90	0.268	0.951	952832545	8245644	0.918	0.289	1.207	0.004847
0.65	9729	6.26	1.00	0.298	1.056	1058702828	9161826	1.424	0.284	1.708	0.006857
0.70	10477	6.19	0.50	0.149	0.528	529351414	4580913	0.656	0.317	0.973	0.003950
0.70	10477	6.19	0.60	0.179	0.634	635221697	5497096	0.678	0.307	0.985	0.003999
0.70	10477	6.19	0.70	0.209	0.739	741091980	6413278	0.713	0.299	1.012	0.004108
0.70	10477	6.19	0.80	0.238	0.845	846962262	7329461	0.829	0.292	1.121	0.004551
0.70	10477	6.19	0.90	0.268	0.951	952832545	8245644	1.102	0.286	1.388	0.005635
0.75	11225	6.13	0.50	0.149	0.528	529351414	4580913	0.673	0.314	0.987	0.004047
0.75	11225	6.13	0.60	0.179	0.634	635221697	5497096	0.699	0.304	1.003	0.004114
0.75	11225	6.13	0.70	0.209	0.739	741091980	6413278	0.790	0.296	1.086	0.004454
0.75	11225	6.13	0.80	0.238	0.845	846962262	7329461	1.082	0.289	1.371	00.05624
0.80	11974	6.07	0.50	0.149	0.528	529351414	4580913	0.705	0.311	1.016	0.004206
0.80	11974	6.07	0.60	0.179	0.634	635221697	5497096	0.745	0.301	1.046	0.004332
0.80	11974	6.07	0.70	0.209	0.739	741091980	6413278	0.924	0.293	1.217	0.005040
0.85	12722	6.01	0.50	0.149	0.528	529351414	4580913	0.730	0.308	1.038	0.004337
0.85	12722	6.01	0.60	0.179	0.634	635221697	5497096	0.846	0.298	1.144	0.004781
0.85	12722	6.01	0.70	0.209	0.739	741091980	6413278	1.172	0.290	1.462	0.006110
0.88	13171	5.98	0.50	0.149	0.528	529351414	4580913	0.753	0.306	1.059	0.004449
0.88	13171	5.98	0.60	0.179	0.634	635221697	5497096	0.938	0.297	1.235	0.005185
0.88	13171	5.98	0.70	0.209	0.739	741091980	6413278	1.425	0.289	1.714	0.007197

Table 7.4 (*continued*). Moor's average single screw transformed from ©F to C_{fm}, C_{ts} '57 and C_{ts} '58.
Ship: L = 122 m, B = 16.76 m, T = 7.32 m, $L·B·T$ = 14957. Model: L = 5 m, L/B = 7.32, T/L = 0.06

C_{fm} ITTC'57	C_r ITTC'57	C_{fs} ITTC'57	C_{ts} ITTC'57	$1+K$ ITTC	$C_{fm}(1+K)$ ITTC78	C_r ITTC78	$C_{fs}(1+K)$ ITTC78	C_{ts} ITTC78	C_{ts}'57/ C_{ts}'78
0.003338	0.000491	0.001621	0.002111	1.110	0.003705	0.000123	0.001799	0.001922	1.098
0.003246	0.000585	0.001589	0.002174	1.110	0.003603	0.000228	0.001764	0.001992	1.091
0.003169	0.000697	0.001563	0.002259	1.110	0.003517	0.000348	0.001735	0.002083	1.085
0.003103	0.000871	0.001540	0.002411	1.110	0.003444	0.000530	0.001709	0.002239	1.077
0.003046	0.001231	0.001520	0.002751	1.110	0.003381	0.000896	0.001687	0.002583	1.065
0.002996	0.001576	0.001502	0.003078	1.110	0.003326	0.001246	0.001667	0.002914	1.056
0.003103	0.001005	0.001540	0.002545	1.117	0.003466	0.000642	0.001720	0.002362	1.077
0.003046	0.001684	0.001520	0.003203	1.117	0.003403	0.001327	0.001698	0.003025	1.059
0.002996	0.002011	0.001502	0.003513	1.117	0.003346	0.001660	0.001678	0.003338	1.052
0.003338	0.000530	0.001621	0.002150	1.120	0.003739	0.000129	0.001815	0.001944	1.106
0.003246	0.000648	0.001589	0.002238	1.120	0.003635	0.000259	0.001780	0.002039	1.097
0.003169	0.000773	0.001563	0.002335	1.120	0.003549	0.000392	0.001750	0.002143	1.090
0.003103	0.001143	0.001540	0.002683	1.120	0.003475	0.000771	0.001725	0.002496	1.075
0.003046	0.002139	0.001520	0.003659	1.120	0.003412	0.001773	0.001702	0.003475	1.053
0.002996	0.002464	0.001502	0.003966	1.120	0.003355	0.002104	0.001682	0.003786	1.047
0.003452	0.000456	0.001659	0.002115	1.140	0.003936	−0.000027	0.001891	0.001864	1.135*
0.003338	0.000582	0.001621	0.002203	1.140	0.003805	0.000115	0.001847	0.001692	1.123
0.003246	0.000722	0.001589	0.002311	1.140	0.003700	0.000267	0.001812	0.002079	1.112
0.003169	0.000923	0.001563	0.002486	1.140	0.003612	0.000480	0.001781	0.002261	1.099
0.003103	0.001743	0.001540	0.003283	1.140	0.003538	0.001309	0.001755	0.003064	1.071
0.003046	0.003811	0.001520	0.005331	1.140	0.003473	0.003384	0.001733	0.005117	1.042
0.003452	0.000498	0.001659	0.002157	1.160	0.004005	−0.000055	0.001924	0.001870	1.153*
0.003338	0.000661	0.001621	0.002282	1.160	0.003872	0.000127	0.001880	0.002007	1.137
0.003246	0.000862	0.001589	0.002452	1.160	0.003765	0.000343	0.001843	0.002186	1.121
0.003169	0.001382	0.001563	0.002945	1.160	0.003676	0.000875	0.001813	0.002688	1.096
0.003103	0.002532	0.001540	0.004072	1.160	0.003600	0.002036	0.001786	0.003822	1.065
0.003452	0.000595	0.001659	0.002254	1.195	0.004126	−0.000078	0.001982	0.001904	1.184*
0.003338	0.000776	0.001621	0.002396	1.195	0.003989	0.000125	0.001937	0.002061	1.162
0.003246	0.001208	0.001589	0.002797	1.195	0.003879	0.000575	0.001899	0.002474	1.131
0.003169	0.002455	0.001563	0.004018	1.195	0.003787	0.001837	0.001867	0.003704	1.085
0.003452	0.000754	0.001659	0.002413	1.240	0.004281	−0.000075	0.002057	0.001983	1.217*
0.003338	0.000994	0.001621	0.002614	1.240	0.004139	0.000193	0.002010	0.002202	1.187
0.003246	0.001794	0.001589	0.003383	1.240	0.004025	0.001015	0.001971	0.002986	1.133
0.003452	0.000884	0.001659	0.002543	1.300	0.004488	−0.000151	0.002157	0.002005	1.268*
0.003338	0.001443	0.001621	0.003064	1.300	0.004339	0.000442	0.002107	0.002548	1.202
0.003246	0.002864	0.001589	0.004453	1.300	0.004219	0.001890	0.002066	0.003956	1.126
0.003452	0.000996	0.001659	0.002655	1.350	0.004661	−0.000212	0.002240	0.002028	1.310*
0.003338	0.001847	0.001621	0.003468	1.350	0.004506	0.000679	0.002188	0.002867	1.210
0.003246	0.003951	0.001589	0.005540	1.350	0.004382	0.002815	0.002145	0.004960	1.117

figures at the lowest F_n figures were in many cases higher than those quoted for the next three F_n values. It is uncertain whether this came about because of experimental error, noting that such low speed data was of little interest at the time these tests were made, or because the data was cross faired for presentation purposes. It was for this reason that this data was not used for a Prohaska analysis of $(1 + K)$ described in §7.1.2.

After a lot of thought the author came to the conclusion that the only plot that would avoid all these pitfalls is one of C_{tm} — and that this would have the further advantage that it can equally well be used for calculations based on the Grigson friction line and $(1 + K)$ formula described in §7.1.3 which is probably the most accurate powering method currently available.

To standardise the C_{tm} plots, these have all been based on a model length of 5 m and have an L/B ratio of 7.28 and a T/L ratio of 0.06.

The fact that the lines are plotted for odd Froude numbers is because these numbers equate exactly with the $V/\sqrt{L}$ values used in the original presentations and this avoids introducing inaccuracy from the cross fairing/interpolation necessary if the curves were to be drawn at single digit Froude numbers — and at the end of a designer's calculations the answer will in any case be a power/speed plot.

As in the Parsons paper, plots are given of both Moor's optimum Fig. 7.6, and average Fig. 7.7, together with the BSRA standard series, Fig. 7.8. Designers will find some merit in using all three plots when making a power estimate. On the one hand, a designer will always want to be certain of achieving the specified speed and for this reason will wish to calculate an "Average Modern Attainment" power. On the other hand, there will always be pressure to aim for the "optimum" so this power should also be calculated.

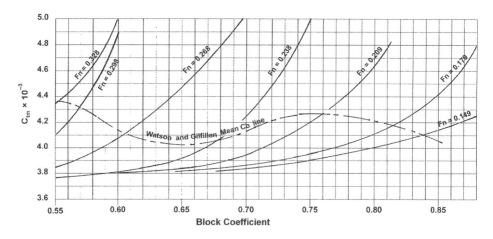

Fig. 7.6. Powering of single screw ships. Moor's optimum. Model $L = 5$ m, $L/B = 7.28$, $T/L = 0.060$.

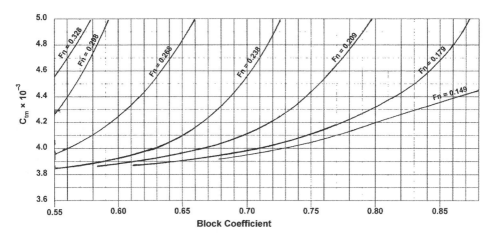

Fig. 7.7. Powering of single screw ships. Moor's average. Model $L = 5$ m, $L/B = 7.28$, $T/L = 0.060$.

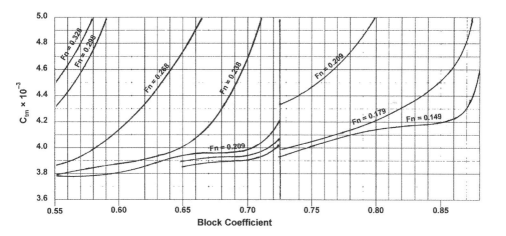

Fig. 7.8. Powering of single screw ships. BSRA standard series. Model $L = 5$ m, $L/B = 7.28$, $T/L = 0.060$.

A comparison between the average and optimum values indicates that almost throughout the range of F_n and C_b, the optimum values are about 8% better than the average. When fixing the installed power a compromise between "optimum" and "average" may be prudent.

A superimposition of the BSRA methodical series on Moor's "average" indicates that the BSRA forms are well up to "average" standard throughout the range of F_n

and C_b and are appreciably better (5% or so) at low F_n (<0.18) and high C_b (>0.75) values. The great virtue of the BSRA data to a designer is that it not only provides a power estimate but also gives a method of developing a body plan which will match the power estimate.

Moor's results on the other hand are derived from a large number of unrelated models and there is no guidance as to the characteristics of the lines which will match the power estimates.

In Fig. 7.6 the Watson/Gilfillan block coefficient line is shown as an indication of the area of this figure which is of most practical importance.

7.2.3 C_{tm} for twin screw ships

The same procedure was applied to Moor's twin screw data and is presented in Fig. 7.9. A superimposition of the twin screw data on any of the single screw figures indicates the generally better naked resistance of the former, but this advantage is of course reversed when appendage resistance is taken into account.

The C_{tm} values for twin screws ships are also presented for a standard model length of 5 m. The L/B ratio remains 7.28 but T/L is reduced to 0.045, a figure more appropriate to twin screw ships.

7.2.4 C_{tm} for frigates, corvettes and high speed ferries

Most warships of the corvette and frigate type are required to have maximum speeds which result in Froude numbers ranging from 0.40 to 0.60, with most of them between 0.45 and 0.55.

Little warship powering data escapes the security net, so it can be helpful to fill out whatever data is available with merchant ship data and the twin screw data given in the previous paragraph can be used in this way, provided care is taken.

In any attempt to do this it is essential to note is that the block coefficient of most warships is much finer than would be dictated by powering considerations.

This fine C_b results from the requirement that these ships should have a high rise of floor to avoid, or at least minimise, slamming which is a very important consideration on these relatively small and lightly constructed ships which are required to maintain high speeds in rough seas. The fine C_b also helps to increase the draft which again improves seakeeping.

A corollary of the fine C_m that these ships have is that C_b ceases to be as good a parameter for C_t as it is for merchant ships and warship designers tend to use C_p. The author would have liked to follow this practice but unfortunately lacked the necessary information on C_p or C_m values to do so.

Anyone wishing to use C_p as a base and having only a C_b value available may find it helpful to use the rough guide to C_m values given in the following formulae (see also Fig. 8.9).

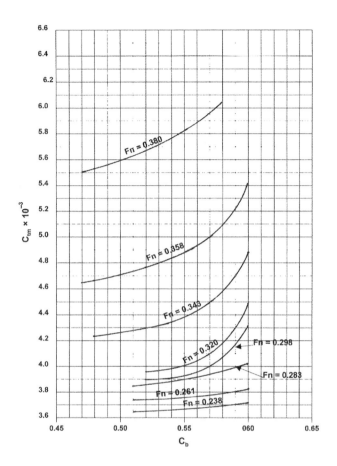

Fig. 7.9. Powering of twin screw ships. Moor's optimum. Model $L = 5$ m, $L/B = 7.28$, $T/L = 0.045$.

For merchant ships $C_m = 0.935 + 0.05\ C_b$

For frigates and similar vessels $C_m = 0.58 + 0.4\ C_b$

Figure 7.10 presents some data from warship tank tests, sticking to block coefficient as the base. The data is again corrected to a standard model length of 5 m but in this case L/B is set at 8.71 and T/L at 0.035 as more representative of warships than the proportions used for twin screw merchant ships.

7.2.5 Calculation of C_{ts}

The calculation of C_{ts} from C_{tm} starts with the calculation of C_{fm} and of $(1 + K)$ using whichever friction line and associated $(1 + K)$ formula is preferred and a spreadsheet similar to Fig. 7.2.

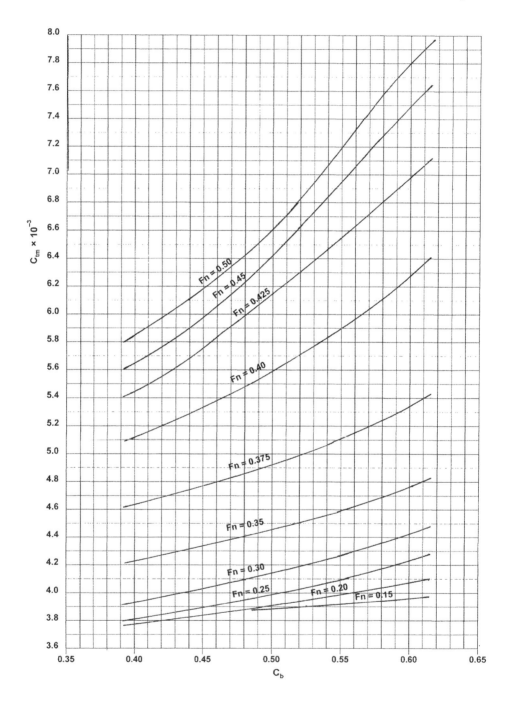

Fig. 7.10. Powering of warships and similar high speed craft. Model $L = 5$ m, $L/B = 8.71$, $T/L = 0.035$.

Using this data, C_r is then established and to this is added $(1 + K) C_{fs}$ and roughness, appendage and air resistance elements all as in eq. (6.7).

7.2.6 Calculation of P_e

The calculation of P_e from C_t uses the formula:

$$P_e = 0.0697 \, C_{ts} \cdot S \cdot V^3 \text{ (in kW)} \tag{7.18}$$

All the C_t values must of course be corrected for the ship's dimensions. If the process outlined has been followed that for length will have already been made, but corrections for beam and draft will still be necessary. These should be made using the ratios of the ship's beam and draft, proportioned to the length of the basis ship divided by the beam and draft of the basis ship together with the appropriate Mumford indices (use twin screw ship also for warships). Although these indices were originally intended for use with ©, they apply to equally to C_t as they are multipliers.

Using L/B and T/L ratios for both the model (1) and the new design (2), the ratios required for the Mumford indices are easily derived as follows:

$$\frac{L_1 / B_1}{L_2 / B_2} = \frac{B_2 \times L_1 / L_2}{B_1} \text{ and } \frac{T_2 / L_2}{T_1 / L_1} = \frac{T_2 \times L_1 / L_2}{T_1}$$

7.3 APPENDAGE RESISTANCE

To complete the resistance (or effective horsepower) calculation it is necessary to add the resistance of any appendages that are fitted to the ship, but were not fitted to the model.

In power estimates appendage resistance is generally added as a percentage of the naked resistance, although there is little logic in this as this implies that an appendage behind a resistful hull has more resistance than an identical appendage behind a less resistful ship.

7.3.1 Most common appendages

The most common appendages are:
- Twin rudders. A single centreline rudder is normally included in the "naked" model, but twin rudders are treated as appendages.
- Twin screw bossing and/or shaft brackets. 'A' brackets, in general, appear to have significantly less resistance than enclosed bossings on twin screw ships, but both need careful alignment with the flow pattern.

- Large unfaired hubs of C.P. propellers. These can have quite a large drag, but are not fitted when the resistance is measured and the effect may appear as a reduction in propulsive efficiency
- Bilge keels used to be assumed as included in the naked power, but as mentioned in §6.2 the resistance of these is now added as an increase in the wetted surface area. The importance of taking great care to align these appendages with the stream line flow in the region in which they are fitted must be emphasised.
- Fin stabilisers
- Bow rudder and/or bow propeller. Bow or stern thruster openings. These must be carefully married into the lines, and indeed the lines may need to be modified to accommodate them. It is worth noting that the additional resistance of a bow propeller can be significantly less behind a bulbous bow than behind a conventional bow.
- Sonar domes on warships and research vessels
- Cathodic protection anodes
- Moon pools on oil exploration and offshore production vessels
- Hopper doors and the ship side slides for the overboard suction pipe trunnions on dredgers. All these can cause a considerable increase in resistance which can, however, be minimised by careful design.

7.3.2 Some approximate figures for appendage resistance

Some approximate figures (regretfully as factors based on the naked resistance) which should be used with care are:

Twin rudders	0.015
Twin screw bossing	0.10
Shaft brackets and open shafts	0.06
Fin stabilisers	0.02
Sonar dome	0.01

If there is more than one appendage the appropriate factors should be added.

An alternative treatment of appendage resistance is suggested by Holtrop and Mennen in their papers to which reference has already been made. This calculates appendage resistance based on their wetted surface area using the C_f value for the ship according to the 1957 ITTC formula.

$$R_{app} = 1/2 \cdot C_f \cdot r \cdot S_{app} \cdot V^2 \cdot (1 + K_2) \tag{7.19}$$

Approximate K_2 values are quoted as follows:

Rudder behind skeg	1.5–2.0
Rudder behind stern	1.3–1.5

Twin rudders	2.8
Shaft brackets	3.0
Skeg	1.5–2.0
Strut bossings	3.0
Hull bossings	2.0
Shafts	2.0–4.0
Stabiliser fins	2.8
Dome	2.7
Bilge keels	1.4

The equivalent $1 + K_2$ value for a combination of appendages is:

$$(1+K_2) \equiv \frac{\sum (1+K_2)S_{app}}{\sum S_{app}} \tag{7.20}$$

The two sets of values for appendage resistance appear to be in reasonable agreement with one another.

There are a number of other items which may need to be treated as appendages, notably bow and stern thrust tunnels. Great care should be taken with the design of the intersections of these with the normal form, particularly for bow thrusters in lines without a bulbous bow.

Moon pools and dredge pipe slides are two items which can be extremely resistful and should be the subject both of very careful design and of special tank tests.

7.4 TYPES OF PROPULSORS

7.4.1 General discussion

It is desirable that a decision on the number and type of propulsors is made before powering calculations are carried out as this decision will frequently influence the design of the lines and therefore the resistance. It might, therefore, have been logical to put this section near the start of Chapter 6, but the convenience of keeping it adjacent to the sections on propulsive efficiency finally determined its position.

For the majority of ships the propulsor will be chosen with the intent of attaining a high propulsive efficiency at an acceptable cost, thereby minimising both the machinery power and cost and the fuel consumption in service. In most cases this will lead to the choice of a single, fixed pitch, propeller, but there are a large number of alternative propulsors which may be preferred for a variety of reasons and these are now considered in turn.

7.4.2 Single or twin screw

If the speed and power are high in relation to ship size and in particular to the maximum allowable draft, it may be necessary to have twin propellers (or triple or quadruple) to enable the required power to be absorbed by propellers whose diameter can be accommodated within the draft.

Twin or multiple screws are generally chosen for ships requiring a high degree of reliability and where the cost of immobility from a breakdown for even a short time is high or verges on being unacceptable — cruise liners and ferries.

Multiple screws are also chosen when there is a particular likelihood of a set of machinery being put out of action — warships by enemy action or ice breakers by ice (although see §16.5.2), with in both cases there being the possibility of either damage to a propeller or the flooding of a machinery compartment.

Twin screws used to be chosen for the better manoeuvrability they provide but the provision of a bow thruster and/or a high performance rudder can now so improve this aspect of the performance of single screw ships that this is no longer the case in relation to slow speed manoeuvring in docks canals and rivers. Where there is a particular requirement for high speed manoeuvring as applies to a warship, twin screws continue to be the best choice.

7.4.3 Controllable pitch propellers

These are often selected in spite of their slightly poorer efficiency because of the contribution they can make to the ease of manoeuvring particularly in ships which have to operate frequently in confined waters. They can also be chosen to match the characteristics of an engine which is best run at one speed all the time and in other cases may be chosen for fuel efficiency reasons if the ship is intended for operation at more than one significantly different speed or displacement. On warships where the cruising endurance speed is often less than half the full speed, a controllable pitch propeller avoids the over-torquing of the cruising engine that might be caused by a fixed pitch propeller.

An alternative to a C.P. propeller in such a case would be a two speed gearbox.

It has been suggested that the hub drag mentioned in §7.3.1 if correctly included would further reduce the efficiency of these propellers.

7.4.4 Highly skewed propellers

Propellers with highly skewed blades can be either fixed pitch or controllable. The skewing makes little difference to the efficiency, but reduces the propeller-induced forces on the hull. Propellers of this type are quite often fitted on cruise liners and ferries on which there appears to be a possibility of vibration and on which the avoidance of any vibration is of course absolutely vital if the passengers are to be kept happy. They are also, not infrequently, retrofitted as a cure for vibration. The tip clearances for a propeller of this type can be less than for a conventional propeller permitting a larger diameter propeller to be fitted with a small gain in efficiency.

The astern thrust of a fixed pitch highly skewed propeller has to be limited by blade strength problems, but there is of course no problem with the astern power of a controllable one.

7.4.5 Self-pitching propellers

At present these are limited to relatively small powers. The concept is highly attractive, particularly when they are used as an auxiliary on sailing yachts in place of a folding propeller but increasing mechanical problems for large sizes seem likely to prevent the application to large ships. An interesting feature is the fact that the open water efficiency, not allowing for the wake, is the same astern as it is ahead.

7.4.6 Other propeller types

Tip Vortex Free (TVF) and Balanced Thrust Loading (BTL) propellers are notable amongst the many attempts made by propeller manufacturers to improve efficiency, but the claims made are not generally regarded as fully substantiated.

The advantage of a contra-rotating propeller is based on the improvement in efficiency which reducing the rotational losses in the propeller race should bring. Quite considerable claims have been made for the improvement in efficiency on the relatively small number of ships so far fitted and these seem credible. The mechanical complexity and high cost of the gearbox, shafting and propellers limit applications at present but developments in hand may result in this type becoming much more widely accepted. This type of propeller would appear to have particular application to direct drive electric propulsion systems where the gearbox problem can be avoided.

The same principle is used when a rotating thruster on the rudder operates in the propeller wake. This is probably only suitable as the provider of a small augmentation of power and when the improved steering performance is the most important factor.

7.4.7 Nozzles

Nozzles shrouding the propeller, notably the Kort nozzle, have as their primary aim the improvement of the efficiency of heavily loaded propellers such as those on tugs when towing or of fishing vessels when trawling. In this condition they are very effective but usually at the expense of the free running speed.

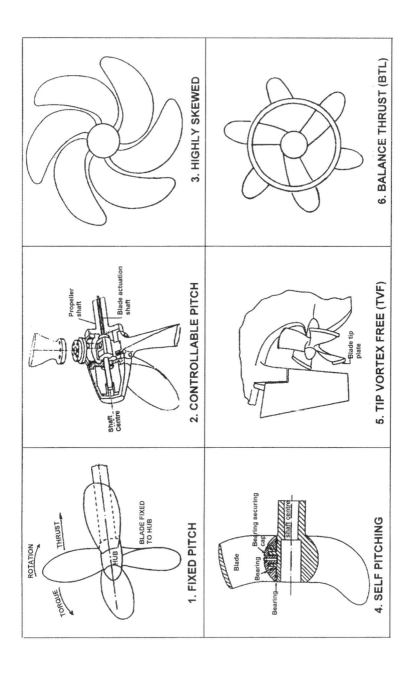

Fig. 7.11 (1–6). Various types of propulsors (*continued opposite*)

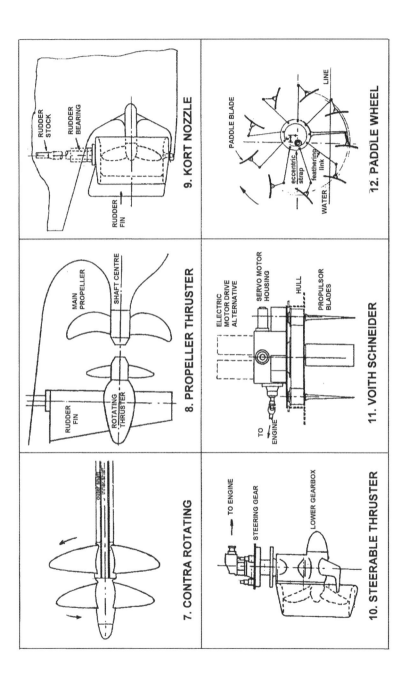

Fig. 7.11 (6–12). Various types of propulsors *(continued overleaf)*.

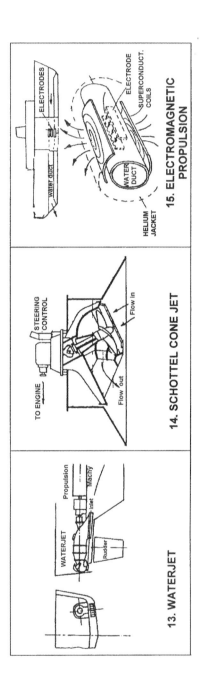

Fig. 7.11 (13–15). Various types of propulsors.

There was a good deal of interest some years ago in the use of nozzles on large tankers and a number of ships had installations of this sort, although its use never really caught on.

More recently with a general acceptance of the advantage which can be obtained from a slow revving propeller, nozzles are now only likely to be considered when for any reason the propeller diameter must be limited. e.g., Great Lakes Bulk Carriers with a small draft in relation to their size.

Both full nozzles and half nozzles can help to reduce propeller excited vibration by improving the flow into the propeller.

Steering nozzles, with or without fins allow the propeller to be moved aft increasing the usable length of the ship but can be liable to maintenance problems.

7.4.8 Propeller position

All the propellers discussed so far can only be fitted at the stern of the ship. This is not necessarily the best place for the thrust to be developed. For tugs there is a very great advantage in the thrust being developed forward, so that the tug is pulled into line with the tow rope, and the risk of capsize thereby greatly reduced or indeed eliminated. Two types of propeller can be fitted in this way: the Voith Schneider and the Steerable Thruster. The former was the first in the field and has the advantage that it does not project so far below the ship to which it is fitted. It has been used in many double ended ferries with considerable success. The latter is cheaper, but projects well below the hull of the vessel to which it is fitted increasing the draft to an extent that may be unacceptable on the shallow draft craft to which it might be applied.

Propellers which can be fitted forward totally within the ship's hull are the Schottel cone jet and the Gill axial flow propellers. Most uses of these types are as a bow thruster or a get-you-home auxiliary as the greatest powers so far made would only be suitable as a main propulsion unit for small/slow ships.

7.4.9 Paddle wheels

A propulsion unit from the past is the paddle wheel. Paddle wheels can be fitted at the sides or at the stern. A feathering paddle can have an efficiency under certain conditions as high as that of a screw propeller, but its vulnerability to damage and its limitation to operation at a fixed draft mean that it is no longer a serious competitor.

7.4.10 Water jet propulsion

A comparative newcomer for serious consideration at quite high powers is water jet propulsion. This system has been used for many years for small craft, but generally has had low efficiency except at high ship speeds. Recent developments have greatly improved the efficiency at lower speeds and at the same time there has been a big increase in the power to which these units are manufactured and an application to warships of frigate size is now a possibility.

The success of "Destriero", crossing the Atlantic at 56 knots will no doubt boost interest in water jet propulsion.

7.4.11 The future

Looking into the future (possibly to quite a distant future) electromagnetic propulsion, a radically different type of propulsor now the subject of theoretical work and small-scale experiments, may turn out to have both the higher efficiency and the good control characteristics being sought.

7.4.12 General

Most of the types of propulsors mentioned above are shown in outline in Fig. 7.11. These by no means exhaust the whole field of propulsors for marine vehicles, but most of the others do not apply to displacement ships. Transcavitating, super-cavitating and surface-piercing propellers apply to fast planing craft; air propellers, fans or jets have their application on hovercraft. There are a few other types which might potentially be used for displacement ships, but these have hardly progressed beyond small scale prototypes.

7.4.13 Open water efficiency

Curves of the open water efficiency of a number of propulsors are shown in Fig. 7.12 which indicates quite clearly the superiority of the conventional propeller from a straight propulsive efficiency point of view, although as already stated this may sometimes be overridden by other considerations.

7.5 PROPULSIVE EFFICIENCY

In the last section in which various types of propulsors were discussed it was shown that a conventional propeller provided the highest propulsive efficiency, although other propulsors might have other advantages that outweighed this. In this section attention will be confined to conventional propellers.

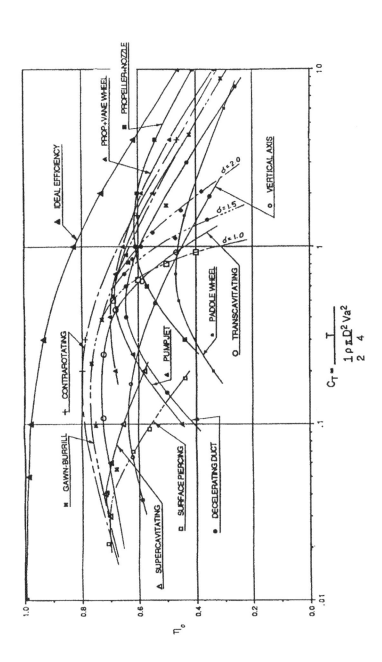

Fig. 7.12. Open water efficiency of a variety of propulsors.

7.5.1 Slow revving propellers

Whilst it is possible to design, and optimise the open water efficiency of, a propeller for any required power output, propeller diameter and revs/min, there is no doubt that the starting point of a search for efficiency should be the use of the largest propeller diameter that the ship can be designed to accommodate without un-acceptable adverse consequences in association with the lowest revs/min that suits this propeller and can be obtained from suitable propulsion machinery.

An indication of the gains to be obtained by using low revs/min is provided by Emerson's approximate formula for QPC which with a slight increase in the constant to bring it into line with modern propeller design, is:

$$\eta_d = 0.84 - \frac{N\sqrt{L}}{10,000} \tag{7.21}$$

where

N = rev/min
L = length BP in metres

A plot of this formula is given in Fig. 7.13. In this figure Emerson's formula has been extended to values of N smaller, and values of L larger than the data from which it was originally derived. Remarkably, in spite of this extrapolation it continues, in the author's experience, to give reasonably accurate answers, although caution must be advised in its use at low N and large L values.

Figure 7.15 shows the improvement in propeller efficiency that is obtained by today's slower revving engines. On a 300 m ship the gain from a change from 110

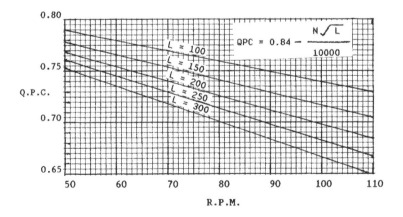

Fig. 7.13. QPC versus rpm and ship length.

rpm, typical of diesel engines of some 15 years ago, to 80 rpm, typical of today's large ships is 7.6%.

An ability to use a lower rev/min depends on whether a larger propeller, which for a given power, must accompany a reduction in rev/min, can be accommodated within the stern aperture, which in turn depends on the load draft.

A relationship which can be used for quick exploration of propeller diameter and rev/min is:

$$d = 16.2 \frac{Pbs^{0.2}}{N^{0.6}} \text{ metres} \tag{7.22}$$

Pbs = service power in kilowatts. A plot of this is given in Fig. 7.14.

This relationship was derived from the general equation

$$P_d/\rho N^3 D^5 = \psi (J, F_n, \omega, R_n)$$

$$\text{so } D^5 = P_d/K_t N^3$$

$$\text{or } D = C P_d^{0.2} \cdot N^{0.6}$$

The constants were derived for a medium range of rev/min, power and diameter and caution is advised in extrapolating the use of this formula to very high or very low rev/min values.

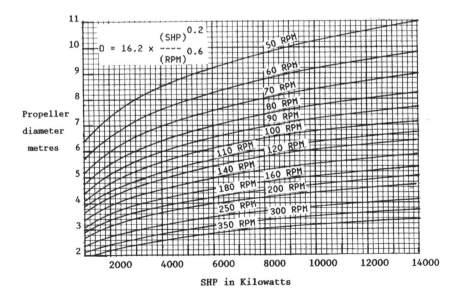

Fig. 7.14. Propeller diameter versus SHP and rpm.

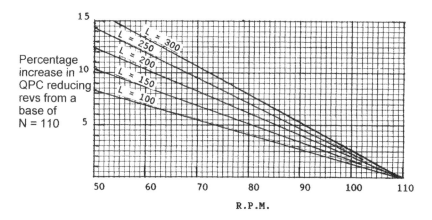

Fig. 7.15. Increase in QPC obtained by reducing rpm.

As well as the problem of accommodating the propeller within the load draft, which has already been mentioned, a larger propeller may make a deeper ballast draft desirable. This may mean an increase in the power required in ballast increasing the fuel consumption in the ballast voyage and cutting into the gain made in the load voyage.

The other factor which governs the use of lower rev/min is, of course the type of machinery used. If this involves the use of gearing, then the rev/min can be chosen to suit the propeller. On the other hand, the propellers of ships fitted with slow speed diesels must be tailored to the engine rev/min.

7.5.2 Slow revving propellers on twin screw ships

The advantage gained by adopting as low revs/min as possible applies equally to twin screw ships, although here the reduction may be from 300 revs/min to 250; or from 250 to 200. Either of these changes gives a very worth while gain in propulsive efficiency.

Again using Emerson's formula, the gain in the first case would be about 8.5% and in the second close to 10%. Once again this is a major extrapolation in the use of Emerson's formula that Emerson himself would almost certainly find quite unscientific, but the author is quite unrepentant in this extension to its use having frequently found tank tests giving better confirmation of the approximate values obtained by this method than of those calculated by more detailed methods.

7.5.3 Optimising open water efficiency

Having obtained the potential for a high propeller efficiency by choosing an advantageous rev/min, the next step is to achieve the best open water efficiency for the propeller's operating regime. The factors which influence this include:

- The choice of material used for the blades; a strong material permits the use of slender blade sections giving a higher efficiency.
- The number of blades; propellers with fewer blades generally have higher efficiency; those with more blades have the advantage of producing smaller pressure pulses and less vibration.
- Blade area; blade shape
- Distribution of pitch;
- Blade rake and skewback.

A well known statement about propellers that is both comforting and disturbing, is that it is very difficult to design a really bad propeller but equally difficult to design a really good one. In practical terms this means that a naval architect can expect to get a propeller whose efficiency is within about 2% of the currently accepted best possible for the design conditions — but will find it very difficult to get that further 2% that will give a really fuel efficient ship.

Probably the most common fault with propellers is that they are not matched to the engine and are either too heavily or too lightly pitched (see the next section).

7.5.4 Specifying propeller design conditions

The design of a fuel efficient ship can only be achieved if amongst other things the propeller design conditions favour fuel efficiency, a question that will generally go back to the specification and contract.

If these require a high trial speed using full power there is bound to be a tendency to design the propeller specifically to meet this. This is likely to result in the propeller being over pitched for the service speed which in turn may mean that the engine will not be able to develop full power within the limit of permitted cylinder pressure and is almost certain to result in to the engine developing excessive pressure and with it excessive cylinder wear.

A form of specification which, whilst recognising that the speed must be measured on trial, gives the designer the best incentive to design for service conditions is:

- On trials in deep water and in fair weather conditions, with the ship newly dry docked and loaded to a draft corresponding to a deadweight of "D" tons, the service speed of "K" knots is to be obtained with the machinery developing not more than "H" S.H.P., thereby demonstrating that when operating at the service power "S" there is a margin of power of $M = (S - H)/H$ to maintain the same speed "K" in service conditions of weather and fouling.
- If trials are run in ballast, the speed "K" is to be obtained on a reduced power calculated from the tank test results to represent an equivalent performance.

For many ship types acceptance trials can only be run in ballast.

7.5.5 Propeller design

The author wanted to include in this book approximate formula that would enable a naval architect deprived of any other data to make a complete initial design, but has decided to admit defeat with propeller open water efficiency. The subject appears to be too complex for any simplified treatment and readers are referred to the standard text books on this subject.

Fortunately, there is rarely any need to consider open water efficiency at the initial design stage as Emerson's QPC formula seems to fill the immediate need in power estimation with reasonable accuracy, enabling open water efficiency to be dealt with by propeller designers at a later stage.

7.5.6 Efficiency of a controllable pitch propeller

The efficiency of a controllable pitch propeller is generally 1–2% less than that of a fixed pitch propeller for the same design condition — this being mainly due to the much larger boss diameter.

If, however, there is a requirement for extensive operation at either a speed or a displacement significantly different from the design condition a controllable pitch propeller can offer substantial advantages in fuel economy and/or speed at the second operating condition.

7.6 HULL EFFICIENCY

If it is impracticable to include in this book a satisfactory way of estimating open water efficiency, there is no point in giving formulae for hull efficiency as it is only the product of these — the QPC — which is of real interest. It is, however, worth discussing the components of hull efficiency to try to identify whether these can be influenced in the design process in a way that will improve powering efficiency.

It will be recalled that:

$$\eta_d = \eta_o \times \eta_h \times \eta_r$$

where

$$\eta_h = \frac{1-t}{1-w_t} \quad \text{and} \quad t = \frac{T-R}{T} \quad \text{and} \quad w_t = \frac{V-V_a}{V}$$

To optimise η_d, the thrust deduction t should be minimised, the Taylor wake fraction w_t should be maximised as should the relative rotative efficiency η_r.

There are a number of formulae for t, w_t and η_r, most, if not all, of which were derived by regression analysis. An examination of these shows:

(i) That thrust deduction t is reduced if the water flow to the propeller is good and the clearances from the hull are relatively large; aspects of design which should in any case be observed to avoid the possibility of propeller induced vibration. Favourable factors are:

relatively small propeller diameter
relatively large L/B ratio
relatively fine C_b
LC_b relatively far forward

In practice a designer is unlikely to be influenced by these considerations.

(ii) That wake fraction w_t depends primarily on the block coefficient with which it increases linearly, indicating that water is entrained to a greater extent by a full ship than by a fine one, and on the ship's length and the B/T ratio.

(iii) That relative rotative efficiency η_r increases primarily with increasing propeller diameter relative to ship length and to a lesser extent with increase in block coefficient.

Although all these formulae may enable a reasonably accurate prediction of hull efficiency to be made, they give disappointingly little guidance on how to improve hull efficiency. There does not for example appear to be any size effect such as is given by the $\sqrt{L}$ in Emerson's QPC equation. This is made even more surprising when the same reference then goes on to give a formula for QPC which has the N $\sqrt{L}$ term plus a number of other refinements.

7.7 TRIAL AND SERVICE ALLOWANCES

7.7.1 Trial power

It is important to remember that power estimates cannot be relied on to be 100% accurate. Grigson's statement that the errors in the ITTC friction line can cause an underestimation of power by as much as 7% certainly suggests caution. There is also the question of shell finish and what standard can be relied on.

The margin of power (over the best estimate) which it is wise to provide should depend partly on whether the ship is reasonably similar to ships which the shipyard has already built and partly on the penalty invoked for non-compliance with the trial speed. The author generally allowed 5% initially in the power given to the marine engineers but was not sorry when a bigger margin arose, as it generally did, when the main machinery was selected.

7.7.2 Weather and fouling — historical treatment

In the past, naval architects used to lump allowances for weather and fouling together as an addition to be made to the power required under trial conditions to enable the trial speed to be maintained in service.

Typically, the service power allowed for an increase of 20% over trial power, although this addition was almost certainly much lower than the actual increase required as was demonstrated by a number of investigations. However under the commercial pressure of competitive tenders 20% remained the general guide line for shipyard naval architects.

The allowance for service should take into account the weather conditions that may be expected on the vessels trade route. Guldhammer and Harvald in the paper quoted in §6.8 suggested average percentage service allowances for different routes as follows:

	Summer	Winter
North-Atlantic Eastwards	15	20
North-Atlantic Westwards	20	30
Pacific	15	30
South-Atlantic and Australian routes	12	28
East-Asiatic route	15	20

These figures predate the improved anti-fouling discussed below but it remains the case that the percentage which must be added to the trial power for weather can depend significantly on the shipping routes for which the ship is intended and on the importance attached to the maintenance of the schedule.

7.7.3 Fouling with modern anti-fouling paints

The 7% power reduction that can be gained by achieving an 80 μm finish on a new ship compared with the 165 μm standard has been referred to in §6.2.6. Even more important, is the fact that, with an advanced constant emission system, the finish will actually improve with service. This is because this type of paint self-polishes in service often actually improving the finish, in marked contrast to conventional paints whose roughness increases, not only with service but also with further paint applications.

With conventional paints the poison leaches out from the interior and the rate of release decreases with time in an exponential way. The use of deeper layers of paint results in the release being slower. With these paints, fouling started to affect the performance of a ship after as little as 20 days out of dock — a fact acknowledged when this was made one criterion of a "best trial".

With the SPCs the paint ablates or dissolves steadily (6 μm/month is a typical figure) and the new surface exposed by this process is as poisonous and active as the original one. If 60 months' protection is required, painting to a thickness of 360 μm will provide this. Not only is little or no fouling allowance required as a ship can now be expected to operate with only minor deterioration in performance between dry dockings but the intervals between these are being progressively lengthened, with three or even five years becoming common.

When repainting becomes necessary a wash with high-pressure fresh water is all that is required and gives a smooth surface for repainting. This contrasts with the high cost in both money and time of the scraping required by conventional paints and with the build-up of roughness from old leached out paint even when this is done.

7.7.4 Weather allowance

With fouling now very nearly eliminated, attention is being increasingly directed to items which can influence the effect which the weather can have on a ship:
(i)   the ship size, small ships being much more affected than large ones;
(ii) wind resistance forces on the hull and superstructure;
(iii) added displacement and windage due to ice accretion;
(iv) additional resistance due to the application of rudder angle to maintain a course in spite of wind and wave forces;
(v)   additional distance travelled due to course instability;
(vi) increase in ship resistance due to ship motions;
(vii) decrease in propeller efficiency due to ship motions.

The design corollaries of these weather effects are that steps should be taken to:
(a)   minimise the windage area of superstructures.
(b)   arrange the superstructure, if possible, in a manner that minimises the amount of weather or lee helm (this is almost certainly a counsel of perfection as in most ships the extent and disposition of the superstructure will be dictated by other considerations);
(c)   pay attention in the design of the lines to the provision of a reasonable area of deadwood aft to help course stability;
(d)   provide good, but not excessive, flare forward together with good free-board to reduce pitching motions and keep the decks forward reasonably free of water;
(e)   provide sufficient water ballast capacity to ensure by a combination of sinkage and trim that there is good propeller immersion in the ballast condition;
(f)   provide modern navigational aids,such as the auto-pilot which was claimed to provide substantial fuel savings by reducing helm movements to a minimum, thereby reducing rudder resistance, by allowing the ship to yaw and then correct itself as it will often do without the application of rudder.

Assuming that reasonable attention has been paid to these design features and that modern anti-fouling paint has been used a figure of between 10 and 15% for the weather allowance would seem to represent general modern practice, although a smaller figure could be used for some routes whilst other routes and relatively small ships may still demand a larger allowance.

7.8 DEVICES TO IMPROVE PROPULSIVE EFFICIENCY

The fact that even a well designed slow revving propeller is unlikely to have a QPC greater than about 0.75 and that the QPC of the higher revving propellers which must necessarily be used on many ships can drop to about 0.50 or lower, has caused a lot of attention to be devoted to examining where the lost energy goes and trying to avoid this loss or reclaim it.

Much of it appears to go into rotational energy and two different approaches have been adopted to avoiding/reclaiming this.

7.8.1 Pre-rotating

A number of devices try to avoid the loss by pre-rotating the water flowing into the propeller in the opposite direction to the "twist" given by the propeller so that the ultimate wake is partially straightened.

Possibly the most radical way of doing this is to change the whole aft end lines of the ship, making these asymmetrical as shown in Fig. 7.16. Apart from some extra loft work this appears to involve minimal extra cost and no extra resistance.

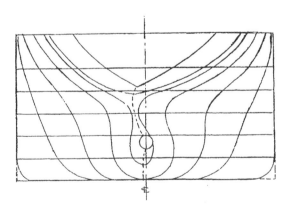

Fig. 7.16. Asymmetrical lines.

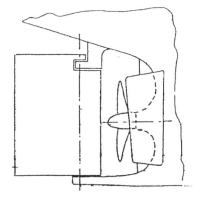

Fig. 7.17. Mitsui duct.

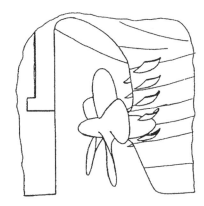

Fig. 7.18. Hull vanes – Grothues spoilers.

Although it took naval architects a long time to think of this idea, or at any rate to use it in practice, it seems only sensible to shape the ship ahead of a propeller designed to rotate in one direction in a way that takes this into account.

An alternative to asymmetrical lines is to move the propeller a little off centre (to starboard for normal clockwise rotation) when the water flow reaching it will be affected in much the same way as it is by asymmetrical lines.

Other devices, all of which aim to induce a more uniform flow into the propeller, include the Mitsui duct shown in Fig. 7.17, the Schneekluth wake distribution duct and Grothues spoilers (Fig. 7.18). All of these devices necessarily add some extra resistance as well as cost and weight. In addition to any improvement in propulsive efficiency, these devices can reduce propeller excited vibration forces and have been used in retrofits for this purpose.

The action of these devices is of course much more local than that of asymmetrical lines and there seems no reason why one of them should not be combined with asymmetrical lines.

7.8.2 Post propeller recovery

An alternative to avoiding or reducing rotational energy loss in the way just mentioned is to seek to reclaim some of the rotational energy from the water abaft the propeller and this is the principle of the Grim wheel. A Grim wheel, shown in Fig. 7.19, has a larger diameter than that of the propeller in whose wake it is placed so that it can also reclaim some tip vortex energy. There seems little doubt that this device can be effective in its energy reclaim role, but the break-up of some wheels poses the question of whether the forces involved are fully understood.

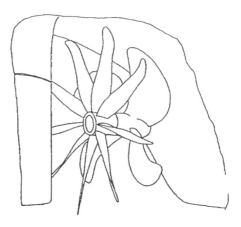

Fig. 7.19. Grim vane wheel.

7.8.3 Efficiency claims

Major claims have been made for each of the devices mentioned, but none of them appear, so far, to have won general acceptance. A question which must be asked in relation to each claim is the extent to which the device in question can improve a good conventional design — or is its best use in rectifying a poor design?

Another interesting question is the extent to which the savings claimed can be additive, where the devices themselves are physically compatible.

7.9 DESIGN OPTIMISATION FOR POWERING

Having dealt with powering methods in this and the previous chapter, it now seems appropriate to consider some of the principal features which determine whether a ship will be power efficient or not.

7.9.1 Block coefficient

The Froude number–block coefficient relationship which is probably the most important factor in optimising a design from the powering point of view has already been discussed in §3.4 and the derivation of the Watson/Gilfillan line has been given. This line is also shown in Fig. 7.6 and it may be noted that this plot reasonably confirms the near optimum nature of the line as its intersections with the F_n lines occur fairly near to the points at which the C_t values start to increase rapidly with increasing C_b.

The parts of the line between $C_b = 0.55$ and 0.65 and above $C_b = 0.80$ take the form that might be expected but the "hump" between $C_b = 0.70$ and 0.80 is not so readily understandable. One possible explanation may be that less attention has been paid to the development of models in this region.

7.9.2 Length

Length affects powering in several ways, the most important being the effect it exercises on Froude number, the significance of which is clearly shown in the C_t figures presented in §7.2 with increasing length reducing F_n for any required speed and thereby reducing C_t.

For a required displacement the effect does not end there since an increase in length as well as reducing the Froude number will also result in a reduction in the other dimensions of B, T and C_b, and reductions in the first and last of these will also reduce the resistance.

A third effect, or possibly another way of looking at the same thing, dealt with in §6.2, is the reduction in the frictional resistance coefficient with length.

At high Froude numbers, a long slender ship shows to advantage, but the smaller wetted surface of a short, beamy, deep ship can be advantageous at lower Froude numbers where frictional resistance predominates.

The reduction in EHP obtainable by increasing length is offset to a small extent by the reduction in propulsive efficiency which is caused by an increase in length.

7.9.3 The effect of $L/\nabla^{1/3}$

Another way of looking at the effect of length is provided by assessing the effect that $L/\nabla^{1/3}$ has on the resistance.

In §6.8 it was noted that plots of resistance against $L/\nabla^{1/3}$ were used by Guldhammer and Harvald in association with a standard B/T.

A feel for values is not easy to acquire, but if the values of the ratios L/B, B/D and T/D suggested in §3.3 are used, it is instructive to see the values of $L/\nabla^{1/3}$ which result.

For a fast ($F_n = 0.30$) long slender fine lined ship with a B freeboard having $C_b = 0.55$, $L/B = 7.5$, $B/D = 1.65$, and $T/D = 0.69$, the value of $L/\nabla^{1/3}$ is 6.24.

For a slow ($F_n = 0.14$) short beamy full ship with an A or B-60 freeboard having $C_b = 0.85$, $L/B = 5.5$, $B/D = 1.9$, and $T/D = 0.77$, the value of $L/\nabla^{1/3}$ is 4.44.

Diagrams in a 1966 RINA paper "The BSRA Methodical Series" by Lackenby and Parker indicate that there is no benefit in increasing $L/\nabla^{1/3}$ above about 5.2 for ships with block coefficients of 0.75 or more.

For C_b values between 0.75 and 0.60 the value should increase from 5.2 to 5.6.

For C_b values of less than 0.60 the value should increase to about 6.0.

Further reductions in resistance can be obtained by increasing towards 7.0, although the gain by doing so must be assessed against the added constructional weight and cost that this will entail.

7.9.4 B/T ratio

Diagrams in the Lackenby paper can also be consulted to evaluate the effect of changes in *B/T*.

At a value of 2.4, which appears to be about the average value for most ships, the resistance values seem, in general, to be only 2% or so above the optimum values which generally correspond to a *B/T* value of a little less than 2.2 — a figure which incidentally it would be almost impossible to achieve in most ships because of the constraints which stability and freeboard rules apply to ship proportions.

In the case of a slow speed ($F_n < 0.14$) and full bodied ship ($C_b > 0.85$), these diagrams show that a change of *B/T* from 2.0 to 3.0 causes an increase in resistance of 3%.

In the case of a fast ($F_n > 0.30$) and fine lined ship ($C_b < 0.55$) a similar change in *B/T* increases the resistance by 5%.

It may be noted that the adoption of a low *B/T* value or, since the breadth is more likely to be fixed in relation to the depth than to the draft, the adoption of as large a draft as possible, is doubly advantageous as, for a fixed displacement it will reduce the capital cost.

Much the same lessons can be drawn from the Mumford Indices which have already been mentioned.

7.9.5 B/T ratio of the ballast condition

In the ballast condition there is a major change in the *B/T* ratio from that applying in the load condition. Typically for a large tanker or bulk carrier *B/T* laden is about 2.2 but in the ballast condition this changes to 5–6. This results in a major change in the hydrodynamics, which is rarely given any consideration although these vessels will spend 40–50% of their time in ballast.

Ships with controllable pitch propellers can adapt their rpm to suit either regime but there may be a case for more consideration being given to the ballast condition in the design of conventional propellers.

Other aspects of the design of the lines to minimise the power required are dealt with in Chapter 8.

Chapter 8

Design of Lines

8.1 OBJECTIVES

In previous chapters the design has progressed to the point at which the main dimensions, the block coefficient and possibly the LCG and LCB have been decided.

A lines plan is now needed for a number of reasons.

(i) So that a General Arrangement plan can be drawn. Whilst this may not be essential at the initial design stage for large, full lined ships, it is an early requirement for fine lined ships, where the arrangement must be tailored to a major extent to suit the space and shape of the deck lines which contain it. This applies very strongly to the design of warships, research vessels and smaller passenger ships.

(ii) So that quantities can be taken off and used for weights, centres of gravity and cost estimates.

(iii) So that cargo spaces and tanks can be arranged and their capacities checked.

(iv) So that hydrostatics can be calculated and trim and stability checked.

(v) To send to the Tank for use as the basis for model tests.

Later, after any modifications found necessary as a result of tank testing have been made, the lines plan becomes the basis for the offsets, loft work, numerical coding and all the structural and arrangement drawings.

The lines plan has to meet a large number of different objectives all of which will have been set by the stage in the design at which it is drawn:

(i) Required displacement at the load draft.

(ii) Required cargo space and tank capacities.

(iii) Required deck areas to accommodate all aspects of the arrangement.

(iv) Features conducive to minimising the powering requirements; low resistance, good hull efficiency and an ability to accommodate the propeller with clearances that make vibration unlikely.

(v)   Features conducive to good seakeeping and good manoeuvrability.
(vi) An LCB position at the load draft which in association with the weights and centres of gravity of the ship and its deadweight items enables the ship to be loaded in a way that will result in satisfactory trim.
(v)   KM values at operating drafts which will ensure satisfactory stability when the ship is loaded as intended.
(vi) The avoidance of discontinuities that may have adverse structural consequences.
(vii) If possible the lines should be production kindly, with as much flat plating as can be arranged and with the minimum amount of double curvature in the shell plating.
(viii)Whenever possible, and particularly on passenger ships, the lines should have an aesthetic appeal.

8.2 THE BOW AND STERN

8.2.1 Normal or bulbous bow

The first decision to be taken in relation to the bow is whether to fit a "normal" or a "bulbous" bow. A normal bow is cheaper to manufacture and a bulbous bow should only be fitted if doing so will reduce the resistance and thereby either increase the speed or reduce the power required and with it the fuel consumption. Figure 8.1 shows the range of Froude numbers and block coefficients at which such an improvement is likely to be obtained when operating at the load draft.

The superimposition of the Watson/Gilfillan C_b line on this diagram indicates the area which is of practical concern and it can be seen that bulbous bows:

(i)   are advantageous for fast ships with C_b values less than 0.625 and F_n greater than about 0.26;
(ii) present no advantage for ships with C_b values between 0.625 and 0.725 — unless these are "over driven" according to the Watson/Gilfillan criterion;
(iii) are again advantageous for C_b values between 0.725 and 0.825, but probably not for C_b values over 0.825.

It is worth noting that at all block coefficients, bulbous bows show to best advantage on over driven ships and are often disadvantageous on ships which are relatively fine for their speeds.

It must be emphasised that this analysis refers only to the load draft condition. It is generally accepted, however, that bulbous bows can offer their greatest advantage in the ballast condition, particularly on full lined ships with block coefficients in excess of 0.75. This being so it would have been nice to make a similar plot for the ballast condition, but unfortunately suitable data to do this does not seem to be available.

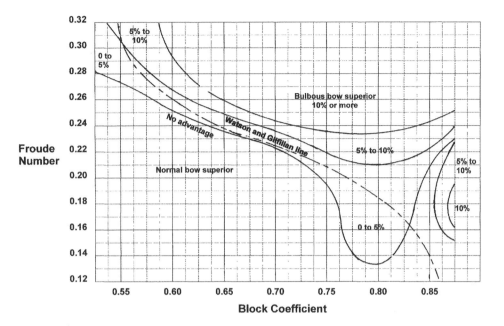

Fig. 8.1. The combination of Froude number and block coefficient at which a bulbous bow is likely to be advantageous.

Some deductions can, however, be made from Fig. 8.1 and in general it appears that if a bulbous bow is not advantageous at the load draft, it will only become advantageous in ballast if the ship is operated at or near its full power giving a speed in ballast at least 10% or say 2 knots or so, more than the loaded service speed.

In the past, when tankers and bulk carriers making lengthy ballast voyages used their full power on this leg, the gain in ballast speed was a clinching argument for fitting a bulbous bow. Today fuel economy often keeps ballast speeds down to, or lower than, the loaded service speed, and the argument for a bulbous bow is reduced. It is worth emphasising that overall economy may require a balance between designing for optimum performance fully loaded and in the ballast condition.

A bulbous bow will generally help to reducing pitching, but on the other hand it is more likely to cause slamming.

8.2.2 Bulbous bow shapes

Bulbous bows come in a variety of shapes and sizes, as shown in Fig. 8.2. One main division is that between a fully faired bulb and one in which there is a sharp knuckle line between the bulb and a normal bow configuration. Claims have been made for the advantages of each type, but the "added" bulb is generally simpler to

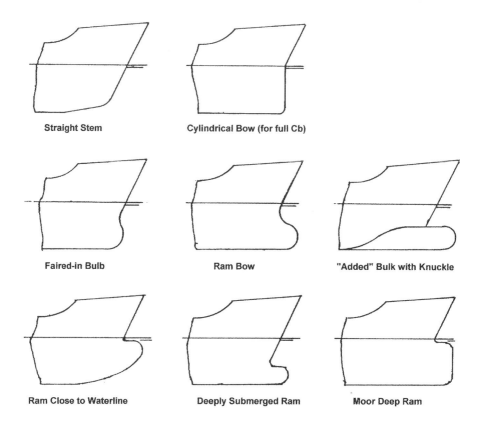

Fig. 8.2. Various bow configurations.

manufacture and seems, on full lined ships, to give at least as good results as the faired bulb.

The next division is between bulbs which project as rams significantly forward of the fore perpendicular, and those with little or no such projection. Ram bows also vary in the vertical positioning of the forward projection, which in some designs commence near the waterline and in others are well submerged.

One of the principal criteria applied to the design of a bulbous bow is the relationship that its sectional area at the fore perpendicular bears to the midship section area.

8.2.3 Bows above the waterline

Above the waterline, bows are raked forward largely to conform with the flare of the adjacent sections. Both rake and flare have as one of their objectives reducing both pitching and the amount of water shipped on the fore deck. Appearance and the minimisation of damage caused to the other vessel in a head on collision are further advantages of bow rake.

Care should be taken not to exaggerate flare too much as waves hitting one side of a heavily flared bow can give rise to torsional vibrations and stresses. This was first noted on some early container ships and Classification Society rules now require additional strength to be provided if the flare is thought excessive.

The severity with which the forces generated by the sea can impact on flare has been shown in a number of accidents, mainly off the South African coast, in which complete bows have broken off, and of course in the recent tragic event in which the bow visor was ripped off the Ro-Ro ferry Estonia.

See also §8.8.3 on the use of knuckles.

8.2.4 Sterns

Sterns have to be considered in relation to the following roles:
- (i) the accommodation of the propeller(s) with good clearances that will avoid propeller excited vibration problems;
- (ii) the provision of good flow to the rudder(s) to ensure both good steering and good course stability;
- (iii) the termination of the ships waterlines in a way that minimises separation and therefore resistance;
- (iv) the termination of the ships structure in a way that provides the required supports for the propeller(s) and rudder(s) plus the necessary space for steering gear, stern mooring and towage equipment etc. and is economical to construct.

8.2.5 Flow to the propeller

Where the propeller diameter (D) on a single-screw ship is of normal size in relation to the draft, i.e. D/T is approx. 0.75, the main consideration is ensuring good flow to the propeller, with a figure of between 28 and 30° being about the maximum acceptable slope of a waterline within the propeller disc area.

Keeping to such a figure tends, of itself, to force the LCB forward on a full bodied ship.

Lloyds' recommended minimum clearances as a fraction of the propeller diameter for a four-bladed propeller are:

Tip to sternframe arch	$= 1.00\ K$
Sternframe to leading edge at $0.7\ R$	$= 1.50\ K$
Trailing edge to rudder at $0.7\ R$	$= 0.12$
Tip to top of sole piece	$= 0.03$

where

$$K = \left(0.1 + \frac{L}{3050}\right)\left(\frac{2.56 C_b \cdot P}{L^2} + 0.3\right)$$

where P = power in kW.

The recommended clearance for a four-bladed propeller on a twin-screw ship, is 1.00 K. Other values are given in the rules for three, five and six-bladed propellers.

8.2.6 Large propellers

Where the propeller is large in relation to the draft of the ship, a number of options exist:

(i) The propeller can be fitted in such a position that the lower tip is below the line of the keel. This is common practice on warships, but merchant ship owners have been reluctant to allow this because of possible damage to the propeller in shallow water and possible additional dry docking problems and costs. With a better understanding of the gains that can be obtained by the use of large diameter propellers it is possible shipowners may be more willing to consider this in the future, although even when a clear water stern is used, most owners demand a substantial rise of both the propeller tip and the bottom of the rudder above the base line.

(ii) The ship can have a designed trim or a raked keel. This is commonly used, and for precisely this purpose, in small ships, notably tugs and fishing vessels. It is also used for the same reason on warships, even large twin-screw vessels. So far it has not been adopted on large merchant ships, partly due to a wish to limit the extreme draft of these ships, and partly because of the increase in structural complexity which is an unfortunate corollary. Winters 1997 R.I.N.A. paper "Application of a large propeller to a container ship with keel drag" merits study and may lead to a greater adoption of this simple and effective way of improving propeller efficiency on large ships.

(iii) A Mariner type rudder, supported by a skeg, can be fitted eliminating the sternframe solepiece and thus permitting a small increase in propeller diameter, but see comments under (i).

(iv) A tunnel type form can be used. The design shown in Fig. 8.3 was used very successfully on shallow draft river craft. It may be noted that the propeller tip can come right up to the static waterline with the tunnel configuration ensuring that it is kept fully immersed. The use of much the same technique on single-screw vessels was introduced by Burmeister and Wain on their fuel economy vessel, and a body plan showing two body plans with lines of this sort used by Port Weller is illustrated in Fig. 8.4. This type of stern brings an incidental gain in displacement and deadweight.

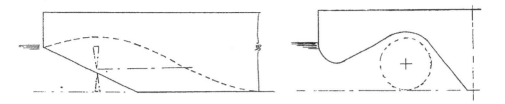

Fig. 8.3. Tunnel form to permit use of a larger propeller.

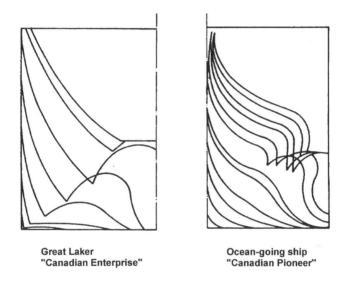

Great Laker
"Canadian Enterprise"   **Ocean-going ship**
                                    "Canadian Pioneer"

Fig. 8.4. Two Port Weller bulk carriers with semi-tunnel single-screw forms to permit use of a larger propeller.

A glance at the designs shown in the 1990 and 1991 numbers of *Significant Ships* shows that the "Mariner" or clear water type of stern mentioned in (iii) above and shown in Fig. 8.5 is now almost universally adopted.

8.2.7 Stern lines above the propeller

It is very desirable from a resistance point of view that the stern lines above the propeller should be continued to form a cruiser stern which is immersed at the operating drafts. As Fig. 8.5 shows, a cruiser stern should extend aft sufficiently to cover the rudder but there is no need for there to be any significant immersion at the end of the waterline; indeed, significant immersion at this point is likely to cause eddies particularly if the cruiser stern is terminated by a flat transom as has become

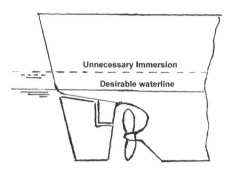

Fig. 8.5. Open water cruiser stern.

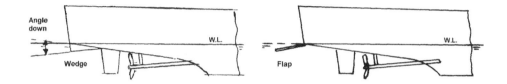

Fig. 8.6. Transom wedge/flap to improve powering performance.

fairly general practice in recent years. The top of the rudder should follow the lines of the stern with only the necessary clearance.

Keeping the stern immersion to the "desirable" WL position has the added advantage of permitting the greatest possible propeller diameter for a given draft.

In merchant ships transoms were initially adopted for cost saving reasons, but once adopted the flat transom concept was progressively developed to provide more deck area for mooring equipment, to provide stowage for a tier of containers or to facilitate moving the accommodation further aft. It was also found that a considerable gain in KM could be obtained by the wider waterlines in the stern, and there did not seem to be any adverse consequences, at any rate on large ships.

The proviso "on large ships" takes note of the loss of a number of stern trawlers due to broaching in severe stern seas which may have been due at least in part to the fact that they had transom sterns.

A transom stern can greatly improve the statical stability of a ship by increasing the KM but if advantage is taken of this to permit more top weight, the ship may have inadequate stability when it suffers the big loss in KM which can occur when the stern comes out of the water when the ship is pitching in a seaway.

Possibly the main argument for retaining a traditional cruiser stern is an aesthetic one and probably for this reason this type of stern is still featured on some cruise liners. Another, but rather unusual reason for retaining a cruiser stern applied to the fishery inspection ship the design of which is described in Chapter 16, §16.5. In this case it was preferred to a flat transom in case the ship had to go astern in ice.

In warships the transom stern was introduced not for cost cutting reasons but because it improved the hydrodynamic performance giving a less turbulent wake particularly at high speeds. As in merchant ships, the resulting increase in KM was appreciated for stability reasons and the additional deck area because it improved the arrangement. In fact in present warship practice the full midship beam is often maintained right to the transom and from upper deck level to very nearly the waterline.

A further development in the sterns of high-speed ships is the transom wedge or flap illustrated in Fig. 8.6. This reduces the high stern wave that used to build up at the stern and thereby reduces the resistance.

8.3 DESIGNING LINES TO MINIMISE POWER

8.3.1 The LCB position

The next item to be considered is the location of the centre of buoyancy. In some ship types this is dictated by the disposition of weight and the need to achieve a satisfactory trim, but in most ships it should be governed by a wish to minimise power requirements.

A very experienced tank superintendent who read this commented that he wished this were so, but had found designers almost always saying that the LCB position had been dictated by trim requirements.

The author believes that the LCG position is closely linked to the LCB position (see, for example, Fig. 4.7) and that unless a ship is being designed to have a particularly heavy local weight, it will trim satisfactorily almost automatically provided reasonably careful thought is given to the disposition of tanks. Throughout his career he has therefore positioned the LCB of his designs where he thought best from a powering point of view.

While it may be wrong to speak of an optimum position of the LCB, it is certainly correct to think in terms of an optimum range. The optimum range of LCB position depends mainly on the Froude number and block coefficient, which as has been shown, are themselves linked.

The range differs for ships with normal and bulbous bows, as the LCB on a form with a bulb will be anything from 0.5–1% further forward than that of an otherwise very similar form with a normal bow.

The range also differs for twin-screw ships for which the optimum range is further aft than it is for single-screw ships, reflecting the fact that the lines of a

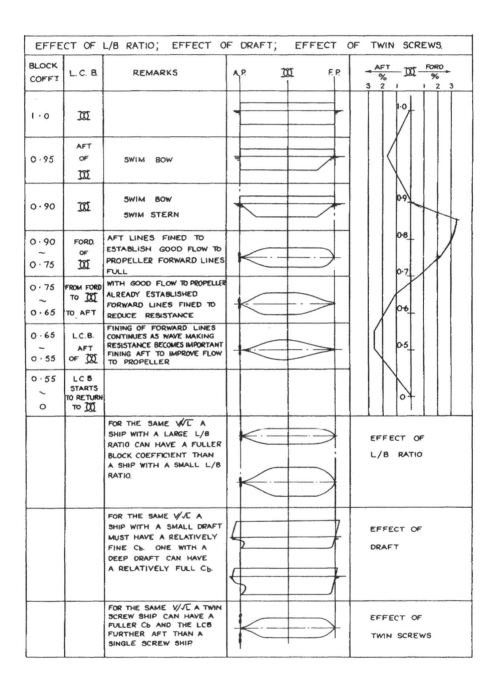

Fig. 8.7. Relationship between block coefficient and longitudinal centre of buoyancy.

twin-screw ship can be optimised almost entirely on resistance considerations with little need to consider the flow to the propellers which, as has been seen, plays a major part in the design of the stern of a single-screw ship.

In thinking about the LCB position and indeed about the value of the block coefficient, it is important not to be mesmerised by the figures themselves but to keep clearly in mind what they mean in terms of the ships lines.

Figure 8.7 tries to illustrate this point and shows how and why the LCB moves as the C_b changes from unity to a very fine form.

(i) At $C_b = 1.00$ the LCB must of course be at amidships.

(ii) For a barge the first essential is a swim bow, so at about $C_b = 0.95$ the LCB moves aft to say 1.5% A.

(iii) The next improvement to be made to ease movement of the vessel is a swim stern, so at about $C_b = 0.90$ the LCB moves back to amidships.

(iv) For the slowest self propelled shipshape vessel the bow is now generally very full-spoon shaped and this coupled with the need for good flow to the propeller(s), requiring fining aft means that for a C_b of between 0.90 and 0.75 the LCB is well forward, say about 2.5–3.0% or even 3.5% The use of big "outboard" type propellers reduces the problem of propeller support on a full ship and to some extent the problems of flow to the propellers and enables very full block coefficients to be used.

(v) Once the run has been made such that it provides a satisfactory flow to the propeller, it is only necessary to fine it very gradually as the block coefficient is further reduced for ships with higher speeds and powers. The forebody, on the other hand, is where reductions in wavemaking resistance can best be effected and from being markedly fuller than the aft body, the forebody changes to being much finer, with the result that the LCB progressively shifts to a position well aft of amidships.

(vi) Finally for very fine ships there is a tendency for the LCB to return towards amidships.

Figure 8.8 gives a plot of the optimum range of LCB position for both normal and bulbous bow forms against C_b, on the assumption that the block coefficient is related to the Froude number generally in accordance with the Watson/Gilfillan line. It will be seen that there is an appreciable range of choice without incurring an excessive penalty — but it is worth remembering that the penalty for a "too far forward" position is usually much worse than that for a position "too far aft".

8.3.2 The sectional area curve

The sectional area curve is one of the principal factors which determines the resistance of a ship and careful attention should be paid to its form. The first step

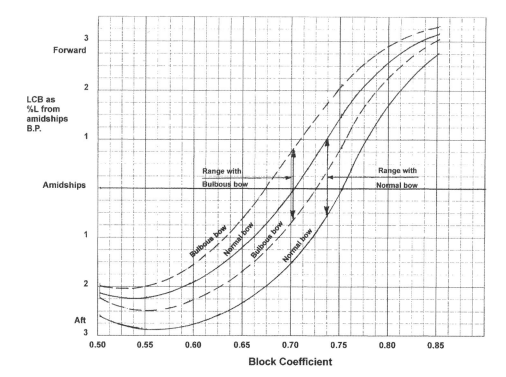

Fig. 8.8. LCB position v. block coefficient. Graph shows the range of LCB positions within which it is possible to produce lines with resistance close to optimum. Assumes F_n–C_b relationship of Watson–Gilfillan mean line.

towards drawing a sectional area curve is the determination of what its maximum ordinate, the midship area coefficient (C_m) should be.

A formula for C_m in terms of the bilge radius and rise of floor is:

$$C_m = 1 - \frac{F[(B/2 - K/2) - R^2/(B/2 - K/2)] + 2R^2(1 - \pi/4)}{B \times T} \tag{8.1}$$

where
F = rise of floor
K = width of keel

With no rise of floor this reduces to:

$$C_m = 1 - \frac{2R^2(1 - \pi/4)}{B \times T} \tag{8.2}$$

Both of these formulae can be transposed to give formulae for R if C_m is known.

$$R = \left[\frac{B \cdot T \cdot (1 - C_m) - F(B/2 - K/2)}{2\{(1 - \pi/4) - F/(B - K)\}} \right]^{1/2} \tag{8.3}$$

and

$$R = \left[\frac{B \cdot T \cdot (1 - C_m)}{2(1 - \pi/4)} \right]^{1/2} \tag{8.4}$$

Whether C_m or R should be fixed first is a matter for debate and there may need to be an interactive process.

There seem to be three motives for keeping the bilge radius small:
  (i) the greater resistance to rolling provided by a "square" bilge;
  (ii) the easier cargo stowage of a squarer hold; and
  (iii) for a given C_b, the finer C_p associated with a larger C_m will generally, but not always, reduce the resistance.
On the other hand, the radius should be sufficiently large to be production-kindly, which probably means about 2.5 m for ships with a beam greater than about 16 m. On fine lined ships it may be desirable to increase it above this figure to assist in marrying it in to the fore and aft lines. Generally however if a fine C_m is desired for any reason — say to increase the draft of a "volume" type ship — this is usually better achieved by the use of a high rise of floor.

For ships with a beam of less than 20 m and no rise of floor an approximate empirical formula for the bilge radius, in metric units, is:

$$R = (1.7 - C_b) \times (B/3.3)^{1/2} \tag{8.5}$$

On ships with rise of floor the bilge radius may be somewhat reduced.

A bilge radius to the above formula and with no rise of floor will result in a C_m of:

$$C_m = 1 - \frac{(1.7 - C_b)^2}{7.7T} \tag{8.6}$$

Although C_m is usually best determined as the product of practical decisions on the dimensions of the bilge radius and the rise of floor, it is sometimes convenient for powering calculations (see §6.9) to have a quick method of estimating a reasonable value in terms of the block coefficient and an approximate C_b–C_m relationship is given as Fig. 8.9.

The big difference between the lines which appear to apply to most merchant ships and that which applies to most warships confirms the view that C_m is best determined by deciding on the bilge radius and the rise of floor rather than *vice versa*.

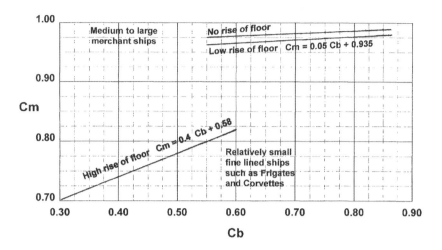

Fig. 8.9. Approximate relationship between C_b and C_m.

The area of the sectional area curve is $C_p = C_b / C_m$

The area curve is divided into three parts:

(i)   the entrance,
(ii)   the parallel middle body,
(iii)   the run.

The parallel middle body should be made as long as possible without distorting either the entrance or run and avoiding hard shoulders at the junctions with these.

The length of parallel middle body is determined largely by the block (or prismatic) coefficient, whilst the length of entrance and run depends on both the block coefficient and the LCB position. In some respects it is more logical to reverse this statement and say that the LCB position is the outcome of decisions on the shape of the sectional area curve taken to minimise resistance.

8.4 DRAWING LINES USING A BASIS SHIP

With both C_b and C_m fixed, a sectional area curve, or for that matter a body plan, can be drawn for the new ship based respectively on the sectional area curve or the lines plan of a "basis" ship of known good performance.

The advantage of working with the lines of a basis ship is that a body plan is the direct result. This should not only be completely fair but its sections should have the general characteristics of the basis ship albeit distorted by differences in the *L/B* and *B/T* ratios of the basis and new ships.

It is, of course, rare for an initial lines prepared in this way to meet all the designers wishes, but it does provide a good basis on which the modifications required can be sketched and eventually faired. If the form derived in this way is not suitable for the new ship it is however, probably better to work with the sectional area curve and avoid being influenced by the form of the sections of the basis ship.

8.4.1 Retaining the same LCB position

The method used is essentially one of adding parallel middle body if an increase in C_b is required or deducting it if a finer C_b is wanted. If:

L_1 is the length of the basis ship, and
C_{b1} is the block coefficient
C_m the midship section coefficient
C_{b2} is the required block coefficient of the new ship
X is the parallel middle body to be added/subtracted
Then

$$(L_1 + X)\, C_{b2} = L_1\, C_{b1} + X\, C_m \tag{8.7}$$

and

$$X = \frac{L_1(C_{b2} - C_{b1})}{C_m - C_{b1}} \tag{8.8}$$

Starting at the F.P. and A.P. respectively new stations are then drawn on the lines plan at a spacing of $(L_1 + X)/10$.

The waterline offsets read at these stations are multiplied by the ratio of the new beam to the basis beam and are plotted on waterlines whose spacing is adjusted from the basis spacing by the ratio of the new load draft to the basis load draft to give a body plan on displacement sections for the new ship.

A comprehensive treatment of ways of varying ships forms may be found in "On the Systematic Geometrical Variation of Ship Forms" H. Lackenby R.I.N.A 92 (1950) p. 289. Computer programs using a number of these methods are available from the firms mentioned in §1.2.

8.4.2 Changing the LCB position

If the new ship's LCB position as a percentage of the length is to be changed from that of the basis ship, then the station spacing should be calculated separately for the fore and aft bodies using the respective half body block coefficients.

These half body block coefficients can conveniently be obtained from Fig. 8.10 which shows the addition to one body and deduction from the other which produces a particular LCB position. It is interesting that a midship LCB corresponds to a forebody $C_{bf} = C_b + 0.007$ and a corresponding aft body $C_{ba} = C_b - 0.007$, and not, as might have been expected, of equal fore and aft body C_{bs}.

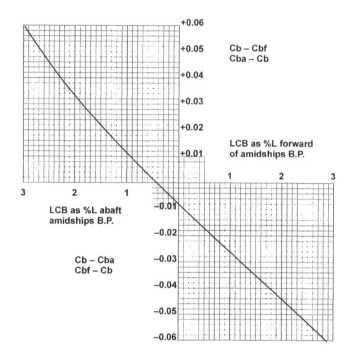

Fig. 8.10. Relative block coefficients of fore and aft bodies for a particular LCB position.

8.4.3 Designing lines direct from a sectional area curve

If the lines are being designed without the benefit of a basis ship, attention should be paid to the following factors which influence the performance:
(i)   the shape of the waterline forward and the angle of entrance;
(ii)   the shape of the waterlines, buttock lines or diagonals aft which guide the flow to the propeller;
(iii) the shape of the sections below water do not appear to have much influence on still water resistance, although there is a view that "U" sections forward and "V" sections aft offer some slight advantage from a propulsive point of view. "V" sections, generally have some advantage in waves and, of course, provide a higher KM value which can be helpful on a design where stability is critical.

8.4.4 Mathematical lines generation

In practice the methods described in this section have largely been replaced by computerised methods based on similar theory.

The author has no practical experience of using lines developed by any of the many mathematical hull generation programs that are available today but a reader of this chapter who had this experience made the comment that most of these programs do not take into account all the factors which have been considered in this chapter, each of which contributes in some way to the overall performance of a design. This can, of course, be overcome by manual intervention at the preliminary body plan stage with mathematical refairing of the resulting sections.

8.5 TWIN-SCREW LINES AND APPENDAGES

The lines of a fine lined twin-screw ship can be designed almost entirely with the objective of minimising resistance, without the need to consider flow to the propeller which necessarily plays a large part in the design of single-screw lines. As a result, twin-screw forms can have a slightly bigger block coefficient (and as already noted a further aft LCB position) for a given Froude number and/or should have better specific resistance of the "naked" hull.

8.5.1 Bossing or shaft brackets

The naked resistance advantage of a twin-screw lines is, however, reversed when the appendage resistance of bossings is added. In the past fully enclosed bossings were the normal fit on twin-screw ships. These had the advantage of protecting the shafts and allowing these to be supported at intervals that avoided problems with whirling vibrations. The additional resistance was, however, high — of the order of 10% of the naked resistance — so more recently the shafts have been left exposed to the sea, supported by one or more "A" brackets. The resistance of this combination can generally, with good detail design, be kept to about 6% of the naked resistance.

These figures are very broad generalisations as appendage resistance can vary widely (see also Chapter 7, §7.5).

8.5.2 Twin skeg forms

A novel approach adopted on several recent twin-screw passenger liners and ferries is the twin skeg form shown in Fig. 8.11. With these forms the bossing in effect becomes part of the main hull. This type of lines appears to increase the wake and certainly provides good support to the propellers, and ought to minimise "bossing vibration", although this is not borne out by experience in more than one ship of this type.

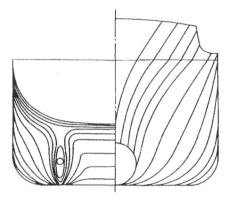

Fig. 8.11. Twin skeg body plan.

The type of form which this design appears to promote has a wide waterline aft giving a high KM value which improves stability.

Claims have been made that the twin skeg form has reduced resistance, but available data does not altogether substantiate this even when appendage resistance is taken into account.

8.6 HIGH STABILITY LINES

8.6.1 General discussion

The motive behind the design of lines as discussed so far has been the desirability of minimising the power required either by minimising the resistance or by maximising the propulsive efficiency.

In some ship types other factors need to be considered and can be so important that the attainment of a worthwhile improvement in them will justify the acceptance of a penalty in powering. On passenger ships and container ships, stability can be such a factor.

Even on other ship types the use of lines which have a high KM for a given beam can help to improve the economics of a design.

In Chapter 4 it was noted that the ratio *B/D* could be reduced if lines designed to give a particularly high KM value were used, and it is worth reiterating that if two designs with different dimensions have same cargo carrying capacity the ship with the smaller *B/D* ratio will probably be the cheaper to build, requiring less steel, and possibly also requiring less power.

If there is a breadth limitation for any reason, the use of high KM lines provides a way to increase cargo carrying capacity by:

(i) permitting an increase in depth;
(ii) permitting an extension to the superstructure; and
(iii) permitting the carriage of additional containers on deck.

The first of these applies to most cargo ships, the second to passenger ships and the third to container ships.

On passenger ships the increase in depth which a bigger KM will permit can, with advantage, be used to increase the freeboard to the bulkhead deck without this having an adverse effect on stability. Such an increase in depth can be made at a modest cost and can be of particular value in the design of car ferries and similar ships where the gain in large angle stability given by increased freeboard can be a great help in improving survivability and the ship's ability to meet damaged stability requirements.

8.6.2 Ways of achieving a high KM

Much of this section is abstracted from a report written by the author as part of a study commissioned by the British Department of Transport into ways of improving the safety of Ro-Ro ships.

The use of high stability forms is by no means a new concept, but the development of these forms has stopped short of what can be achieved without incurring significant penalties in powering or seakindliness.

The ways in which the KM can be increased for given ship dimensions are:
(i) filling out the waterline — increases BM
(ii) adopting V sections — increases KB
(iii) adopting a high rise of floor — increases KB

8.6.3 Filling out the waterline

Filling out the waterline forward tends to increase the resistance and should only be done to a limited extent. The comparison of the waterlines of two ships of similar Froude number shown in Fig. 8.12 is instructive as there does not appear to have been any penalty in the powering of ship II. The change in KM between the two ships which is largely obtained by the more pronounced shoulder on ship II amounted to about 3%.

Filling out the waterline aft can be taken appreciably further than it is wise to do with the fore body. Figure 8.13 shows the body plan of the stern of ship II as built and as it could have been built with the waterline filled out. Although the ship as-built already had quite a full waterline the modification proposed would have increased the KM by 7.5%

The stern of this ship was not immersed at the load draft, so a further gain in KM could be achieved by adopting a transom wedge, as shown in Fig. 8.14 and

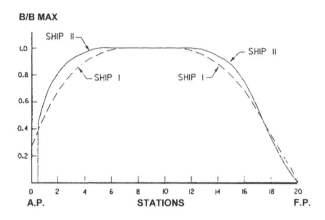

Fig. 8.12. A comparison of LWL shapes.

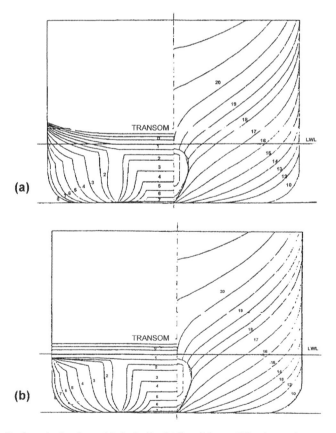

Fig. 8.13. Ro-Ro ferry body plans. (a) As built. (b) Possible modification to improve stability. The modification shown in (b) would increase KM by more than 1 m.

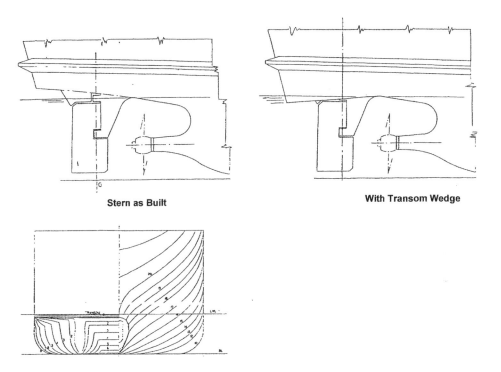

Stern as Built

With Transom Wedge

Fig. 8.14. Ro-Ro ferry stern plans. Transom wedge added to increase KM (by more than 1.7 m).

described in §8.3, giving the propeller a semi-tunnel position. Whether this arrangement would increase or reduce the power required is not known, but the latter seems equally likely. The gain in KM (from the basis) resulting from this change increases to 12%.

8.6.4 V sections and high rise of floor

For various reasons, neither of these modifications were applicable to the ships which were used as the basis of the study. V sections undoubtedly improve stability. If fitted forward they tend to increase the power required; fitted aft they can be good from a powering point of view particularly if associated with "straight" buttocks. They do not always suit the internal arrangement of the ship too well particularly if carried to an extreme.

The use of high rise of floor can improve the stability but almost certainly means increasing the depth and draft of the ship if the displacement and internal volume are not to suffer.

8.6.5 Flared form

In this section attention has so far been focused on the provision of high stability at the load and worst damage condition drafts, but the high GMs which Ro-Ro ships usually have at these drafts are provided not for the intact condition but because they are the corollary of providing the minimum allowable GM after damage.

Flared ship sides provide a way of improving the stability after damage, but in a ship such as a Ro-Ro the freeboard to the bulkhead deck is usually small thus limiting the gain from the flare. On the other hand, the height to the upper deck is usually considerable and if the flare is carried all the way to that deck the increase in the breadth of the ship at this deck will also be considerable: with corollaries of more material and cost and a higher ship VCG.

A variety of ways of arranging flare with and without associated tumblehome are shown on Fig. 8.15. Of the alternatives shown, that in (e) seems to offer most advantages, but unless the freeboard is increased as shown in (f) the gain from flared sides hardly seems worthwhile.

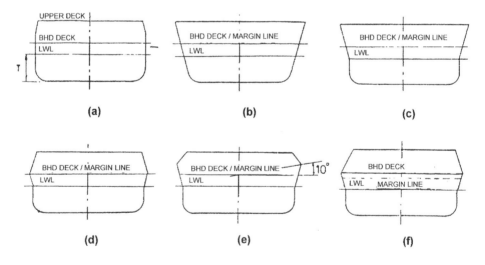

Fig. 8.15. Alternative flare/tumblehome configurations.

(a) Existing ship; margin line 76 mm below bulkhead deck.
(b) 15° Flare faired into sections; some loss of displacement made good by filling lines and/or increasing *B* and/or *T*.
(c) 15° Flare with knuckle at LWL; maximum beam at upper deck increased by 3.8 m.
(d) 15° Flare between knuckles at LWL and margin line.
(e) 15° Flare between knuckles at LWL and 10° angle from margin line.
(f) 15° Flare between knuckles at LWL and new bulkhead deck (freeboard increased by 1 m).

8.7 SEAKEEPING AND MANOEUVRABILITY

Although this chapter is devoted to the design of the lines, the subjects dealt with in this section affect other aspects of design and indeed of ship operation and it seems sensible to deal with these at one time.

8.7.1 Seakeeping

With some exceptions, seakeeping is regrettably low in the design priorities for most types of ship. The principal exceptions to this are warships, research vessels and offshore oil production and storage vessels all of which have to spend long periods at sea and have to provide a platform on which their crews can undertake demanding tasks whatever the weather conditions. Seakeeping is also recognised as being important on passenger ships and particularly on cruise liners, although because these are generally very large ships they can achieve a fairly good sea-keeping performance by virtue of size without requiring much special attention being paid to this feature of their design.

Small ships whose seakeeping ability should undoubtedly have more attention paid to it include fishing vessels, whose broaching problems are mentioned in §8.3.

Deciding what is involved in good, or at any rate acceptable, seakeeping is a difficult task, varying as it does with ship size and speed and the sea areas in which the ship is required to operate. The following features certainly enter into it:
   (i) shipping water on deck;
   (ii) pitching motions;
   (iii) rolling motions;
   (iv) slamming; and
   (v) broaching.

8.7.2 Shipping water on deck

The amount of water shipped on deck is determined primarily by the size and speed of the ship together with the freeboard at the bow (see Chapter 11, §11.2). Within these constraints the practical ways of minimising it are well flared forward sections coupled with well placed knuckles.

8.7.3 Pitching motions

These are mainly determined by ship size and the longitudinal moment of inertia. If a high proportion of the weight of the ship and its cargo are concentrated near amidships the pitching period will be relatively small and the accelerations large; if on the other hand a high proportion of the weight is "winged out" towards the ends of the ship the pitching period will be longer and the accelerations less. Other design constraints generally prevent any significant action to follow this counsel but it should be kept in mind.

The heavily flared sections forward advocated to minimise water on deck can increase pitching (and forward damage) so a compromise between pitching and water on deck may be necessary.

A bulbous bow may help to reduce pitching but is more likely than a normal bow to cause slamming.

8.7.4 Rolling motions

Rolling motions are largely a function of the metacentric height. A ship with a high GM (say 2.0 m or more) will have a short period of roll with uncomfortably high accelerations; a ship with a low GM (say 0.2 m or less) will be much more comfortable with low accelerations but the amplitude of roll may be large.

Stabilisers, either of fin or tank type provide the best way of reducing rolling. The fin type gives the best reduction in roll amplitude when the ship is operating at speed but if the ship may have to operate at a slow speed, tank stabilisers provide the best answer.

Whether the ship has stabilisers or not, bilge keels should be fitted if at all possible and should be carefully sized to their task. On full bodied ships the bilge keels should extend over most of the length of parallel middle body; on ships with no parallel middle body, or where the extent of parallel middle body is limited, the bilge keels must necessarily extend into the entrance and run but should be limited in length and great care should be taken to ensure that the keels runs along streamlines. If streamline tests are not available Isherwood's R.I.N.A. paper provides a good guide to the line to follow.

On fine lined ships where the length of bilge keels are limited this should be compensated by the use of deeper bilge keels to maintain the area. The depth of the keel should be limited so that it does not extend beyond the square of the hull and the keel should be arranged normal to the shell.

An approximate formula for the length of bilge keel that it is usually practical to fit is:

$$\text{length of keel} = 0.6 \times C_b \times L \tag{8.9}$$

A formula for the depth of bilge keel which offsets the reduced length where the C_b is small which the author used for ships of up to about 180 m is:

$$\text{depth of keel} = 0.18/(C_b - 0.2) \tag{8.10}$$

For today's bigger ships this formula should probably be modified to include some small scaling with ship's length.

There are a few ship types, of which icebreakers are one, which should not have bilge keels, in this case because they are too easily ripped off by ice. On some vessels used for scientific research, such as fisheries and oceanographic, scientists tend to argue that bilge keels should be omitted to lessen "noise", but this should only be accepted if an effective tank stabilisation system is being fitted.

The shape of the midship section also influences rolling with a small bilge radius giving a squarer shape which has more inherent resistance to rolling.

8.7.5 Slamming

As slamming occurs when the forebody re-enters the water after having emerged during pitching, all the measures already suggested as ways of reducing pitching help to minimise this also.

In addition, a deep still water draft is an important factor in minimising slamming because it reduces the risk of emergence in a seaway . The advantage to be gained from a deep draft is particularly relevant to the ballast condition.

If it is impossible to stop the forebody emerging from the water the next best thing is to ensure that it re-enters with minimum force, which can be done by shaping the bottom with a pronounced V form in the slamming region so that it acts like a knife.

Frigates and corvettes, having a small draft and the need to maintain speed in very rough seas, would tend to slam very badly if their lines were not very carefully designed with this in mind. They are given a very high rise of floor which both increases the draft and creates a V bottom, extending the full length of the ship, which can re-enter the water without much fuss but, although greatly reduced, some slamming still occurs.

Slamming/pounding or something very like it can still take place even if the forefoot doesn't actually emerge. It is essential that there should not be a large flat area of bottom in the region of maximum relative motion/ acceleration which goes quite a long way aft.

The bottom of bulbous bows should be angled as shown in Fig. 8.14.

8.7.6 Broaching

Broaching occurs when a ship is travelling down wind and/or down the path of the waves in a seaway. Yachts, which broach much more frequently than ships, do so when a particularly strong gust of wind generally from a slightly further forward than the prevailing wind hits their sails. This makes the yacht turn uncontrollably into the wind bringing the relative direction of this from aft round to the beam. With the change in the wind direction there is a broadside force on the sails which causes the yacht to heel very severely often "onto her beam ends".

The best way of avoiding a broach is by a very quick helm response, made at the very first sign that a broach is imminent, aimed at turning the yacht so that the wind is dead astern. The best, indeed almost the only, method of recovering from a broach is to completely free the sheets of all the sails so that these stream out in the wind and cease to develop any force.

In ships, broaching is a much rarer phenomenon and the mechanism which causes it is slightly different, the broaching of ships being caused by the waves rather than by the wind. If, when sailing away from waves the crest of a particularly large wave overtakes the ship and hits one quarter — rather than hitting the stern squarely — the ship will tend to be thrown round into the direction in which the wave is moving. This will bring the ship beam on to the sea and result in a severe list. The fact that at the time the crest hits the stern the preceding trough will be passing along the ship means that the draft at amidships and forward is reduced lessening the resistance to the turning force exerted by the crest. Whilst the crest is still on the quarter, the bow is buried and the stern is raised, the next trough then arrives causing the waterline to drop away from the rudder reducing, or in a severe case eliminating, its effectiveness just when it is most acutely needed.

The operational lessons to be learnt are:

(i) the need for the helmsman to take great care when steering a ship in following waves of significant size and to be ready to turn the ship's stern very quickly into any larger waves approaching from either quarter;

(ii) the need for the Master to decide in good time if the waves seem to be becoming too large for the ship to take comfortably on the stern to alter course immediately and if necessary to heave-to.

This digression into yacht sailing and ship operation has been made because few naval architects appear to understand what broaching is and just how dangerous it can be. It is perhaps worth emphasising that broaching is not limited to small ships but can be a real danger to quite large ships.

Reverting now to the steps that can be taken at the design stage to produce a ship less likely to broach.

As a broach is caused by a wave hitting the stern quarter of the ship the effect will be more severe if the stern presents a big target to the wave. Large flat transoms on comparatively small ships such as fishing vessels appear to be a case in point.

As the mechanism of a broach involves the ship's stern being carried round any thing that helps to resist this sort of motion is helpful, so a reasonable amount of deadwood aft should be retained. On the other hand a bulbous bow and particularly a ram bow may contribute to broaching by digging into the water and becoming the fulcrum about which the ship swings. There may even be something to be said for the preference which some older yacht designers had for having some balance between the two ends of a yacht.

To minimise the proportion of the rudder that comes out of the water in a broach reducing the helmsman's ability to correct the course, rudders should be made as deep as possible.

8.7.7 Ship motion calculations

The calculation of ship motions at the design stage is a comparatively recent practice stemming from the conjunction of the importance which is now attached to minimising motions of ships and particularly those of the types mentioned in §8.7.1. with research work which has developed the necessary calculation methods and the ability of modern computers to handle these fairly considerable calculations with speed and economy.

The calculations can be based on a range of environmental data using three types of sea spectra: Pierson-Moskowitz, ITTC two parameter and Jonswap and modelling the waves either as unidirectional (long crested) or spread with a user-defined wave spreading function.

The calculation method used is based on strip theory with frequency domain computations for motions, added resistance and loads from waves and time domain calculations for hull response to slamming. The output can consist of any or all of the following:
- the ship motions of heave, pitch, roll, yaw and sway;
- the added resistance for a variety of ship headings;
- dynamic loads imposed on the ship;
- total motions at specified places in the ship;
- possible sustained sea speed against a variety of limiting factors;
- structural responses due to slamming.

Figure 8.16, reproduced by courtesy of Kockums Computer Systems, illustrates the required input and the range of output.

8.7.8 Course stability and manoeuvrability

Course stability is put first partly because it is a requirement that applies to all ships and partly because it seems a natural follow-on to the discussion of broaching in the last section. It is of course the quality that a ship should have of not deviating from its set course unless the rudder is put over to initiate a turning movement.

Most ships whose lines have been designed to minimise resistance and promote good flow to the propellers will have good course stability as a natural corollary without any special measures. Ships with a low length/beam ratio especially if this is coupled with a full block coefficient are likely to have poor course stability unless particular care is taken to ensure that this is satisfactory — and course stability may set a limit to the present trend to very low L/B ratios.

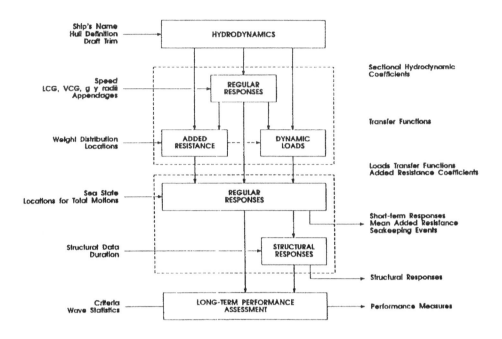

Fig. 8.16. Seakeeping data flow diagram.

The best practical if somewhat simplistic advice would seem to be the maintenance of as much deadwood aft as possible.

8.7.9 Manoeuvrability

The requirement for manoeuvrability varies with ship type and intended service. Whilst all ships should have a reasonable minimum standard a significantly higher performance should be specified for those intended for operation in narrow or crowded waters. Warships whose manoeuvrability provides one of their best defences against approaching torpedoes or missiles and the possibility of avoiding mines seen at the last moment need a particularly good capability.

In most single-screw ships turning ability is provided by a single centreline rudder operating in the propeller race. The performance is largely a function of the rudder area and its relationship to the product of length × draft which provides an approximate measure of the ship's resistance to turning. The waterline shape of the rudder is also important and should be a streamlined aerofoil with a high lift characteristic. Because the flow onto the rudder from the propeller is different above and below the shaft centreline there would appear to be some advantage in shaping the upper and lower parts of the nose of the rudder differently, and there have been patent types of rudder on which this was done, but possibly because of the extra complication in the construction few such rudders were built.

Some rudders on single-screw ships are supported on pintles from a closed arch type of sternframe with in this case the sternframe acting as the nose of the rudder; on other ships the sternframe is open aft with a heel extending to provide a bearing at the bottom of the rudder. On ships with a closed arch sternframe the rudder area can be somewhat less than should be the practice with an open arch sternframe as the sternframe tends to act at least partially as part of the rudder.

For single-screw ships of above 100 m in length, a K_a value (rudder area / $L \times T$) of 0.0125 for a streamlined aerofoil rudder in association with a streamlined closed arch sternframe or of 0.0135 for a rudder without a sternframe providing a leading edge, has given a satisfactory turning circle on a considerable number of ships.

The author has little data on ships below 100 m for which higher K_a values are normally used, with the value of K_a increasing as the length decreases, probably because the smaller the ships are the more confined the waters in which they are required to manoeuvre. Barnaby gives figures for a range of vessel sizes in his "Basic Naval Architecture". Whilst his figures seem high for ships of 100 m, possibly because his data may date back to single plate rudders, a brief summary of his figures for smaller ships is given below in default of other information.

Length (m)	K_a
25	0.024
50	0.021
75	0.018
100	0.016

The most common requirement for performance appreciably better than that obtainable from a conventional rudder applies when the ship has to manoeuvre at slow speed and there are several ways of providing this capability, *viz.*: active rudders of which there are several very effective types; a bow thruster; both bow and stern thrusters.

Twin-screw ships used to be built with a single rudder on the centreline, but it was found that the performance of a rudder operating clear of the propeller race was often poor and virtually all twin-screw ships built today have twin rudders positioned in the propeller wake. Because on twin-screw ships stern shaft withdrawal is almost invariably "out" from the ship twin rudders must be sufficiently displaced either inboard or preferably outboard, because this position gives a bigger turning lever and there is less wake effect, of the lines of shafting to permit this.

Twin rudders are helpful in recovery from a broach as one of them should be well immersed even if the other comes out of the water but the suggestion made in the paragraph on broaching that the rudder should be as deep as conveniently possible is still worth following particularly on shallow draft ships.

A rudder is more effective if its top is arranged with only a small clearance from the shell as this effectively increases the aspect ratio and reduces eddies.

8.7.10 Ship manoeuvring calculations

Like ship motions, ship manoeuvring has only become the subject of calculations in recent years. Considering how important a role a ship's manoeuvring capability can have in the avoidance of collisions or of grounding in coastal waters, it is surprising how long it has taken for statutory and classification societies to become interested in this capability as an essential safety feature in a ship's design.

In the past the interest of these authorities was limited to the time taken by the steering gear to put the rudder hard over and even owners seemed content to record the turning circle and rarely specified a maximum diameter except for ships where manoeuvring was of particular importance. This aspect of design was left almost entirely to the shipbuilder and they, in turn, contented themselves with providing a rudder area based on a coefficient related to the product of the ship length and draft that had given satisfactory performance for a previous ship of similar characteristics for which they had records.

It is not surprising that quite a number of ships, particularly those with high block coefficients, have had poor manoeuvring and/or poor course stability.

Modern computer-based calculation methods take account of a rudder force coefficient based on the aspect ratio of the rudder, the rudder position relative to midships, the flow velocity to the rudder based on the propeller race in which it operates and the rudder/hull interaction based on the turning velocity of the ship. If necessary shallow water effects can also be evaluated.

Results can be presented for the turning circle, for a zig-zag manoeuvre and for a reverse spiral curve, the last of these being used to detect potential directional instability.

8.8 THE LINES ABOVE THE WATER LINE

8.8.1 Section shape

Above the waterline the shape of the sections can be determined by a number of factors. The freeboard at bow, stern and amidships helps to determine the flare. If the freeboard is relatively small, the angle of flare must be large partly to achieve the required deck area and partly to help limit the water shipped on deck. If the freeboard is large both of these objectives can be obtained by a gentle flare.

Flare at the bow and outward sloping sections aft have been practised for many years; more recently, however, the advantage of flaring the midship section has been pointed out. This advantage lies in the gain in large angle stability that flared sections provide. This will improve intact stability in almost all cases but in the often more critical case of damage stability will only do so to the extent of the gain in breadth at bulkhead deck level. Flared sections will add weight and cost and will raise the VCG. Clearances to cranes, dock walls and tugs must also be considered.

Very infrequently used nowadays although once quite common is the reverse of flare namely "tumblehome". The purpose of tumblehome was cost saving and a lowering of the VCG thereby improving static stability. In practice it was found quite helpful in keeping the ship topside, even if the ship was slightly heeled, clear of dockside cranes.

8.8.2 Deck lines

The lines plan shows all decks that extend to the ship's side: the upper deck, forecastle, bridge (in merchant parlance a side to side erection amidships), poop — together with all lower decks.

These decks are, of course, shown in plan view, elevation and sections.

In the past all decks exposed to weather generally had both sheer and camber. Lower decks might follow the same pattern or could have camber and no sheer, or neither.

Sheer was traditionally arranged parabolic on the Upper deck at ship's side. Standard sheers forward and aft are specified in the loadline rules, the aft sheer being half of the forward sheer. The standard sheers were intended to keep the decks reasonably clear of water, but in practice designers often thought it wise to exceed these values. In metric units standard sheers are:

forward $= 0.0166\,L + 0.508$
aft $= 0.00833\,L + 0.254$

Camber was traditionally also parabolic and again a standard was set by the loadline rules as breadth/50.

Modern practice is to eliminate sheer over most of the ship and if any sheer is required to give the height of bow needed, either to meet the rule requirement or thought necessary as a result of seakeeping tests or calculations, to have a straight line sheer forward of an appropriately positioned knuckle.

Straight decks with no sheer have several advantages: their steelwork is cheaper to construct, as is any joinerwork fitted to them; the stowage of containers or modules is simplified; in association with level keel trim they keep the deck in the same relationship with port facilities such as passenger gangways, cranes, coal chutes etc. throughout the ship's length. Unfortunately they lack the aesthetic appeal of the old parabolic sheer.

Camber is now generally also arranged on a straight line basis with a level area extending from the centreline to knuckle points P&S outboard of which the deck is sloped to provide drainage.

This again reduces shipbuilding costs and improves cargo stowage particularly on container ships and car ferries.

8.8.3 Knuckles

The "fair" lines of the shell are sometimes interrupted by a knuckle line or lines. These can be introduced for a number of reasons:

- (i)   to enable a high angle of flare to be used in the lower part of the sections without this carrying on become too extreme in the upper part;
- (ii) to avoid the end of a forecastle deck projecting in a way that might cause contact with dockside cranes or similar;
- (iii) to improve seakeeping (although there is disagreement over this) by the detachment of waves from the shell;
- (iv) to reduce shipbuilding cost by increasing the number of plates that do not need to be rolled in two directions.

For economy in fabrication, knuckles are generally best positioned a short distance above a deck.

Chapter 9

Machinery Selection

9.1 INTRODUCTION AND CRITERIA FOR CHOOSING THE MAIN ENGINE

The selection, arrangement and specification of the main and auxiliary machinery is the province of the marine engineer. In this chapter only those aspects of these tasks which directly affect the naval architect as the overall ship designer are dealt with — and the treatment is necessarily a simplified one. It commences with an examination of the criteria against which the choice of main engines is made, which include:

9.1.2   Required horsepower
9.1.3   Weight
9.1.4   Space
9.1.5   Capital cost
9.1.6   Running costs
9.1.7   The ship's requirement for electrical power and heat
9.1.8   Reliability and maintainability
9.1.9   The ship's requirement for manoeuvring ability and/or for slow-speed operation
9.1.10 Ease of installation
9.1.11 Vibration
9.1.12 Noise and other signatures
9.1.13 Availability

The importance of each of these criteria differs from one ship type to another. In some ships only a few of the criteria need be considered, in others all must be taken into account although with different degrees of emphasis. Each criterion is considered briefly in the following sections.

9.1.2 Required horsepower

The naval architect, when calculating the power to specify to the marine engineer, has to make a number of assumptions. The most important of these assumptions relates to the number and type of propulsors and to the propeller revolutions. All these must be known to enable the quasi propulsive coefficient to be estimated and this has, of course, a major influence on the required power. A secondary influence on the power stems from the effect on the displacement of whatever assumption is made in respect of machinery type with the influence this exercises on machinery and fuel weights etc.

All these assumptions must be relayed to the marine engineer, who should feel free to question them. If by changing any or all of the assumptions the marine engineer can offer a technically better and/or cheaper solution, a dialogue with the naval architect should ensue and the power estimate adjusted to suit what are then agreed as the main technical features of the machinery.

Apart from adjustments of this sort the power is of course the fundamental criterion.

9.1.3 Weight

This is not generally a very important matter for the majority of merchant ships, although it undoubtedly plays quite a significant part in the selection of machinery for ferries and similar relatively fast, fine lined ships, particularly if these are also subject to a draft limitation.

In the design of warships, planing craft and catamarans, the need for a high speed from a relatively small ship makes the power/weight ratio a matter of vital importance.

9.1.4 Space

Much of what has been just been said about weight also applies to space. As far as the main engines are concerned space and weight generally go together, but if a trade-off between weight and space is possible, then ships designed on a dead-weight basis should be fitted with the lighter machinery, even if this takes more space, whilst those designed on a volume basis should be fitted with the less bulky machinery even if this is heavier.

On warships space, like weight, is at a premium and the power/volume ratio is very important.

9.1.5 Capital cost

The cost of the main engine itself must be considered along with any differential costs which may arise from its installation. Such differential costs could include the cost of gearing and/or the need for separate pumps for one alternative whereas another may use direct drive and have engine driven pumps included in the main engine price.

9.1.6 Running costs

Usually, the most important item of running costs is the annual fuel bill. In recent years fuel prices have been very volatile, as Fig 9.1 shows. Whilst at the time of writing (1995) fuel prices are well down from their peak in the period 1979–1985, the lessons learnt then ensure that these costs remain a very major factor in machinery selection.

For a required horsepower there are, in principle, two fundamentally different ways of minimising expenditure on fuel:

(i) by fitting as fuel efficient an engine as possible even if this requires a relatively expensive fuel; or

(ii) by the use of machinery which can burn a cheap fuel even if its specific consumption is comparatively high;

and, of course a compromise between these extremes, with the ideal being an engine capable of achieving a low specific consumption whilst burning a cheap fuel.

Whilst seeking minimum fuel costs, however, it is important not to overlook other running costs, such as lubricating oil, spare gear, annual maintenance and — not least — the cost of manning. A reduction of one in the number of engine room staff may reduce running costs by as much as, or more than, can be achieved by expensive improvements in engine efficiency.

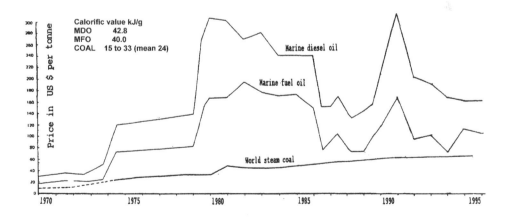

Fig. 9.1. Fuel prices through the years. For recent years the oil prices are those recorded in M.E.R. for January at Las Palmas which is a reasonable mean between the cheapest prices (generally at Rotterdam) and the most expensive (generally in the Far East). The MFO prices relate to IFO 180.

9.1.7 The ship's requirement for electrical power and heat.

Because the main engine will generally be able to burn a cheaper fuel than is required by the generators, the use of the main engine(s) to provide electrical energy and/or heat for engine auxiliary plant and hotel services via shaft driven alternator(s) and exhaust gas boiler(s) respectively can have an important influence on running costs.

On passenger and other ships with high electrical loads this can lead to a preference for diesel electric propulsion.

9.1.8 Reliability and maintainability

These aspects — very dear to all practical seagoing marine engineers — must be considered on all ships, but become of outstanding importance on ships for which the consequences of a breakdown may be particularly severe. Such ships include passenger ships where not only are particularly high costs incurred in dealing with the immediate emergency but future profitability may be prejudiced by attendant publicity. On warships reliability is made the subject of very detailed studies and redundancy is introduced to minimise the consequences of any loss of capability whether this is caused by mechanical breakdown or enemy action.

Some marine engineers tend to favour the use of a slow-speed diesel because this will have fewer cylinders, reducing the parts requiring maintenance, whilst others prefer the lighter and more easily handled parts of a medium-speed engine.

9.1.9 The ship's requirement for manoeuvring ability and/or slow-speed operation

An ability to manoeuvre quickly and accurately can be an important factor in the choice of main engines and, of course, their associated propulsors on ships which berth or use canals or constricted waters frequently.

A need to be able to operate at slow speeds using low power, particularly if this may have to be for protracted periods, can rule out certain machinery options, unless this function is undertaken by an auxiliary system, such as a controlled slipping clutch device or an auxiliary propulsion system such as a thruster.

9.1.10 Ease of installation

This is probably a second-order criterion, but there is no doubt that some engines, particularly of the slow-speed type have much simpler systems than others of the same type and this may be taken into account when the choice is finely balanced.

9.1.11 Vibration

Any vibratory forces or couples that may emanate from a main engine under consideration must be carefully assessed before it is accepted as suitable. An engine which develops even a moderate couple should only be considered if it can be clearly shown that the resulting vibration is within acceptable limits at all parts of the ship where it could affect personnel or equipment. It is worth noting that the relativity of the position in which the engine is to be fitted to the nodes and anti-nodes of the ship's vibration profile can have a significant effect. (See also the next section on noise.)

9.1.12 Noise and other signatures

In some vessels such as fishery and oceanographic vessels and warships operating submarine detection equipment such as towed array, the minimisation of the under-water noise signature becomes a driving factor in the whole machinery installation.

Where the noise targets are stringent, consideration must be given to raft mounting the engine(s) and enclosing them in an acoustic enclosure. Both of these requirements impose limits of both weight and space on the choice of engines. Even if these measures are taken, the required performance will demand the choice of an engine with minimum vibration and noise characteristics.

In mine hunters the magnetic signature becomes so important that all machinery must be constructed of non-magnetic materials.

9.1.13 Availability

This is a warship concept which is discussed in Chapter 14. It was a particularly significant factor when the change from the general use of steam turbine machinery to gas turbines and/or diesels was first being considered. Today, its greatest design influence lies in the general move it has caused towards repair by replacement to reduce out-of-service time (see also §14.6).

9.2 ALTERNATIVE MAIN ENGINE TYPES

9.2.1 Diesel engines

By far the majority of merchant ships in service or under construction today have diesel engines. These are available in three main types:
  (i)   slow-speed diesels with a speed of rotation mainly in the range 60 to 150 revs/min;

At a Power of about 7000 kW

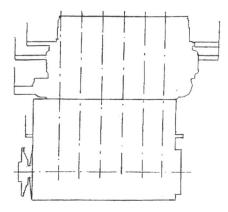

Slow speed diesel:
7000 kW @ 100 rpm
317 tonnes
1.11 tonnes/m³ or 0.90 m³/tonne
0.044 tonnes/kW or 23 kW/tonne

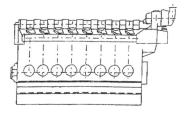

Medium speed diesel:
7650 kW @ 520 rpm
153 tonnes
0.80 tonnes/m³ or 1.25 m³/tonne
0.02 tonnes/kW or 50 kW/tonne

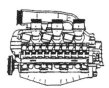

High speed diesel:
7000 kW @ 1300 rpm
21 tonnes
0.84 tonnes/m³ or 1.9 m³/tonne
0.003 tonnes/kW or 330 kW/tonne

Fig. 9.2. A comparison of the size, density and weight per kW of various types of main engines.

(ii)   medium-speed diesels with a speed of rotation mainly in the range 450 to
       800 revs/min;
(iii) high-speed diesels with a speed of rotation mainly in the range 1000 to
       3000 revs/min

There are of course some diesel engines with revs/min in the gaps between the
above ranges, but these are the generally accepted bands.

A comparison of the size, the density and the weight per kW of each of these
machinery types is given in Fig. 9.2, all of the examples chosen being of about 7000
kW — this power being chosen as one at which all three engines compete. A similar
comparison between a slow-speed diesel and a marine gas turbine is given in Fig 9.3.
In this case the power is 22000 kW representing the top end of the gas turbine range.

At a Power of about 22000 kW

Slow speed diesel:
22000 kW @ 100 rpm
770 tonnes
1.04 tonnes/m^3 or 0.96 m^3/tonne
0.035 tonnes/kW or 28 kW/tonne

Marine gas turbine:
22000 kW
22 tonnes
0.35 tonnes/m^3 or 2.8 m^3/tonne
0.001 tonnes/kW or 1000 kW/tonne

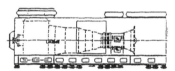

Fig. 9.3. A comparison of the size, density and weight per kW of various types of main engines.

9.2.2 Slow-speed diesels

Slow-speed diesels are today the almost universal choice for all large merchant ships: container ships, bulk carriers, tankers, and gas carriers. These engines, which are all of the two-stroke type, are often quoted at two different powers and associated speeds.

The lower speeds quoted enable the engine to be associated with a large diameter highly efficient propeller bringing an attendant gain in fuel economy. On the other hand, a reduction in speed to say 72% of the full revs will result in the power reducing to about 55% of that developed at full speed and this will make it necessary to fit an engine with about 50% more cylinders and accept a corresponding increase in capital cost.

The lowest engine speed currently quoted is 57 revs/min.

The highest power currently quoted is 51,840 kW for a 12-cylinder. K90MC B&W engine (at 94 rev/min).

The specific fuel consumptions quoted range from about 174 down to 156 g/kWh (with efficiency booster). The fuel used is heavy marine fuel oil — the cheapest oil fuel available (see Fig. 9.1).

Most slow-speed engine models are made in 4–12-cylinder versions, with all cylinders vertical in line.

With powers starting at 2500 kW for a 4-cylinder engine operating at about 150 rev/min, the slow-speed diesel range extends to quite small and relatively slow ships, but in the lower part of the range its merits must be assessed against those of medium-speed engines.

The principal manufacturers are MAN B&W, Sulzer and GMT.

9.2.3 Medium-speed diesels

The slower speed medium engines can be used with direct drive on small single-screw vessels and on slightly larger twin-screw ships on both of which an engine speed of around 450 rev/min may not be much higher than the propeller speed, which would in any case be dictated by the limit on propeller diameter imposed by the lines.

On larger ships medium-speed diesels are fitted with gearing to reduce the propeller speed to the lowest value that can be accommodated by the largest propeller that can be fitted, thereby optimising the efficiency. Before the development of medium-speed engines to the high powers now available, it was quite usual practice on single-screw ships to fit two engines geared together. A similar practice was adopted on larger twin-screw ships with two engines geared to each shaft line.

Until the makers of slow-speed diesels reduced their revs, the low propeller revs and the resulting high propeller efficiency that could be obtained using a geared diesel was one of the best selling points for this type of installation, offsetting the gearing efficiency loss (about 1.5%) and the higher specific fuel consumption of this type of engine when compared with a slow-speed engine. This particular advantage no longer applies, but the medium-speed diesel can still offer considerable weight and space advantages over the slow-speed engine, whilst the redundancy advantage of having two engines rather than one can sometimes be a favourable factor in the choice.

Medium-speed diesels are also frequently chosen as the prime mover in diesel–electric installations.

Medium-speed diesels are mainly of the four-stroke type. A number of models are manufactured both as "in-line" and "V" engines. The in-line models generally range from 6 to 9 cylinders, with the V models taking over in the range from 12 to 18 cylinders, although there are a few 8-cylinder V models.

The highest power currently available from a medium-speed engine is 23,450 kW from an 18 V PC 4.2 Pielstick at 428 rev/min. A twin installation can therefore give almost 47,000 kW.

More typical, however, are powers per cylinder ranging from about 150 kW to about 1000 kW and engine speeds ranging from 450 to 750 revs/min.

The specific fuel consumptions claimed range from about 200 down to 167 g/kWh. In general, the larger engines have the lower specific fuel consumption. The fuel used is again a heavy marine fuel oil, but the cruder versions of this are generally to be avoided and these engines require to be changed to diesel oil when manoeuvring.

There is a far wider range of manufacturers than for slow-speed diesels with Daihatsu, MAN B&W, NEI Allen, SEMT Pielstick, Wartsila, Sulzer and Stork being amongst the principal firms.

9.2.4 High-speed diesels

The use of high-speed diesels as main propulsion prime movers is confined, as far as merchant ships are concerned, to small vessels and ships with diesel–electric installations.

In warships where a high power-to-weight ratio is essential, high-speed diesels are used either on their own or in conjunction with gas turbines.

As high-speed diesels are generally manufactured on a production line basis, models tend to be available in a limited number of cylinder options.

The highest power currently available from a high-speed engine is 7400 kW from a V20 M.T.U. at 1300 rev/min (370 kW per cylinder).

More typical, however, are powers per cylinder of from about 20 to 200 kW and speeds of from 1200 to 2100 revs/min.

The specific fuel consumptions claimed range from about 250 down to 187 g/kWh. In addition to these engines having a higher specific fuel consumption than medium-speed engines, consideration must be given in any economic comparison of the types to the higher cost per tonne of diesel oil (about 2.0 × cost of heavy fuel, see Fig. 9.1) and the reduced time between overhauls.

Manufacturers at the top end of the power range are M.T.U. and SEMT Pielstick. In the middle and lower range, Caterpillar must be one of the best known names but there are many others.

9.2.5 Gas turbines

The main advantages that gas turbines have over competing engines, which are principally high-speed diesels, are their extremely good power/weight and power/volume ratios. Their main disadvantage has been their much higher specific fuel consumption. This disadvantage — already quite significant at full power operation power — becomes much worse for most gas turbine types when they are run at part load and if operation at part load power is a frequent requirement this has generally eliminated them from consideration.

A few merchant ships were built with gas turbine propulsion in the late sixties/early seventies and were successful technically, but became quite uneconomic when fuel prices rose dramatically in 1973 and the ships concerned have since either gone out of service or been re-engined with diesels.

The situation with warships is quite different with the high power/weight ratio of the gas turbine making this very attractive on a ship where any additional weight increases the power required so much that it becomes easy to get into a vicious spiral. The poor specific fuel consumption of a gas turbine does not matter as much on a warship as it does on a merchant ship because warships tend to use full power infrequently and for relatively short bursts. It has, however, become quite usual practice for warships to have two separate sets of machinery — one for maximum speed and one for the cruise regime. These two sets of machinery can consist either of two gas turbines, a high power set for the high-speed regime and a lower power set for the cruise regime, or more frequently a combination of high-power light-weight gas turbines for maximum speed with high or medium-speed diesels with good specific fuel consumption for the cruise regime. The latter combination clearly making very good sense.

Both of these combinations can be arranged in two ways with the two machinery fits as alternatives (COGOG) or (CODOG) or with them so linked that both are used to develop the maximum power (COGAG) or (CODAG).

Although the latter combination offers more power there are complications in the gearing required and at present the former is the generally preferred option. On frigates there are usually either two gas turbines or a gas turbine and a diesel geared to each of the twin shafts; on corvettes one gas turbine and one diesel may be fitted with a central gearbox dividing the power between the two shafts.

Gas turbines are generally arranged as modules suitable for repair-by-replacement with the machinery casings sized to suit.

As most gas turbines for marine use have been developed from aircraft engines, the number of models is quite limited and come principally from Rolls Royce and General Electric. The powers of those currently being fitted are given in Table 9.1.

A return to the use of gas turbines on merchant ships may not be far away as it seems likely that developments presently in hand will reduce the specific consumption of a gas turbine to much the same as that of a diesel engine and go a long way towards eliminating the reduction in efficiency at part power.

In addition, as discussed later under electrical propulsion, the low weight and particularly the low space requirements of these prime movers is starting to look attractive for cruise ships.

Table 9.1

Currently fitted gas turbines

	Make	Power (kW)	SFC (Full) (kg/kWh)	SFC (at 25 %) (kg/kWh)
Rolls Royce	Tyne	4000	0.29	0.45
	SM1A	14000	0.235	0.34
	SM1C	18000	0.23	0.33
	Olympus	21000	0.30	0.46
General Electric	LM2500	21000	0.23	0.33
	LM500	4000	0.27	0.40

The increase in specific fuel consumption at 25% power should be noted. The fuel used in gas turbines is marine diesel oil.

9.2.6 Steam turbines

Using oil as a fuel, even the very advanced steam turbine plants which have been proposed cannot compete in fuel economy with diesel engines, but they do provide the easiest route by which coal and nuclear fuels can be used.

Apart from the possibilities with these fuels, it is worth noting that oil companies' refining methods have resulted in recent years in lower grades of fuel being sold on the marine market and that this is a trend which may continue. So far, diesel manufacturers have managed to keep improving their engines' ability to use these poorer fuels, but there may be a limit to this and if a sufficient price differential develops between the cost of the cheapest oil and that which a diesel can tolerate, there may again be a role for the turbine.

Another way in which the steam turbine may return is in a combination system with a gas turbine. In such a system, advantage is taken of the large amount of heat in the exhaust gas of a gas turbine, which is a corollary of its relatively poor thermal efficiency. The system involves fitting a very large and efficient exhaust gas boiler, steam from which is led to a steam turbine in an arrangement along the lines of that shown in Fig. 9.4. The overall efficiency of such a system could be quite attractive, but the capital costs do not at present make it economical.

9.3 PROPULSION SYSTEMS

9.3.1 Direct drive

For the sake of completeness, this section must start with direct drive which is, of course, much the most common propulsion system and is the almost invariable choice with slow-speed diesels. The components of this system consist of shafting

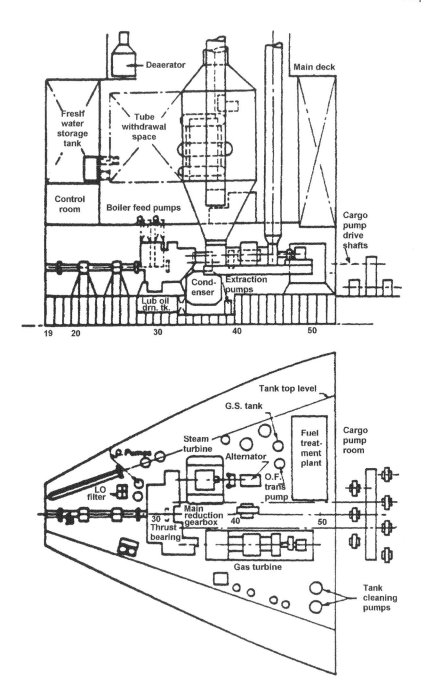

Fig. 9.4. Combined gas turbine/steam turbine arrangement for a 30,000 kW tanker machinery.

and a conventional propeller, with all manoeuvring being done by adjusting the engine speed and direction of rotation.

9.3.2 Geared drive

Geared drive can be associated with most of the other prime movers. Quite frequently, gearing has more than one function, although the most common requirement is the reduction of the revs from the engine output figure to that which is required for the efficient operation of the propeller. It can, however, also be used to combine the power of two prime movers onto one shaft or alternatively to divide the power reaching it between two shafts or between shafting connected to the propeller and a drive to a shaft alternator.

Reversing is a further function that gearing can be called upon to provide, although this is an infrequent requirement as most diesels can be reversed fairly easily whilst the reversing of ships with gas turbine machinery is generally provided by fitting them with controllable pitch propellers.

9.3.3 Electrical propulsion

Electric propulsion has been used for many years, dating back to such famous pre-Second World War passenger liners, as "Viceroy of India", "Normandie" and many others. The prime movers on these ships were all steam turbines and electrical drive was adopted for a number of reasons, with a mistrust of the reliability of large reduction gearboxes certainly figuring in a number of the decisions.

More recently, most electrical propulsion systems have had diesel engines, either medium or high speed, as their prime movers.

The merits of electrical propulsion include the ease of control which it provides giving an excellent manoeuvring capability together with an ability to operate economically and for lengthy periods at reduced speed and power.

The principal disadvantage of electrical drive has always been that it is much more expensive in first cost than the geared alternative. This economic disadvantage is compounded by the fact that the mechanical efficiency is lower, leading to increased fuel consumption and cost.

In early electrical propulsion systems D.C. motors were used and the ships invariably had completely separate electrical systems for propulsion and other purposes.

The development of marine-type thyristor converters has now made it possible to equip ships on the power station concept with propulsion, engine room auxiliaries and ship's hotel load all drawing from a common energy pool, which is in turn fed by whatever number of generators is needed with all engines therefore operating at near peak efficiency.

Bringing all the electrics together in one system — along with some reduction in the relative cost of electric propulsion systems — have combined to reduce the extra cost of today's type of electrical propulsion and it is now the favoured system for large cruise liners, on which its many operational advantages outweigh any residual extra cost.

A system of this sort has also become widely accepted for specialist ships such as research vessels, ice breakers, cable ships, fish factory ships, oil production vessels.

In general, electric propulsion is attractive either:
 – where there is a large non-propulsion electrical load as on a passenger ship, or
 – where there are a number of propulsion devices spread throughout the ship, such as the azimuth thrusters of a dynamically positioned offshore vessel.

Amongst the advantages which electrical propulsion has for a cruise liner are:
 (i)   the possibility of maximum attenuation of noise and vibration;
 (ii)   the ease of providing the large, but very occasional, electrical demands of a bow thruster without the need of a special system;
 (iii) an ability to operate at very low ship speeds;
 (iv)   a high degree of redundancy giving good reliability.

A good description of a modern diesel electric system which explains in some detail the reasons for its adoption and the alternatives considered is given in a paper entitled "Fantasy and reality" presented by J.W. Hopkins to the Institute of Marine Engineers in 1991.

The prime movers for a diesel–electric installation of this sort can be either medium- or high-speed diesels, the former being the preferred choice on cruise liners and the latter on smaller specialist vessels.

The number and size of generators must be arranged to suit a scenario of different loadings and there will generally be advantage in having two different powers of engine to match these as closely as possible. "Fantasy" for example has four 12-cylinder engines, each developing 7920 kW and two 8-cylinder versions of the same engine, each developing 5280 kW.

The advantage of having all the engines on the ship of the same type starts with the price advantage from bulk buying and continues with such benefits as the need for only one fuel, a reduced requirement for spares, easier maintenance and repairs. If medium speed engines are chosen there is the advantage of using a cheaper fuel for all purposes instead of this being used only for propulsion.

The high power/weight and more importantly power/space ratio of gas turbines is now starting to be appreciated by the owners of large cruise liners who can see a change to this type of prime mover helping them to achieve a worthwhile increase in the number of passengers carried in a particular size of ship.

9.4 FUELS

9.4.1 General

All the propulsion machineries considered so far use oil fuels, albeit of a variety of different grades. At times when the price of oil has peaked (see Fig. 9.1), considerable attention has been given to alternative fuels and in particular to the use of coal and nuclear energy and the two sections which follow look at these alternatives —although at today's fuel prices neither is currently attractive.

9.4.2 Coal burning ships

When the cost of oil fuel increased by a factor of eight in the decade 1970–1980, that of coal increased by a factor of about three. This led to a renewed interest in coal as a marine fuel, particularly amongst Australian shipowners, who ordered a number of ships when the ratio of the cost per tonne of oil/coal in Australian ports was about 5:5. Since 1980, oil prices have fallen and there is at present no likelihood of more coal-fired ships being ordered in the near future, however the design problems involved are interesting and a brief look at them may not be out of place.

Recent coal-burning ships have had mechanical chain grate stokers serving their boilers, but fluidised bed combustion seems likely to take over in any future ships. Unfortunately the thermal efficiency of a boiler/turbine combination is low: 25% being typical of a medium-sized present day installation, although this should rise to about 40% in a large next generation installation with reheat; but even this compares poorly with the efficiency of a modern design of diesel which may attain 50%

Coal has a much lower calorific value (24 kJ/g) than oil fuel (40 kJ/g). When this is taken along with the lower efficiency, the weight of fuel required for a coal burning ship becomes 2 to 2.5 times that needed by a diesel ship. This is not the end of the difficulties, however: coal requires more storage space since it stows at 1.15–1.35 m^3/tonne, increasing, effectively to about 1.7 m^3/tonne when allowance is made for the space required for conveyors and the "self trim" empty space at the top of the bunkers, as compared with about 1.05 m^3/tonne (SG 0.96) for oil fuel.

Furthermore, coal cannot be stowed in double bottom and wing tanks that are so conveniently used for oil fuel and instead requires space free of structure, which in most ships means space that could have been used for cargo. Because of the large consumable weight of coal, considerable care must be taken in its fore and aft disposition or, alternatively, provision made for substantial water ballast capacity if trim problems are to be avoided.

It can be readily seen why there must be a big cost differential between the price of oil and that of coal before the latter becomes an attractive alternative.

9.4.2 Nuclear power ships

Nuclear power has been used extensively in US, British and USSR submarines, in a few US aircraft carriers and in a number of USSR ice breakers. In each of these uses, nuclear power has particular advantages and all these installations appear to have been technically successful. Costs have, of course, hardly mattered in these uses and in most of them there really was no other technical solution giving anything like the same performance.

The use of nuclear power in merchant ships reached the stage of three different prototype vessels and was undoubtedly shown by these to be technically feasible. It could even be economically attractive, assuming a high oil price of the order of that prevailing in 1980, although it would be necessary to build a considerable number of ships to a standard design to defray the very high development costs which would be involved. However, the resistance which the prototype vessels met from environmental groups worldwide was so great that their entry into ports was denied and the ships were laid up without completing their planned trial periods.

In all the nuclear installations to date, the nuclear reactor acts as a boiler supplying steam, generally at moderate steam conditions, to one or more steam turbines.

9.5 AUXILIARY POWER

9.5.1 Electricity generation

Every ship has many systems, equipment and machinery requiring auxiliary power. The requirement is generally for electrical energy, but heat energy can also be used directly via steam, hot-water or hot-oil systems, whilst mechanical energy is used directly to drive pumps on oil tankers and dredgers.

In the interest of economy, a designer should start by looking for ways of minimising the electrical power and heat demands. The next priority is to find ways of meeting these demands with the minimum consumption of the cheapest fuel, taking into account in doing so the associated capital and maintenance costs and, of course, also keeping a wary eye on the system complexity.

9.5.2 Waste heat utilisation

In many merchant ships, the amount of waste heat available from exhaust gas is such that there is no great need to economise in "other heat" demands, but on the other hand the amount is insufficient to justify its use for generating electricity.

In other cases, notably on large bulk carriers, the exhaust waste heat is sufficient to make its use in generating electricity worthwhile as it can meet the comparatively

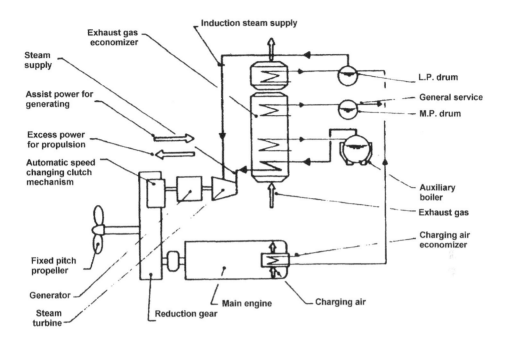

Fig. 9.5. Schematic for electrical generation and propulsion power augmentation by waste heat utilisation.

small sea load of this type of ship. Under these circumstances the exhaust waste heat may become a scarce resource and "other heat" demands should, if possible, be met by the use of the heat in cooling water or lubricating oil systems.

When the electrical power which can be developed from waste heat is more than that needed to meet the normal sea load, it becomes possible to use this power either to increase the ship's speed or to reduce the power drawn from the main engine, thereby reducing the fuel consumption. A schematic for this is shown in Fig. 9.5.

It may be worth noting that improvements in the efficiency of diesel engines have reduced, and will no doubt continue to reduce, the waste heat available and will in general also tend to reduce its quality by increases in the mass flow and reductions in the temperature, which can now be as low as 150°C. That there is still a large quantity of energy available is shown in Fig 9.6.

9.5.3 Shaft-driven alternators

If the electrical power required cannot be produced from waste heat, the next best thing is to produce it by burning fuel oil rather than diesel oil, but unfortunately the

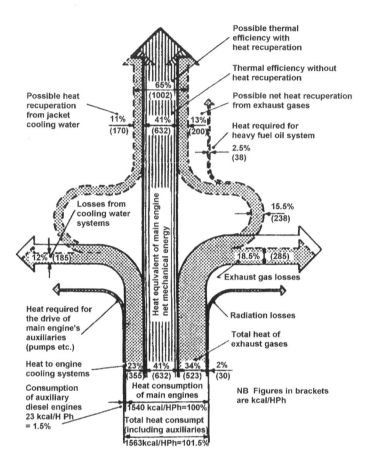

Fig. 9.6. Typical heat balance diagram for a Sulzer RND diesel engine. (Abstracted from "The Modern Diesel Engine and the New Trend of Transport and Energy Systems" by J.A. Smit (Sulzer) 1976.)

heaviest and cheapest grades of fuel oil are not acceptable fuels for auxiliary diesel engines within the power bracket required to drive the size of generators fitted to most ships.

The main engine(s) of most medium to large ships do, however, burn fuel oil, so the use of a shaft- or gear-driven alternator provides a means of generating electricity in a fuel efficient manner.

It is an added bonus that the use of such an alternator can provide a capital saving since the addition of one cylinder to a slow-speed main engine will provide the necessary additional power and generally cost quite a lot less than the saving which can be made by reducing by one the number of diesel alternators.

The economics are, of course, more complicated than this, since means of meeting the port electrical load and of providing standby generation must also be considered. In addition, a shaft-driven alternator may introduce problems of frequency control necessitating the use of a controllable pitch propeller or the use of static frequency converters, but in suitable circumstances a shaft-driven alternator can make a useful contribution to both fuel and overall economy.

The number of generators fitted on most merchant ships is determined by the need to have one on standby at all times. On warships and specialist merchant ships the number should allow for one generator being under repair in addition to the one on standby. On most merchant ships with a relatively light sea load it is usual to fit three generators with two sharing the sea load. On ships with a higher sea load four generators are the usual fit, but sized so that three of these share the sea load. In some cases the economic argument between three large generators and four smaller ones can be closely balanced.

9.6 OTHER ENGINE ROOM AUXILIARIES AND EQUIPMENT

9.6.1 Items specified and arranged by the marine engineer

The engine room accommodates, and the marine engineer supervises, a wide range of machinery and equipment. These may be conveniently be divided into six groups.

(i) Items associated with propulsion such as:
 – couplings,
 – gearing,
 – thrust block,
 – shafting,
 – bearings and sterntube,
 – propeller.

(ii) Items associated with auxiliary energy such as:
 – main and emergency generators,
 – steam generating plant,
 – heat exchangers.

(iii) Other major items of machinery such as:
 – pumps,
 – air compressors,
 – oily water separators,
 – incinerators,
 – water purification plant.

(iv) Piping systems in the engine room consisting of the following systems:
 – fresh water,
 – sea water,

- fuel oil,
- lubricating oil,
- compressed air,
- steam,
- condensate and boiler feed,
- bilge and ballast,
- general service,
- fire fighting,
- together with the associated valves, fittings, lagging, etc.

(v) Engine room structure and fittings:
- uptakes, silencers and funnel,
- downtakes, where fitted,
- floorplates, gratings, ladders,
- ventilation fans and trunking,
- workshop and storeroom fittings,
- engine room fire extinguishing systems and equipment,
- engine room painting and insulation,
- lifting gear,
- spare gear and tools.

(vi) Engine room controls:
- instrumentation and alarms.

It is suggested that a standard grouping of this sort should be used for all detailed machinery weight and centre of gravity estimates and it may be remarked that a very similar grouping appears in Chapter 17 as specification headings, whilst a simplified version is suggested in Chapter 18 as a basis for cost estimation.

A naval architect cannot be expected to know about all these items in any detail, but he should aim to know enough to be able to discuss any problems that may arise in relation to their arrangement and operation in a ship.

9.6.2 Items specified by the naval architect

In addition to the above list there are a number of items which are generally the responsibility of the naval architect to specify and arrange, but which are very often fitted in the engine room. These consist of items such as:
- cargo and stores refrigeration machinery,
- air conditioning refrigeration machinery,
- sewage plant,
- fin stabilisers,
- stern thrusters.

A good estimate of the space requirements for such items should be passed to the marine engineer at an early stage in the design and it is obviously very important that weight and cost estimates by both naval architect and marine engineer are prepared to a standard demarcation.

A wise naval architect will ensure that his marine engineering colleagues are closely involved in the specification and arrangement of any major items of machinery even if these are to be fitted well away from the engine room. Experienced engineering advice on such items as bow or stern thrusters can be invaluable.

9.7 OTHER PROPULSION DEVICES

The high fuel prices of the eighties stimulated studies into the use of so-called "free" energy sources and this chapter seems the most appropriate place in this book to discuss these.

9.7.1 Wind power

Proposals for the use of wind power have varied from fairly conventional designs derived from the sailing ships of former days, with either fore and aft schooner or square rigs such as Dyna ship, solid aerofoils such as those designed by Walker Wingsail, revivals of the Fletner rotor ship, the use of wind turbines, or even kites.

Some proposals have been based on the use of wind power as the main propulsive force when there is a satisfactory wind blowing; in other proposals the wind is used to assist a conventional main engine, either increasing the ship's speed or reducing the power delivered by the engine and therefore improving the fuel consumption.

One of the most pleasing uses of wind power has been on a number of small cruise ships, where as well as saving fuel costs the fact that the passengers get the exhilaration of being under sail is a major sales point to those who cruise on these ships. That these vessels can also have an attractive appearance must be a further plus point for a cruise liner.

Some guidelines for successful wind power designs are:
(i)   It must be possible to build the wind powered ship to a size not too different from that of the conventional vessel with which it must compete. If it cannot, it will lose out on the economics of size.
(ii) The design must be such that cargo handling is as easy as it is on a competing ship. For general cargo it has therefore either to be suitable for containers, which seems likely to be very difficult, or have a competitive cargo handling system, which may be equally difficult to achieve. For bulk

solids, it must have large hatches which will not be easy to marry in with a sail rig. On the other hand, the carriage of bulk liquids does not present the same difficulty and a small tanker with auxiliary sail has now been operating successfully for a number of years.

(iii) The crew required must not be significantly more than that needed on a conventional ship. To achieve this a high degree of automation of rig handling must be provided. Speedy reefing (or its equivalent) must be possible for safety when strong winds are encountered.

(iv) The automation must not have such a large auxiliary power demand nor incur such high costs as to negate the savings made by the use of "free" energy.

An interesting problem associated with the "wind assist" mode is the need for the ship's propeller to be able to adapt to negative slip, with a controllable pitch propeller providing the answer.

Wind power does not seem likely to return as the major energy source on ships. Its use as an auxiliary may be another matter and a number of recent installations appear to be providing satisfactory economies in trades in which the wind spectra is suitable.

9.7.2 Wave power

Wind is not the only free energy source available to ships and a system invented by a Norwegian engineer utilises wave power. This device is said to be limited to ships of up to 50 m in length and it must be admitted that its extension to larger ships would seem unlikely. The device consists of a moveable foil placed horizontally on an axis beneath the vessel and arranged to have an angle of attack which results in its vertical movement as the ship moves in the waves providing a horizontal force imparting forward movement.

9.8 FUEL ECONOMY

Much of this chapter has been concerned with fuel economy and it seems right to conclude it with the summary of the various contributors to fuel economy given in Table 9.2.

Whilst it is not possible to add together all the fuel savings attainable by the various methods mentioned in this and previous chapters, there is no doubt that large aggregate savings can be made.

Some savings reduce the possibilities of further savings; for example, it has already been noted that the improved efficiency of diesel engines leaves less waste heat to be reclaimed. On the other hand, there is the odd case where one saving may

Table 9.2

Contributors to fuel economy

The transportation need, ship size, speed, routing	Shipowner and transportation economist
Design optimisation, lines and propeller	Naval architect
Shell finish and minimising fouling	Shipbuilder and paint manufacturer
Fuel efficient engine, use of waste heat	Marine engineer
Reduction in hotel load and other electrical demands	Naval architect

Table 9.3

Savings in fuel costs per cargo tonne-mile 1975–1995

	Change	Reduction in fuel costs (%)
1.	Speed reduction of 15% to a more economical speed (e.g. 12 knots in lieu of 15 knots or 23 knots in lieu of 27 knots)	25
2.	Increase in deadweight by use of a fuller C_b together with weight saving due to improved structural design	2
3.	Improved lines and proportions of ship main dimensions reducing resistance	8
4.	Improved shell finish reducing frictional resistance and fouling	6
5.	Slower revving propeller increasing propulsive efficiency	12
6.	Asymmetric lines, Grimm wheel or reaction fairings reducing or recovering rotational energy	10
7.	Improved specific consumption of main engine (170 g/kWh v. 210 g/kWh)	19
8.	Shaft-driven alternator reducing cost of fuel used for electricity generation	2
9.	Reductions in electric load due to insulation etc.	1
10.	Better utilisation of waste heat	2
	Cumulative saving	56

help to increase another, as happens when a reduction in EHP makes it possible to lower the propeller revs and thereby gain in propulsive efficiency.

The extent to which the economy of a typical modern ship built in the last five years has improved compared with a similar ship built about twenty years ago is shown in Table 9.3. On this basis, the fuel consumption of a good modern ship per tonne-mile would be about 44% of that of a ship built twenty years ago. The cumulative saving is, of course, obtained by multiplying the percentage savings and not by addition.

This table is open to criticism and the author would not attempt to defend it in detail, but believes it gives a feel for the savings that have been made. A naval architect should always be on the look-out for new ideas that will provide improvements but should assess these critically before accepting the claims made!

Chapter 10

Structural Design

10.1 FACTORS INFLUENCING STRUCTURAL DESIGN

Structural design decisions have as their primary objective the evolution of a structure that will withstand all the forces acting on it. The most important of these forces are the bending moments and shear forces which stem from the waves which the ship encounters and the loading applied by the cargo carried. As the structure must continue to meet these forces throughout the ship's life, the scantlings must include allowances for the corrosion and wear which can be expected.

Theoretical approaches to strength calculations are described in textbooks on naval architecture, whilst up-to-date practical methods which should be used are given in the rules of the classification societies and there seems no point in repeating them here.

Regrettably, classification rules are so complicated nowadays that they provide little guidance to designers towards the best structural configuration, however excellent they may be for checking a design once this has been completed.

Fortunately, computer programs such as LRPASS take much of the work out of determining the scantlings for a new design, but the optimising tricks that a good designer used to learn in the course of his work no longer come so easily.

Classification societies verify the scantlings required by a proposed structural design, but unless the design has serious shortcomings they do not usually suggest changes in the main features of the structural design, although such changes might improve the reliability of the structure, reduce the steelweight and/or improve the ease of construction.

It is on such features of structural design that this chapter's attention is focused.

10.1.2 Redundancy

Whilst a designer aiming for economy will usually try to minimise structural redundancy, recent bulk carrier casualties suggest that a measure of redundancy is desirable so that the loss, or a severe reduction in the strength, of some structural members can be absorbed without catastrophic failure. Particular attention should be paid to this in the design of parts of the ship which seem likely to be susceptible to corrosion and/or fatigue.

In some parts of the structure the design should be governed by local strength and/ or vibration considerations; in other areas it may be important to limit deflection.

In a 1992 I.E.S.I.S. paper entitled "Safety of bulk carriers" J.M. Ferguson of *Lloyds Register* gave a useful reminder of the many factors which may influence the safety of a ship and this is reproduced as Fig. 10.1. The paper lists the main types of defects and their locations as:

(i)   cracking at hatch corners;
(ii)   plate panel buckling of cross deck strips and stiffening structure;
(iii)   cracking of hatch coamings;
(iv)   cracking at the intersection of the inner bottom plating and the hopper plating;
(v)   grab and bulldozer damage to the side frames lower brackets;
(vi)   grab damage to the inner bottom plating, hopper and lower stool plating;
(vii)   cracking at main frame bracket toes;
(viii) both generalised and local corrosion of main frames and brackets;
(ix)   cracking at fore and aft extremities of topside tank structures;
(x)   corrosion within topside tanks.
(xii)   general corrosion of transverse bulkheads.

Although this list refers directly to bulk carriers the importance of good detail design and good operational practice is equally applicable to all ships, and is worth emphasising at the start of this chapter on structure.

10.1.3 The variety of structural calculations

Although longitudinal strength is the most important strength consideration in almost all ships with both the vertical bending moment and the vertical shear forces requiring investigation, a number of other strength considerations must be considered. Prominent amongst these are transverse, torsional and horizontal bending strength, with torsional strength requiring particular attention on "open" ships with large hatches arranged close together.

In later more detailed scantling calculations, watertight bulkheads must be designed to meet the hydrostatic loads that they will receive if one of the adjoining compartments becomes open to the sea, whilst bulkheads of large tanks which may

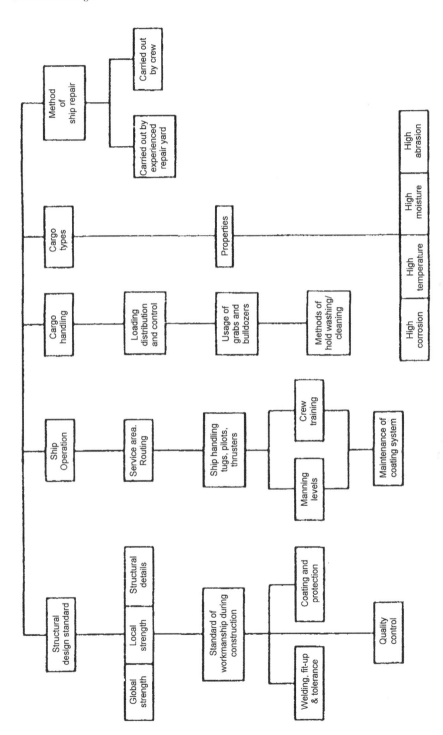

Fig. 10.1. Factors which influence the safety of a ship.

be partly filled with liquids must be designed to meet the sloshing loads caused by the ship's motions in a seaway.

10.2 LONGITUDINAL STRENGTH

Although as already stated, most Classification Society rules tend to be complex, some remain quite simple and it is possible to give one brief extract from rules now agreed by all the Classification Societies as I.A.C.S. requirement S11 which can be used in advance of detailed longitudinal strength calculations to give a first approximation to the required midship section modulus.

The minimum midship section modulus Z about the transverse neutral axis at the deck or at the keel is not to be less than:

$$Z = f_1 \cdot k_L \cdot C_1 \cdot L^2(C_b + 0.7) \times 10^{-6} \text{ m}^3$$

The wave bending moment is to be taken as:

$$M_w = f_1 \cdot f_2 \cdot M_{wo}$$

In these formulae

f_1 = Ship service factor = 1 for unrestricted service
  0.8 for short voyages, and
  0.5 for sheltered water

$f_2 = -1.1$ for the sagging moment

$\quad = \dfrac{1.9 C_b}{(C_b + 0.7)}$ for the hogging moment

k_L = steel strength factor
  = 1 for mild steel
  = 0.78 for 315 N/mm^2 higher tensile steel, and
  = 0.68 for 315 N/mm^2

C_1 varies with ship length, but typically for the main range of ship sizes (90–300 m) is:

$$C_1 = 10.75 - \frac{(300 - L)}{100^{1.5}}$$

C_2 is a factor depending on the position along the ship's length = 1.0 for amidships

$$M_{wo} = 0.1 \, C_1 \cdot C_2 \cdot L^2 \cdot B \cdot (C_b + 0.7) \text{ kN m}$$

Taking the sagging moment case, the allowable stress built into these formulae is:

$$\sigma = \frac{M_w}{Z} = \frac{0.1(-1.1) \cdot C_1 \cdot C_2 \cdot L^2 \, B(C_b + 0.7)}{f_1 \cdot k_L \cdot C_1 \cdot L^2 \, B(C_b + 0.7) \times 10^{-6}} \; \text{kN/m}^2$$

which reduces to:

$$\sigma = \frac{-0.11 \cdot C_2}{f_1 \cdot k_L \times 10^{-6}} \; \text{kN/m}^2$$

For unrestricted service, the amidships stress and mild steel structure, all the above constants are unity and the formula corresponds to a wave bending moment stress $\sigma_w = 110 \, \text{N/m}^2$

It should be noted that this stress is based on the sagging wave bending moment with the corresponding calculation for the hogging moment being a little more complicated.

The permissible combined stress for still water plus wave bending moments is given by: $\sigma = 175/k_L \, \text{N/mm}^2$.

Built into a modulus derived using the wave bending moment only therefore is an allowance for the still water bending moment being

$$\frac{175 - 100}{100} \text{ or 59\% of the wave bending moment.}$$

If the actual still water bending moment is in excess of these figures, the modulus must be adjusted accordingly.

10.2.2 Structural decisions based on longitudinal strength

One of the first decisions which must be taken in structural design is whether to use longitudinal or transverse framing.

For large ships (over about 200 m) longitudinal framing will generally be a classification requirement, but even if this is not the case its use will usually be desirable on economic grounds because it results in a lighter steelweight.

For small ships (under about 65 m) longitudinal strength is of secondary importance and longitudinal framing brings no advantage in steelweight, whilst the greater complexity of this system of construction increases fabrication costs.

For medium-sized ships — between these limits — the choice lies with the designer, who can decide whether it is more advantageous to minimise steel material weight or steel work man-hours.

It is worth noting that it need not be a straight choice between longitudinal and transverse framing however as a combination of these methods can have advantages in some ship types/sizes. Such a combination will generally use longitudinal framing for the bottom framing and for the strength deck, i.e., for the two flanges of the hull girder, whilst transverse framing is used for the ship side (the girder web) and also for the supporting structure of any decks near the neutral axis.

10.3 SPECIAL STRENGTH CONSIDERATIONS FOR PARTICULAR SHIP TYPES

Apart from the structural considerations already mentioned which affect all ships to a greater or lesser extent, there are some special considerations which are applicable to particular types, such as fast cargo ships, passenger ships and warships.

10.3.1 Fast cargo ships

The fine lines of a fast cargo ship tend to result in there being a lack of material in the upper deck in way of the forward hatch and particularly at the forward corners of this. There have been a number of incidents in which the structure in this vicinity has been damaged with in some cases the whole bow of the ship being lost.

As a consequence, Classification Societies now have special strength requirements for the strength of this area, depending on the ship's speed and the shape of the cross section — but wise design will try to avoid the problem.

An over-heavy flare forward should be avoided as this may result in bow flare slamming, which has led to considerable damage on some ships.

10.3.2 Large passenger ships

Whereas on most ship types there is a clearly defined deck which forms the upper flange of the hull girder with the superstructures above this level being relatively short and therefore not contributing to the overall longitudinal strength and consequently fairly lightly constructed, passenger ships tend to have a mass of superstructure decks which in most of today's designs extend for almost the complete length of the ship.

In passenger liners built before World War II, attempts were made to relieve the superstructures of stress by fitting these with expansion joints. In a number of liners built after World War II, aluminium superstructures were used. Whilst the main reason for the use of this material was its light weight and the greater extent of superstructures which could therefore be built within the limit imposed on the VCG by stability considerations, it was hoped that the fact that aluminium has a much lower Young's modulus than steel would enable the junction of the two materials to provide much the same effect as the expansion joints of earlier ships.

In both cases, however, cracking seems to have been a frequent problem indicating that the superstructures were taking stresses for which they had not been designed.

Expansion joints, whilst relieving the superstructure of stress, caused stress concentrations at their lower ends and this often led to fatigue cracking in this area.

The use of aluminium was discontinued some years ago mainly due to a recognition of the increased danger of structural collapse in the event of fire, which is a consequence of the material's low melting point, but partly also due to a realisation of its poor fatigue properties in a marine environment.

In modern cruise liners, designers in search of economic efficiency usually want to fit extensive superstructures — both in length, which often extends from a short distance abaft the bow to very near the stern, and in the number of tiers fitted. On a number of these ships the superstructure is stepped in from the ship's side at the upper deck to accommodate lifeboats at this level, designers finding this advantageous for a number of reasons:
 – to improve the launching of the lifeboats as a contribution to passenger safety;
 – to reduce top weight enabling more accommodation to be fitted;
 – to improve the amenity of the top decks for passengers.

In conjunction with these design decisions, designers then wanted to get the best possible contribution from the superstructure to the longitudinal strength of the ship and fortunately found a new design tool to hand in finite element calculations.

Finite element calculations can be used to solve the complex problems posed by openings in the deckhouse sides, the stiffness of the deck on which the house sides are supported and the three dimensional effects interrelating these.

Three-dimensional F.E.M's cannot be used at the all-important initial design stage, so naval architects involved in this type of work owe a considerable debt to Professor Caldwell for his 1957 R.I.N.A. paper on the subject and to J.W. Fransman for his 1988 R.I.N.A. paper "The influence of passenger ship superstructures on the response of the hull girder", in which analytical methods of calculation are developed. Some appreciation of the approach adopted in these papers may be given by Fig. 10.2 abstracted from the latter paper.

10.3.3 Warships, and more especially, frigates and corvettes

Although individual navies have their codes for structural design, none of these rules for warships are as detailed or as freely available as merchant ship Classification Society rules. There are, however, a number of very good papers on the subject and most of the following is abstracted from one or other of the papers mentioned in the bibliography.

For an introduction it is hard to better the statement which J.D. Clarke of A.R.E. (the British Admiralty Research Establishment) made in his 1986 paper "Wave loading in warships" to the effect that "the most important loading exerted by the sea on the usually slender hull of a warship (this mainly refers to frigates and

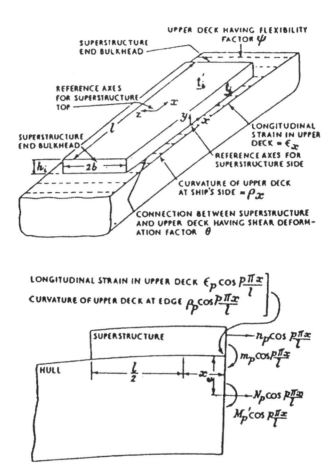

Fig. 10.2. An analytical method of calculating the forces and moments in hull and superstructure.

corvette types) is vertical bending of the hull girder. This results in alternating compressive and tensile stresses in the deck and bottom structure which must be limited to avoid buckling, fracture or fatigue failure."

The paper goes on to suggest that longitudinal strength calculations should be based on bending moments and shear forces which have a 1% probability of exceedance in the estimated life of the ship taking into account the sea areas in which the ship is expected to operate.

In warships of frigate or corvette size the strength of the upper deck in compression is usually the critical factor. The fact that it is compression in the deck rather than in the bottom that is critical arises from the fact that the bottom structure has to be of heavier construction than the deck to meet hydrostatic loads.

The bottom can therefore resist the compressive loads from a hogging bending moment better than the deck can resist the compressive loads from a sagging moment. In addition, the hull form generally causes the sagging wave bending moment to be greater than the hogging moment, whilst slamming can add another component to the sagging bending moment.

The still water bending moment of a warship is usually relatively small when compared with the wave bending moment, but if there is any choice in the matter it is marginally advantageous to design so that there is a hogging still water bending moment.

The compressive strength of the upper deck must therefore be evaluated as accurately as possible if minimum structural weight is to be achieved. The most probable form of failure of the deck is a column-like collapse of the longitudinals and the thin plating to which it is attached between transverses. The well known "Euler strut" equation is used for calculations of this sort in the elastic regime, but in practice the material reaches yield point and there is an interaction between buckling and yielding which depends on the standard of fabrication of the structure with imperfections and "as built" stresses having a significant effect. The behaviour is usually quantified in terms of a "column curve" relating failure load and stiffener size, several versions of which are given in specialist books on the subject.

Superstructures can play an important part in the strength of warships if required to do so although the possibility of severe damage to the superstructure in action together with the possibility of it having to be modified during service life causes some prudent designers to ignore its contribution. Superstructures intended to contribute to longitudinal strength should be made as long as possible, but there is much to be said for the alternative philosophy of making the hull as large as possible and reducing the superstructure to a minimum and not asking it to contribute to longitudinal strength at all.

Superstructure contributions to longitudinal strength should be calculated on the lines discussed for passenger liners.

Because of the light scantlings of warships, particular attention should be paid to the suggestions made later on ways to minimise vibration and stress concentrations. Great attention should also be paid to the design of special strengthening at the end of superstructures to marry these into the hull strength.

One difference between warship and merchant ship structural design is the need to investigate and detail several structural sections in the former as opposed to the "midship section" which has generally sufficed for the latter. This need is partly due to the variation in the sections caused by the hull form of these fine lined ships and partly to ensure that the strength is being maintained in way of large openings and/or the ends of superstructures.

The relative absence of large deck openings on warships as compared with merchant ships means that torsion rarely presents any difficulties. The number of decks and bulkheads prevents lateral strength being a problem.

The high speed of these ships and in particular the frequent need to maintain this in adverse weather, can result in severe slamming in spite of measures which are usually taken in the design of the lines to minimise this. The structure in almost the whole fore body and in the stern must be designed with this in mind.

Underwater explosions can result in intense shock effects, which may induce whipping of the hull structure. The damage done by shock to machinery and equipment is more important than that done to the structure of a well designed ship, but measures are nevertheless taken to improve the structural resistance to shock. These measures include the use of symmetrical rather than asymmetrical sections for stiffeners, the design of connections in a manner that helps to ensure continuity of strength and stiffness and sometimes involve the fitting of additional tripping brackets. There is usually a requirement for resistance to air blast due to nuclear or other explosions. This affects the above water structure including the super-structure. Specialist publications should be consulted for details of the methods used which make use of the ultimate strength of the structural material allowing permanent deformation but avoiding complete collapse.

10.4 OTHER STRENGTH CALCULATIONS

10.4.1 Torsional strength

Although torsion is not usually an important factor in ship design for most ships, it does result in significant additional stresses on ships, such as container ships, which have large hatch openings. These warping stresses can be calculated by a beam analysis which takes into account the twisting and warping deflections. There can also be an interaction between horizontal bending and torsion of the hull girder. Wave actions tending to bend the hull in a horizontal plane also induce torsion because of the "open" cross section of the hull which results in the shear centre being below the bottom of the hull. Combined stresses due to vertical bending, horizontal bending and torsion must be calculated and must meet classification society rules. A "closed" structure such as side tanks can add torsional stiffness and should generally be incorporated in this type of ship.

10.4.2 Fatigue

Fatigue can result in the growth of cracks under the cyclic loads to which a ship is subjected by the bending moments imposed by waves and vibration. The cracks may not be dangerous in themselves but can lead to brittle fracture if let go too far.

They also have a nuisance value due to leaks that they permit and can necessitate expensive repairs. Cracks almost invariably start from welded joints and the avoidance of cracks demands that these are given a high standard of detailed design.

Fatigue endurance appears to be independent of the type of steel employed, so the higher stress levels used with higher tensile steels mean that there is a lower fatigue life with these steels. As the fatigue endurance/stress relationship is governed by a cubic law, a small increase in stress can lead to a major reduction in fatigue life — a fact demonstrated in recent years by the problems found on large tankers built using a major proportion of higher tensile steel.

The poor fatigue properties of aluminium in a marine environment have already been mentioned. If the stresses in an aluminium deckhouse are to be so reduced that it will have an equivalent fatigue life to that of a steel deckhouse, the scantlings have to be so increased that there is little weight saving.

10.4.3 Brittle fracture

Brittle fracture causes the sudden propagation of a crack which can extend for a considerable distance and has led to the loss of a number of ships. It is triggered by the rate of application of stress and is greatly influenced by ambient temperature and steel thickness.

Ordinary mild steel becomes prone to brittle fracture at temperatures approaching 0°C, and Classification Societies require the use of notch tough steels in areas of ships which are subject to high stress levels, particularly when these require thick plating, and a more general use in ships, such as ice-breakers and refrigerated cargo ships, in which the steel will be subject to sub-zero temperatures.

The quality of steel is usually indicated by its "Charpy" value which is a measure of the energy required to propagate a crack. Charpy "J" values together with the temperatures at which these tests are made for different grades of steel are given in Table 10.1.

In specialist ships such as warships a more general use of notch tough steel may be wise.

10.5 MINIMISING STEELWEIGHT AND/OR STEELWORK COST

Minimising steelweight is of particular importance in deadweight carriers, in ships required to have a limited draft, and in fast fine lined ships.

On other ship types it is still desirable to minimise steelweight to reduce material cost but only when this can be done without increasing labour costs to an extent that exceeds the saving in material costs. On the other hand, a reduction in structural labour cost achieved by simplifying construction methods may still be worthwhile even if this is obtained at the expense of increasing the steelweight.

Table 10.1

Strength of various higher tensile steels

Lloyds steels	Yield stress		Tensile stress		Charpy J
	N/mm^2	tons/in^2	n/mm^2	tons/in^2	
Normal A,B,D,E	235	15.23	400–490	26–32	27 (ex A)
Higher tensile					
DH, EH 32	315	20.43	440–590	28–38	31
DH, EH 34S	340	22.04	450–610	29–39	34
DH, EH 36	355	23.00	490–620	32–40	34

Charpy temperatures: A –; B, AH 0°C; D, DH –20°C, E,EH –40°C.

10.5.1 Spacing of stiffeners

The spacing of frames and longitudinals can have a significant effect on steel weight. Relatively close spacing of stiffeners — longitudinals, frames, beams, web frames, bulkhead and casing stiffeners — reduces the overall weight but as doing so will add more stiffeners with more cutting and welding the labour cost inevitably rises.

10.5.2 Choice of type of sections

The use of rolled sections, rather than the construction of fabricated ones, can be a significant labour cost saver, and if a somewhat closer spacing of web frames, longitudinals or frames enables the required modulus to be obtained in this way this is usually worth doing.

Some special sections such as Admiralty long stock T-bars cost significantly more per tonne than more standard sections and should only be used where their use is sufficiently advantageous in other ways to justify the extra cost. As well as the extra material cost, the shape of these sections makes them difficult to join together whereas asymmetrical sections with a flat face can be connected with lap welded beam knees or brackets and are therefore generally to be preferred from a construction cost point of view. On the other hand, asymmetrical bulb plates and angles can be subject to premature tripping under compressive loads and the use of symmetrical sections is therefore often preferred by warship designers.

Cutting two T-bars out of an I-beam using a castellated cut that increases the depth of the resulting sections whilst providing scallops can be an economical production process.

10.5.3 Spacing of bulkheads

In bulk carriers and tankers the spacing of main watertight bulkheads as well as meeting any subdivision requirement should, if possible, be so arranged that the lengths of holds or tanks are such that they are multiples of the web frame spacing, thereby minimising the number of web frames..

10.5.4 Hatch arrangement

On ships with large cargo hatches, only the deck plating and associated longitudinals outside the hatch coamings contribute to the upper flange of the ship girder. Purely from the point of view of efficient structure and economy in steelweight, hatches should not therefore be any wider than is necessary for efficient cargo handling — although whether a cargo handling enthusiast would be willing to settle for anything less than a 100% spotting ability is open to question.

Recent bulk carrier casualties suggest that more importance should be attached to the structural strength of hatch covers and their ability to maintain watertight integrity in extreme conditions, than has been the case in the past. This applies particularly to the foremost hatches.

10.5.5 Alignment of structure

Other deck openings such as stairwells and access hatches should, if possible, be kept inside the line of the main hatch or engine casing openings and their longest dimension should, if possible, be in the fore and aft direction. If there are a series of such openings as on a passenger ship or a warship, these should be lined up so that a minimum number of longitudinals have to be cut.

10.5.6 Use of higher tensile steel

The use of higher tensile steel is — certainly on larger ships — one of the best ways of reducing weight and although the material cost per tonne is higher, the reduced tonnage usually means the total material cost is reduced and in some cases there can also be a reduction in labour cost. Table 10.1 shows the respective strengths of mild steel and the various higher tensile steels.

When higher tensile steel is used, Lloyds hull girder section modulus can be multiplied by a factor k_L, where

$k_L = 245/\sigma$ or 0.72, whichever is greater.

For local scantling requirements of plating, stiffeners etc the corresponding multiplier is:

$k_L = 235/\sigma$

σ in both cases being the minimum specified yield stress in N/mm².

Whether a change to higher tensile steel brings a reduction in labour cost depends on how much of the steel used is of a grade and thickness which requires special heat treatment.

The thinner scantlings used with higher tensile steel means that a structure constructed in this material inevitably carries a higher fatigue and corrosion risk.

10.5.7 Corrugated construction and swedged stiffeners

The use of corrugated construction for transverse bulkheads provides a good way of reducing both steel weight and labour costs but this is achieved at the expense of the cargo capacity unless the cargo is a liquid or a fairly finely divided solid, which will stow in the corrugations.

The use of swedging to provide the stiffening of steel casings saves both weight and labour cost but may introduce arrangement complications, whilst fatigue at the end connections is possible.

10.5.8 Other ways of minimising steel work labour cost

(i)   Reducing the number of parts and the complexity of fit-up by paying great attention to the detail of the design.

(ii)   Coordinating the design with the intended production methods to maximise downhand welding and work done in the assembly shop and minimise overhead welding and that done on the ship.

(iii)   Rationalising plate thicknesses and stiffener sizes used can help to reduce the steel stock maintained in a shipyard, and possibly reduce buying costs by increasing bulk buying. It may also simplify construction, but the advantages are regrettably at the expense of a somewhat higher steel-weight.

(iv)   Taking account of production considerations in the structural design. For example arranging stringers and girders in a tanker in positions where they can to act as platforms to aid the fabrication process, rather than positioning them entirely to minimise steel weight.

10.6 OTHER FACTORS WHICH SHOULD INFLUENCE STRUCTURAL DESIGN

Apart from structural strength, weight and labour cost, there are a number of other factors which ought to be considered in well thought out structural design, but which do not always get as much attention as they deserve:

(i)   avoiding or minimising structural discontinuities,

(ii)   avoiding or minimising vibration,

(iii) reducing corrosion and facilitating maintenance and repairs.

10.6.1 Avoiding, minimising or compensating for structural discontinuities

Some of the best possibilities of minimising structural discontinuities occur when the outline general arrangement is being drawn as decisions taken at this time determine the type, extent and positioning of erections, the ends of which will form some of the most severe discontinuities.

Whilst a forecastle is likely to be necessary from a seakeeping point of view on all except a very large ship, a poop is only likely to be required on a fairly small vessel. However, the ends of both forecastles and poops are well away from amidships and do not generally present any significant structural problem.

A side-to-side erection near midships such as a bridge should be avoided if possible, but if it is a necessary arrangement feature then it should be made as long as possible for two reasons:

(i)   so that its ends are as far away from amidships as practicable thus minimising the stress at the discontinuity, and

(ii)   the maximum value is obtained from its contribution to longitudinal strength.

The junction of the strength deck and the ship side plating occurs at the point of maximum stress due to longitudinal bending. On large ships local stresses at this connection are now greatly reduced by the practice of fitting a radiused gunwale plate. The danger of cracking at this point is recognised and reduced by the requirement of Classification Societies for construction in this region to be of notch tough steel.

The use of notch tough steel of course goes well beyond this immediate area with Classification Societies requiring the use of the different grades "D" and "E" in various stressed positions depending partly on ship size and partly on plating thickness.

It is also most important that welded attachments to stressed plating of this sort of such items as stanchions and fairleads are kept to a minimum (preferably none).

A similar situation arises at the bilge which is again a highly stressed area, and great care should be taken with the detail design of the attachment of the bilge keel to the hull plating.

Local discontinuities come from many other features; sometimes these discontinuities must be accepted and compensated for, in other cases they can be eliminated by wise design.

10.6.2 Avoiding or minimising vibration

One of the best ways of avoiding or minimising vibration is to eliminate cantilever construction if this can reasonably be done.

If for any reason cantilever structure is desirable it should be stiffened to take its natural frequency well clear of that of any possible exciting force. Structures most likely to need consideration in this way include bridge wings, masts and derrick posts.

Another most desirable way of minimising vibration and indeed of avoiding stress concentrations is by ensuring that the ends of superstructure deckhouses land on steel casings or bulkheads.

It is also desirable to have a number of transverse casings in the superstructure lining up with bulkheads in the main hull, whilst deckhouse sides should if possible line up with longitudinal bulkheads, girders or longitudinals supporting the upper deck.

Openings should be avoided near the ends of deckhouse sides.

Modern "tower" like superstructures tend to have low natural frequencies and these should be evaluated including their interaction with the main hull girder, whilst the recommendations for the provision of good support to superstructures are particularly important.

10.6.3 Reducing corrosion and facilitating maintenance and repairs

Reducing corrosion is best achieved by eliminating any confined pockets within which water can lie.

The measures which should be taken to facilitate maintenance and repairs are in general also measures which will ease construction, although the ability to turn units upside down during construction may mean that something which is quite easy to build is by no means easy to repair. The sheer size of large tankers and bulk carriers makes inspection by surveyors a matter of great difficulty, and it looks as though provisions to ease this problem are going to be essential in the new generation of double hull tankers.

10.6.4 Detailed structural design

The importance of good detailed design throughout but particularly in the primary structure can hardly be over-emphasised. In terms of potential ship loss, the cost of repairs and incidence of fatigue cracks detail design has been shown by accident statistics to be more important than hull girder strength *per se*.

Although cost considerations have in the past made shipbuilders strive to eliminate redundancy, recent casualties suggest that a measure of redundancy should be introduced in any area where the loss of a member due to corrosion or fatigue might be calamitous.

Classification Societies have in recent years issued some very useful guidance on good and bad details and these should be carefully studied by all designers.

10.7 STRUCTURAL STRENGTH UNITS

Because the author learned his theory of structures many years ago, he found he needed an *aide memoire* on modern units and their equivalents whilst writing and this is included as Table 10.2.

Table 10.2

Structural strength units

Unit	S.I. and metric equivalents		Old British
Mass	1 tonne = 1000 kilograms $= 10^6$ grams	1 tonne = 0.9842 tons 1 ton = 1.016 tonnes	1 ton = 2240 lbs
Acceleration due to gravity	9.81 metres per sec^2		32.2 feet per sec^2
Force	1 Newton = force to accelerate a mass of 1 kg at 1 metre per sec^2	1 Newton = 7.233 Poundals	1 Poundal = Force to accelerate a mass of 1 lb at 1 ft/sec^2
	1 tonne force = force due to gravity on 1 tonne mass		1 ton force = force due to gravity on 1 ton mass
	1 tonne (F) = 9.81 kN		1 ton (F) = 7.2128×10^4 poundals
Stress and pressure	1 N/mm^2 = 1 MN/m^2	1 tonne/mm^2 = 635 tons/in^2	tons/in^2
	1 tonne(F)/mm^2 = 9.81×10^3 N	1 ton/in^2 = 0.001575 tonnes/mm^2	tons/ft^2
		1 ton/in^2 = 15.42 N/mm^2	lbs/in^2

Metric multiples:

10^9 = giga 10^{-3} = milli
10^6 = mega 10^{-6} = micro
10^3 = kilo 10^{-9} = nano

Chapter 11

Freeboard and Subdivision

11.1 INTRODUCTION

This chapter and the two following try to present as compact a summary as possible of the major statutory rules which govern aspects of merchant ship design. These rules contain many detailed provisions and condensing these has required both a simplification of the treatment of some items and the omission of some less important items.

For 100% accuracy the originals must be used, but most important matters are covered and calculations based on the abbreviated data given in these chapters should be accurate enough for most design purposes; it is hoped that the summaries make easier reading than the detailed statutory rules. In some places guidance towards desirable design options has been added.

11.2 FREEBOARD

Rules requiring a statutory freeboard apply to all merchant ships of 24 metres length or more[1], although in practice the freeboard of most passenger ships is set by the more severe requirements of the subdivision rules, and the same may now apply to cargo ships following the introduction of subdivision requirements for these in part B1 amendment to SOLAS 1974, which is dealt with in Section 3 of this chapter.

There are no equivalent rules for warships although the standards of seakeeping and damage survival demanded for these vessels ensure that they have adequate freeboard.

---

1   Some countries, including the U.K., extend the requirement for a statutory freeboard to ships less than 24 m in length

The freeboard rules currently in force are embodied internationally in the provisions of the International Convention on Load lines 1966 and for British ships in The Merchant Shipping (Load Line) Rules 1968. The freeboard rules are divided into two sections:

(i)   a section which lays down ship construction requirements which deal with structural strength, stability, the watertight integrity of the ship, the safety of the crew, and in some cases the ability of the ship to withstand flooding of specified compartments; and

(ii)   a section dealing with the calculation of freeboard, based on the geometry of the ship.

It is not proposed to elaborate on the first section here, although the stability requirements and the ability to withstand flooding are dealt with in a later section of this chapter.

11.2.1 Different types of freeboard

There are a number of different freeboards which apply to different types of ship and each of these has some special construction requirements.

Each ship also has a number of different freeboards which apply under different circumstances of sea area, time of year and water density. The basic freeboard for any ship is its "summer" freeboard. "Summer" here is a technical term defined in the rules as covering particular periods of the year in particular sea areas.

Reverting to the different types of freeboard applicable to different ship types, these are as follows:

(1) Type A

This is a reduced freeboard permitted for tankers designed to carry liquids in bulk and which have certain features. Since the rules were written new requirements introduced by IMO for segregated ballast tanks have so changed the design of oil tankers that their design is no longer weight based and currently this type of freeboard is rarely used.

(2) Type B

This is the standard freeboard which applies to the majority of ships.

(3) Type B-60 and B-100

These are reduced freeboards which can be given to ships, which can be shown to have an ability to withstand damage of an extent required by the rules. This type of freeboard is particularly advantageous for large bulk carriers and especially those intended for the iron ore trade.

The "100" in B-100 refers to the difference between a type A and a type B freeboard. The "60" refers to 60% of this difference.

The damage which these ships must be able to withstand is dealt with in §12.4.

(4) Timber freeboards

These freeboards which are also less than type B, are assigned to ships designed to carry timber on deck, in recognition of the fact that this type of buoyant cargo provides a major contribution to the ship's survivability. There are specific requirements for the stowage of the deck cargo.

(5) Dredger freeboards

Dredgers which have open hoppers without hatch covers and also have bottom dump valves, have in the first of these a facility that results in cargo spillage if the ship heels beyond a certain angle and in the second an ability to dump cargo rapidly in the event of an emergency. They are also usually employed in waters close to land.

The increased survivability which these features give is recognised by giving these ships the possibility of a freeboard less than the statutory minimum and relief from the statutory bow height.

Under different conditions the freeboard may be one of the following:

5/8 B; or 1/2 (B-60); or 1/2 (B-100)

subject in all cases to a minimum of 150 mm. The conditions attached to this dispensation are given in the Department of Transport Instructions to Surveyors. They include requirements:
  - that the longitudinal strength is adequate for the corresponding draft,
  - that the ship has operational limits which normally do not exceed 15 miles from land,
  - that the ship complies with special stability and flooding requirements, which are discussed in §12.3,
  - that draft indicators are fitted.

The last requirement may not appear to be of quite the same importance as the others, but it is in fact an important contribution to the safety of a ship which loads at sea.

Hopper dredges/barges with a less than statutory minimum freeboard are also marked with statutory marks. (The non-statutory marks are red in colour and their use is limited to very specific (named) local applications for which they are assigned.)

11.2.2 Seasonal freeboards

The freeboards mentioned so far are in each case the basic or summer freeboard (S) which applies in salt water in summer. Ships are in general also given a number of subsidiary freeboards which apply in different conditions.

These are respectively tropical (T), winter (W), winter North Atlantic (WNA), freshwater (F) and tropical freshwater (TF) and are derived from the basic summer freeboard using the following relationships:

$$T = S - \frac{T_d}{48} \; ; \; W = S + \frac{T_d}{48} \; ; \; WNA = W + 50 \; ; \; F = S - \frac{\Delta}{4 \times \text{Tpc}}$$

TF bears the same relationship to *T*, as *F* does to *S*. All of the above are in mm.

T_d = Summer draft in metres

Δ = displacement in tonnes

Tpc = tonnes per centimetre

The subsidiary timber freeboards have the preface *L*. The calculation of *LT*, and *LF*, are as above. *LWNA* remains as *WNA* and *LW* is calculated from:

$$LW = LS + \frac{LT_d}{36}$$

LT_d is summer timber draft in metres.

11.2.3 Freeboard calculations

The calculation of freeboard is not a difficult matter and most naval architects will have access to a computer program which will perform this task for them very quickly. The treatment that follows is neither completely accurate nor fully comprehensive, but as the approximations involved result in a very small error, it can be used with confidence in preliminary design work when no other reference books are available.

The calculation of a summer freeboard starts with a tabular freeboard read from the rules in which this is tabulated against ship length. In the rules values are quoted for every metre from 24 to 365 m. Linear interpolation from the very abbreviated table given below remarkably introduces a maximum error of less than 25 mm and suggests the rules might have been simplified!

Length of ship (m)	Type A (mm)	Type B (mm)
24	200	200
50	443	443
76	786	816
100	1135	1271
150	1968	2315
200	2612	3264
250	3012	4018
300	3262	4630
350	3406	5160
365	3433	5303

The standard freeboard is then corrected for the following features:

(i) Hatch cover correction

If the hatches on a type B ship are not of the pontoon or similar type (i.e., are old fashioned covers on portable beams)

Approx addition = $50 + 3 [L - 100]$ mm
(max error 25 mm)

The fitting of the type of hatch covers which necessitate this addition is now very rare.

(ii) Ships of L < 100 m where the enclosed superstructures are less than 35% L

Addition = $7.5 (100 - L) (0.35 - E/L)$ mm

(iii) Block coefficient correction

If $C_b > 0.68$

$$\text{Addition} = \frac{SF(C_b - 0.68)}{1.36}$$

where SF is summer tabular freeboard

(iv) Depth correction

If depth exceeds $L/15$

Addition = $R(D - L/15)$ mm

where
$R = L/0.48$ for $L < 120$, and
$R = L/250$ for $L > 120$
If $D < L/15$, the freeboard can be correspondingly reduced, but only if there is an enclosed superstructure amidships of at least $0.6 L$

(v) Correction for efficient erections

A deduction is made for erections based on their location, their breadth and the proportion of the ship's length which these occupy.

Standard heights are specified for erections on a basis of length, and if the actual height is less than the standard or if the erections do not extend to the ship's side, the actual length (S) of these is reduced *pro rata* to give an effective length (E).

Table 11.1

Line	Type of freeboard	E AS % L						
		0	0.3	0.4	0.5	0.6	0.7	1.0
1	A	0	21		41		63	100
2	B (F)	0	15		32	46	63	100
3	B (F + B)	0	19		36	46	63	100
4	Timber	20	53	64	70			100

where F indicates with Forecastle but no detached bridge, and F + B indicates with Forecastle and detached bridge.

Standard heights are:

Raised quarter decks	0.90 m for $24 < L < 30$
	0.90–1.07 m for $30 < L < 122$
	1.07 m for $L > 122$
Other superstructures	1.80 m for $24 < L < 75$
	1.80–2.3 m for $75 < L < 125$
	2.30 m for $L > 125$

The deduction for $E = 1.0\ L$ is:

          350 mm for $L = 24$

          860 mm for $L = 85$

          1070 mm for $L = 122$

with intermediate values by interpolation

The percentage deductions for different types of freeboard and different types of erections are give in Table 11.1, again with intermediate values by interpolation. There are some qualifications to the use of lines 2 and 3.

(vi) Sheer correction

Standard freeboard assumes the ship has standard sheer which is defined as a parabolic sheer with its lowest point at amidships and ordinates at the:

A.P. of 25 $[L/3 + 10]$ mm

F.P. of 50 $[L/3 + 10]$ mm

The deficiency (+) or excess (−) of sheer is given by the difference between the actual and standard mean heights of sheer throughout the ship's length. For parabolic sheers this is:

$$\frac{(S_{fa} + S_{aa}) - (S_{fs} + S_{as})}{6}$$

where

S_{fa} = actual sheer forward
S_{aa} = actual sheer aft
S_{fs} = standard sheer forward
S_{as} = standard sheer aft

The correction for sheer is:

Addition/deduction = defy/excess $[0.75 - S/2L]$

where S = length of superstructure.

There are some detailed rules about the measurement of the actual sheer and in relation to variations from the standard profile.

	Freeboard		Date	
Ship No.				
Dimensions	L	B	D	C_b
	+ mm	– mm		
Standard Freeboard (SF)				
Hatchcovers correction				
Depth correction = $[D - L/15]$ R			If $D > L/15$ +, if $D < L/15$ –	
Cb correction = SF×$(C_b - 0.68)/1.36$			+ if $C_b > 0.68$	
Erections correction = ×				
Sheer correction = + Def ×$[0.75 - S/2L]$			+ def: – excess	
Bow height correction				
Total deductions				
Freeboard				

			Sheer type	
			Actual sheer F	
Moulded depth D		m	Actual sheer A	
Stringer/sheathing		m	Standard F	
Freeboard depth		m	Standard A	
Freeboard		m	Def/Excess	
Moulded depth		m	Erections type	
Keel		m	S	
Full draft		m	E	
			%	

Standard freeboard type A	
Standard freeboard type B	
A–B	
60% (A–B)	
Standard freeboard B-60	

Fig. 11.1. Freeboard calculation table.

(vii) Bow height correction

If the bow height is less than the minimum set by the formula below, the freeboard must be increased by the deficiency.

$$\text{Min bow height} = 56\,L[1 - L/500]\,\frac{[1.36]}{C_b + 0.68}$$

L need not be taken as more than 250 m
C_b not to be taken as less than 0.68

Finally it should be noted that the use of a standard calculation sheet such as that given in Fig. 11.1 eases and speeds the task and, above all, helps to avoid omissions.

11.3 SUBDIVISION — GENERAL

There have been international rules for the subdivision of passenger ships (carrying more than 12 passengers) since the 1929 International Conference on Safety of Life at Sea. The standard of subdivision required and the various factors involved in the supporting calculations have been modified over the years but remain much the same in the present rules, which are as set by the International Convention for the Safety of Life at Sea (SOLAS) 1974 as modified by the 1978 protocol and 1981 and 1983 amendments.

The method of calculation in general use has however changed with the advent of computers which make direct calculations easy and ended the use of the "B.O.T floodable length curves" method.

The requirements of these rules were set out in a deterministic format, but have now been joined by the alternative Passenger Ship Equivalent Regulations adopted as IMO Resolution A.265 (VIII), which employs a probabilistic approach.

Until very recently, the only subdivision requirements for cargo ships were for those with reduced freeboards (B-60) and classification societies requirement for a minimum number of bulkheads, but this changed with the adoption by IMO in 1990 of Resolution MSC 19(58) which added a new part B1 to the 1974 SOLAS.

This follows the general line of, but improves on, the passenger ship probabilistic method and it is expected that the latter will be modified in due course to bring it more in line with the cargo ship method.

The probabilistic approach addresses the probability of damage occurring at any particular location throughout the ship. It considers the likelihood of damage resulting in the flooding of one, two or any number of adjacent compartments and of penetrating or not penetrating longitudinal bulkheads and watertight decks or flats. The probability of the ship having sufficient residual buoyancy and stability to survive in each case of damage is assessed and the summation of all positive probabilities gives an "attained Subdivision Index" which must be greater than a "Required Subdivision Index" which is based on ship's length and complement for passenger ships and on ship's length only for cargo ships.

Some of the pros and cons of deterministic and probabilistic methods can be illustrated by a few examples. These are based on the passenger ships rules, as of course there are no deterministic rules for cargo ships.

In the deterministic rules the statutory maximum transverse extent of damage is set at B/5 measured inboard at any point from the half breadth at the subdivision waterline. This makes the precise positioning of a longitudinal watertight bulkhead critical: if it is inboard of the B/5 criterion by a few millimetres the space inboard of it becomes "intact" buoyancy for the purpose of considering compliance with the prescribed damage stability criteria and may also be used to increase the permissible length; if it is outboard of the B/5 criterion then from a subdivision point of view it does not exist, although the rules require that damaged stability calculations be done on the assumption it is not breached if this results in a more onerous situation.

Under the probabilistic regulations the precise positioning ceases to matter and damage calculations are made with and without the bulkhead being breached.

The probabilistic rules take a somewhat similar approach to the water tightness of decks with calculations being made on the alternative assumptions that these are and are not breached.

The probabilistic method takes into account not only safety against flooding as in the deterministic method, but also safety against capsize. In doing this the probabilistic rules takes account of the fact that it is usually the damaged stability that is the ultimate determining factor in the deterministic approach. A ship can pass the subdivision rules but fail on damaged stability, but a ship meeting the damaged stability requirements will also pass the subdivision ones.

The arguments clearly favour the rational of the probabilistic rules and it is thought likely that the deterministic rules will be phased out in the foreseeable future.

There would seem to be two main objections to the probabilistic rules. The first of these is the extremely large amount of calculations required, which although acceptable in this computer age is scarcely to be welcomed. The other objection is the lack of guidance that it gives to a designer, who may even be driven to continuing use of the deterministic method in initial design, changing to the probabilistic later — and hoping this does not entail major changes! It must be admitted that although the design guidance given by the deterministic rules eased the designer's task, some of its features, as instanced above, did not lead to optimisation in the interests of safety.

Even the deterministic rules do not provide any direct guidance on what freeboard should be provided in a design. Instead, they set a standard of subdivision which must be achieved. If the freeboard has been set "too low", the number of bulkheads necessary will be excessive and the space between them such that the development of a satisfactory design will be difficult or impossible. A wise designer should therefore select an initial freeboard ratio or draft/depth ratio based on that of a ship having approximately the same factor of subdivision and an arrangement reasonably similar to that intended for the new design.

In choosing a basis ship for this guidance, attention should be paid to the block coefficient and the sheer of the bulkhead deck of the basis ship and the corresponding values intended for the new design as these affect the floodable lengths which can be obtained with a given freeboard ratio, particularly for compartments towards the ends of the ship.

It is also wise to think ahead to damaged stability, which not infrequently makes it desirable to have more freeboard than is needed purely to meet the subdivision requirements.

If initial subdivision calculations show an improvement in subdivision is required, the most usual ways of obtaining of achieving this are by the respacing of bulkheads, adding a bulkhead, reducing the draft and/or increasing the depth to the bulkhead deck.

11.4 DETERMINISTIC RULES FOR PASSENGER SHIPS

This section gives a condensed version of the main provisions of these rules but reference to the rules themselves remains essential as there are many secondary provisions.

Reference should be made to a standard textbook on naval architecture for a description of the first principles method of calculating floodable length. Regrettably, there does not seem to be any quick/approximate method suitable for inclusion in this book.

General definitions

The floodable length at a given point is the maximum length of compartment, having its centre at that point which can be flooded without the margin line being submerged.

The margin line is 76 mm below the top of the bulkhead deck at side.

Uniform permeabilities as determined by the following formulae are to be used for the machinery space and the spaces forward and aft respectively of the machinery space:

Machinery space $\qquad U = \dfrac{85 + 10}{v}(a - c)$

Spaces forward and aft $\qquad U = 63 + 35\ a/v$

respectively of machinery space
a = volume of space for passengers within the volume v
c = volume of space used for cargo, stores or coal within the volume v
v = total volume to the margin line of the portion of the ship under consideration

Detailed calculations may be allowed in unusual cases, one of these being the presence of "intact buoyancy" — see §12.5.1. for extent of damage assumptions.

The factor of subdivision (F) depends firstly on the ship's service, i.e., whether the ship is to be certified for international voyages which are dealt with in the remainder of this section or short international voyages, which are dealt with in §11.5.

11.4.1 International voyages

The factor of subdivision depends on the length of the ship and its Criterion of Service numeral. There are three classes based on length:

(i) L (metres) = or < 79
(ii) $79 < L$ (metres) < 131
(iii) L (metres) ≥ 131

The length criterion is based on two factors A and B related to length as follows:

$$A = \frac{58.2}{L - 60} + 0.18 \qquad \text{and} \qquad B = \frac{30.3}{L - 42} + 0.18$$

(L = 131 m or more) (L = 79 m or more)
A corresponds to a ship which is primarily a cargo carrier and has a C_s value of 23 or less ($F = A$ for $C_s = 23$).

For $L = 131$ m, $\qquad A = 1.00$

B corresponds to a ship which is primarily a passenger carrier and has a C_s value of 123 or more ($F = B$ for $C_s = 123$).

For $L = 79$ m, $\qquad B = 1.00$
For $L = 131$ m, $\qquad B = 0.52$.

Criterion of Service Numeral: Interpolation between A and B is carried out by the Criterion of Service Numeral (C_s) which provides a measurement of the extent to which the ship is devoted to the carriage of passengers. There are two alternative formulae for C_s.

Where $P_1 > P$:

$$C_s = 72 \frac{(M + 2P)}{V + P_1 - P}$$

In other cases:

$$C_s = 72 \frac{(M + 2P)}{V}$$

M = volume of machinery space below margin line and between bulkheads
P = volume of passenger spaces below margin line
$P_1 = K \cdot N$
N = number of passengers
$K = 0.056 L$

Factor of subdivision (F): The factor of subdivision for ships of 131 m or more is generally calculated from the formula:

$$F = A - \frac{(A - B)(C_s - 23)}{100}$$

There are the following exceptions to this formula:
   (a)   where $C_s = $ or > 45 and $0.50 < F_f < 0.65$ then F shall be 0.50
   (b)   where $F_f < 0.40$ and it can be shown to be impracticable to meet this in the machinery space F may $= 0.40$ in this compartment.
   F_f = factor from the above formula.

The factor of subdivision for ships where $79 < L$ (m) < 131 is determined by the formula below and lies between 1.00 and the value B with interpolation using the factor S:

where $S = \dfrac{3574 - 25L}{13}$

$\quad\quad\quad$ ($= 123$ for $L = 79$)
$\quad\quad\quad$ ($= 23$ for $L = 131$)

$$F = 1 - \frac{(1 - B)(C_s - S)}{123 - S}$$

For ships where $L < 79$, the factor of subdivision $F = 1.00$, but this may be relaxed, if it can be shown to be impracticable.

A factor of subdivision $F = 1.00$ also applies to ships of any length carrying more than 12 passengers but less than $L^2/650$ or 50 whichever is less (these equate at $L = 180.3$ m).

11.4.2 Other details dealt with in the deterministic rules

The rules contain special provisions relating to:
- steps in bulkheads
- the position of peak bulkheads
- the extent of double bottoms
- openings in watertight bulkheads
- shell openings below the margin line
- machinery and system
- electrical installation, etc.

Many of these requirements have a major influence on important details of the ship, its outfit and machinery, but do not greatly affect the main design aspects which this book addresses.

11.5 DETERMINISTIC RULES FOR SHORT INTERNATIONAL VOYAGES

Because a ship engaged in a short international voyage is never far from a port, the rules allow a relaxation in the life-saving appliances carried, provided the ship meets more severe subdivision requirements, increasing the probability of its staying afloat after damage or at worst increasing the time available for rescue. The improved subdivision required for these ships also takes account of the greater likelihood of a collision or a stranding in the waters in which these ships operate.

The permeability of the spaces forward and aft of the machinery space on these ships is calculated using the following formula.

$$U = 95 - 35 \, b/v$$

where

$b =$ volume below the margin line and above the inner bottom or peak tanks used for cargo, stores, coal, oil fuel, fresh water within the volume v.

The Criterion of Service numerals given in §11.4.1 apply except that K for unberthed passengers $= 3.5$ m^2 and for berthed passengers either this value or that from the formula given in §11.4.1, whichever is greater.

The factor of subdivision formula for ships of 131 m or more is slightly modified from that applying to International voyages, with B being replaced by BB, which is determined as follows:

$$BB = \frac{17.6}{L - 33} + 0.20$$

This applies to $L \geq 55$ m (BB has a value of 1.0 for $L + 55$).

With this change

$$F_f = A - \frac{(A - BB)(C_s - 23)}{100}$$

If $F_f < 0.50$, then $F = 0.50$ or the value by the formula given in §11.4.1, whichever is smaller ($F_f = F$ from formula).
  If $F_f > 0.50$, then $F = 0.50$.

The factor of subdivision for ships $55 < L$ (m) < 131 lies between 1.00 and the value BB with interpolation using the factor S_1. Where

$$S_1 = \frac{3712 - 25L}{19}$$

($S_1 = 123$ for $L = 55$; $S_1 = 23$ for $L = 131$)

$$F = 1 - \frac{(1 - BB)(C_s - S_1)}{123 - S_1}$$

For ships of $L < 55$, $F = 1.00$, but relaxations may be allowed in a limited number of compartments.

11.6 PROBABILISTIC RULES FOR CARGO SHIPS

As the present regulations for probabilistic calculations for Passenger Ships given in IMO Res. A 265 (VIII) are expected to be modified fairly soon to bring them more into line with the regulations for Cargo Ships set out in IMO MSC 19 (58) and become a new Part B1 to SOLAS 74, it seems sensible to deal first, and in more detail with the cargo ship rules and limit treatment of the passenger ship rules to highlighting their principal differences from the cargo ship rules.

  This section goes into the rules for the subdivision and damaged stability of cargo ships in much greater detail than this book has given to any of the other rules. This is done partly because these rules are very new and have yet to take their place in naval architecture textbooks and, partly because they are very complicated and require many lengthy calculations. These are so involved that they will almost invariably be carried out by computer and it will be very difficult for designers to understand the principles involved, which must be a precondition to them being able to optimise their designs effectively. It is hoped that the manual treatment presented here, albeit of a simplified case, will go some way towards helping an understanding of the rules.

  The rules, which apply to ships constructed on or after 1 February 1992, apply at present to cargo ships with a subdivision length L_s of over 100 m, although it is hoped to extend them, probably with some modifications, to ships under this size.

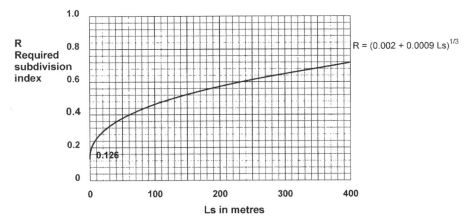

Fig. 11.2. Probabilistic subdivision of cargo ships — Required Subdivision Index.

Subdivision length has a new definition: the greatest length of the ship at or below the deck which limits the vertical extent of flooding.

To assist understanding, graphs have been drawn of each of the factors involved, with the corresponding equations alongside. Figure 11.2 gives a graphical representation of the Required Subdivision Index R. The fact that this is related only to length has been criticised by those who think that the crew numbers should be a factor. An answer to this criticism is that if increasing crew numbers resulted in a higher value of R and therefore in a higher building cost, it would have the undesirable effect of increasing the pressure to reduce crew numbers. In any case cargo ship crew numbers vary little with size of ship.

The next factor is the Attained Subdivision Index A which is:

$$A = \Sigma p_i \cdot s_i$$

where
   i represents each compartment or group of compartments (see Fig. 11.3). This shows a ship with six single compartments, five groups of two compartments, four groups of three, three groups of four, two groups of five and one comprising the whole ship.
   p_i represents the probability that the compartment or group under consideration may be flooded, disregarding any horizontal subdivision;
   s_i represents the probability of survival after flooding of the compartment or group, including the effects of horizontal subdivision.

The calculations are made at level trim. The summation is made for all cases of single compartment flooding plus all cases of flooding of two or more adjacent compartments that the ship can survive.

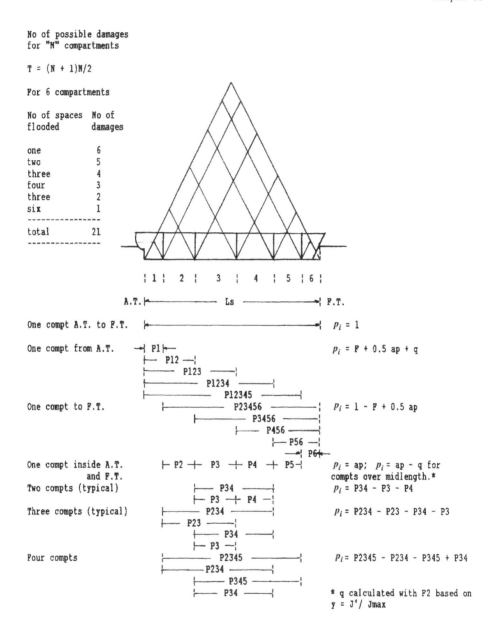

```
No of possible damages
for "N" compartments

T = (N + 1)N/2

For 6 compartments

No of spaces  No of
flooded       damages

one           6
two           5
three         4
four          3
three         2
six           1
------------------
total         21
------------------
```

Fig. 11.3. Probabilistic subdivision of cargo ships. Calculation of p_i.

The contribution of compartments that are subdivided by longitudinal bulkheads are to be adjusted to allow for the probability of these bulkheads being breached or not.

The calculation of the factor p_i is a complex one and with a need to calculate this for many different cases, it will generally be done by computer. To give a feel for the realities behind the complex figuring, Fig. 11.4 presents a manual calculation for a simplistic ship of 100 m. This ship has no longitudinal bulkheads or watertight decks below the bulkhead deck. The first column gives all the formulae, and the second the conditions attached to their use.

Reverting to Fig. 11.3, it will be seen that there are four different formulae for p_i of so-called "single" compartments, which may be made up of one, two, three, four, five or six (max. in this case) actual compartments, depending on whether these:

– extend to both AT and FT,
– extend to AT,
– extend to FT,
– extend to neither AT nor FT.

From these 15 "single" compartment p_i values, the p_i values for the 5 two-compartment cases, the 4 three, the 3 four, the 2 five can be calculated using the method shown in Fig. 11.3. In each case the calculation starts with the p_i figure for the group as a single compartment, from this is subtracted the two p_i values for groups of one less compartment starting from the same end points and finally the bit that this procedure deducts twice is added back.

Some of the symbols used in MSC 19(58) tend to obscure the physical meaning of parts of the calculation, which becomes clearer if it is realised that J is the compartment length non-dimensionalised by expressing it as a proportion of the ship's length and y is therefore the ratio of actual damage to the assumed maximum damage J_{max}.

A plot of J_{max} against length is given in Fig. 11.5(a). Figure 11.5(b) shows a plot of F_1 and F_2 against y. These factors figure in the formulae for p and q as shown in Fig. 11.5, whilst p and q feature in the formulae for p_i. At the top of Fig 11.5(b) there is an approximate indication of which y values apply to single, two, three, four, etc. compartments.

Figure 11.5(c) shows a plot of a and F against E, which is a non-dimensional representation of (2×) the distance of the centre of a compartment from amidships with a negative value indicating an aft compartment and a positive value a forward one.

J' is a non-dimensional representation of (2×) the distance from amidships to the nearest bulkhead of the compartment. The choice of J' as the symbol for this seems unnecessarily confusing with J and J_{max} representing quite different features.

Figure 11.4 presents a spreadsheet illustrating the use of all these formulae. For this purpose, the figure has been set out to cover the six single compartments of the ship, although three separate sheets should be used in practice — one covering 1 to N compartments, each of which includes the A.T., a second covering 1 to $(N-1)$ compartments, each of which includes the F.T. and a third for the $(N-2)$

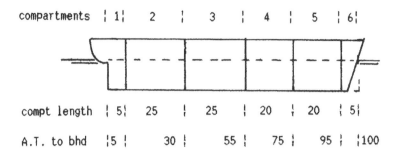

Item	Formula	Condition	Single Compartments					
			1	2	3*	4	5	6
X1			0	5	30	55	75	95
X2			5	30	55	75	95	100
E1	X1/Ls		0.00	0.05	0.30	0.55	0.75	0.95
E2	x2/Ls		0.05	0.30	0.55	0.75	0.95	1.00
E	E1 + E2 − 1		−0.95	−0.65	−0.15	0.30	0.70	0.95
J	E2 − E1		0.05	0.25	0.25	0.20	0.20	0.05
J′	J − E	E ≥ 0				−0.10	−0.50	−0.90
J′	J + E	E < 0	−0.90	−0.40	0.10			
J_{max}	48/Ls	max 0.24	0.24	0.24	0.24	0.24	0.24	0.24
a	1.2 + 0.8 E	max 1.2	0.44	0.68	1.08	1.20	1.20	1.20
F	0.4+0.25 E(1.2 + a)		0.01	0.09	0.31	0.58	0.82	0.97
y	J/Jmax		0.20833	1.04167	1.04167	0.83333	0.83333	0.20833
F1	$y^2 − y^3/3$	y < 1	0.04039			0.50154	0.50154	0.04039
F1	y − 1/3	y ≥ 1		0.70833	0.70833			
F2	$y^3/3 − y^4/12$	y < 1	0.00286			0.15271	0.15271	0.00286
F2	$y^2/2 − y/3 + 1/12$	y ≤ 1		0.27865	0.27865			
p	$(F1)(J_{max})$		0.00969	0.17000	0.17000	0.12037	0.12037	0.00969
q	$0.4(F2)(J_{max})^2$		0.00007	0.00642	0.00642	0.00352	0.00352	0.00007
p_i	F + 0.5(a)(P) + q*	Aft incl. A.T.	0.01270					
p_i	1 − F + 0.5(a)(P)*	Ford incl. F.T.				amid-shp.		0.03582
p_i	(a)(P)*	ex A.T. & F.T.		0.11560	0.17718	0.14444	0.14444	
s_i		compartments reduced by q,						
$p_i s_i$		cal. with F2						
A	sum of $p_i \cdot s_i$	based $y=J'/J_{max}$						

*amidships

Fig. 11.4. Tabular calculation of probabilistic subdivision of cargo ships. Example features ship with six compartments and $L_s = 100$ m.

(a)

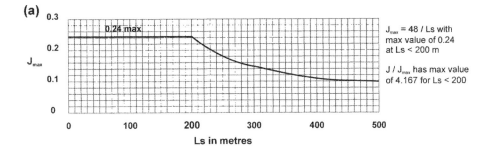

J_{max} = 48 / Ls with max value of 0.24 at Ls < 200 m

J / J_{max} has max value of 4.167 for Ls < 200

(b)

Approximate number of compartments

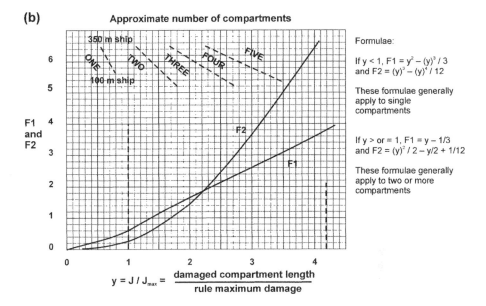

$y = J / J_{max} = \dfrac{\text{damaged compartment length}}{\text{rule maximum damage}}$

Formulae:

If y < 1, F1 = $y^2 - (y)^3$ / 3 and F2 = $(y)^3 - (y)^4$ / 12

These formulae generally apply to single compartments

If y > or = 1, F1 = y – 1/3 and F2 = $(y)^2$ / 2 – y/2 + 1/12

These formulae generally apply to two or more compartments

(c)

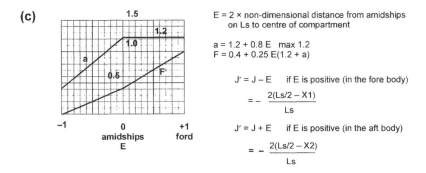

E = 2 × non-dimensional distance from amidships on Ls to centre of compartment

a = 1.2 + 0.8 E   max 1.2
F = 0.4 + 0.25 E(1.2 + a)

J' = J – E   if E is positive (in the fore body)

$= - \dfrac{2(\text{Ls}/2 - X1)}{\text{Ls}}$

J' = J + E   if E is positive (in the aft body)

$= - \dfrac{2(\text{Ls}/2 - X2)}{\text{Ls}}$

Fig. 11.5. Probabilistic subdivision of cargo ships. (a) J_{max} versus L_s. (b) F_1 and F_2 versus J/J_{max}. (c) Factors a and F versus E.

compartments, each of which lie inside both A.T. and F.T. — all as shown in Fig. 11.3. A further spreadsheet can then be used for the calculation of groups of compartments, also as shown in Fig. 11.3. Shown at the foot of Fig. 11.4, but not used in this case, are lines for the tabulation of s_i values leading on to the $p_i \cdot s_i$ values and their aggregation to give the attained subdivision index A.

In the simplistic ship shown in Fig. 11.4 there were neither longitudinal bulkheads nor watertight decks introducing complications in the subdivision. These are probably undesirable features from a subdivision and damaged stability point of view, but other reasons may nevertheless make them essential. Figure 11.6 shows all the possible structural variants that can arise out of combinations of these features and the different damage possibilities.

Longitudinal subdivision is dealt with by the use of a reduction factor r applied to the p_i value of the compartment concerned. Values of r are given in Fig. 11.7.

Horizontal subdivision is on the other hand treated as a factor affecting s — the probability of survival, already mentioned but not yet defined.

The formula for s, together with graphs of its two components s/c and c are given as Fig. 11.8. Both require an input from damaged stability calculations for the compartment concerned; the former the GZ and the range and the latter the final angle of heel.

It is a requirement that calculations of s are made at both the deepest subdivision loadline and at a partial loadline set at 60% of the difference between the light draft and the subdivision draft, with the results averaged.

The permeabilities to be used in the damaged stability calculations are:

stores   0.60   accommodation   0.95
machinery 0.85   void spaces   0.95
dry cargo   0.70   liquid tanks   0 or 0.95, whichever results in the more
                                                          severe requirement.

Horizontal subdivision within the compartment brings in a reduction factor v which represents the probability that spaces above the division will not be flooded. A graph of v_i is given as the third graph in Fig. 11.8 and should be studied in association with the sections on Fig. 11.6.

As was said earlier, the rules are very complicated and demand the use of a computer program.

It has been said that very few existing ships would meet these rules, so some guidance for new designs is clearly desirable. Unfortunately practical designers have so far had little experience of the implications of the rules to guide them and until this is forthcoming must go cautiously and carry out full calculations before a design becomes fixed.

Some good immediate guidance could be obtained by carrying out calculations for one or two existing ships and seeing by how much these fail and how they might best be modified to pass.

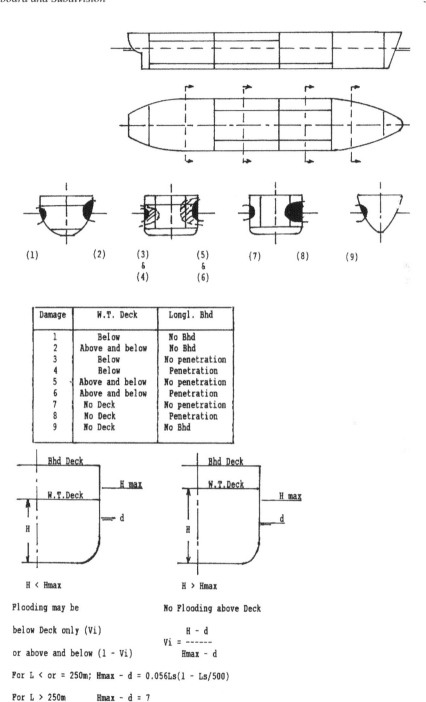

Fig. 11.6. Different types of damage.

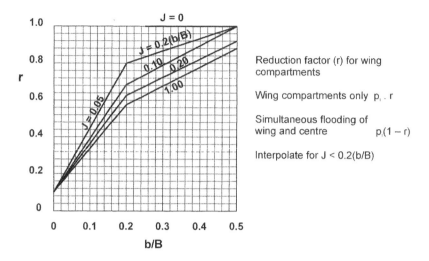

Fig. 11.7. Probabilistic subdivision of cargo ships. Reduction factor for wing compartments (*r*).
Formulae:

$$J \geq 0.2b/B \quad r = b/B\left[2.3 + \frac{0.08}{J+0.02}\right]+0.1 \quad \text{if } b/B \leq 0.2$$

$$r = \left[\frac{0.016}{J+0.02}+b/B+0.36\right] \quad \text{if } b/B > 0.2$$

Some general guidance which can be given from passenger ship experience suggests:
- avoid longitudinal subdivision if possible and where it is essential cross connect the wing compartments with automatically operating ducts or pipes;
- think very carefully before introducing horizontal subdivision as damage above this may result in a loss of waterplane inertia without the benefit of added weight low down;
- choose dimensions that will ensure good stability and range before damage to help to ensure that the *s* value after damage is good;
- make compartments generally of about the same length so that each contributes fairly equally to *A*. If longer compartments are required, try to locate these in the forebody, where they will benefit from a higher *a* value (see Fig. 11.4).

Opposite: Fig. 11.8. Probabilistic subdivision of cargo ships.
(a) *S/C* versus GZ × range.
(b) *C* versus final angle θ.
(c) $(H-d)/(H_{max}-d)$ versus $H-d$.

(a)

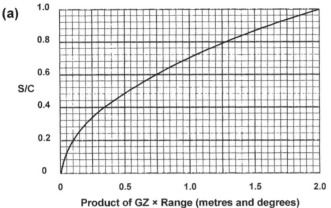

$$S/C = (0.5 \times GZ \times Range)^{1/2}$$
$$GZ_{max} \leq 0.1 \text{ m}$$
$$Range \leq 20°$$

S/C (vertical axis)

Product of GZ × Range (metres and degrees)

(b)

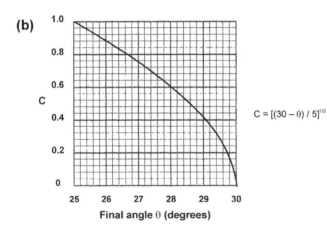

$$C = [(30 - \theta) / 5]^{1/2}$$

C (vertical axis)

Final angle θ (degrees)

(c)

$$V_i = \frac{H - d}{H_{max} - d}$$

(H − d) metres

Factor V_i

Probability of flooding being restricted to below a horizontal subdivision at height H

$1 - V_i$ = Probability of flooding above and below

H_{max} = Maximum damage height
$= d + 0.056 \ Ls(1 - (Ls/500))$ m
$= d = 7$ max for Ls > 250 m

Finally, however, it may be necessary to make the compartments smaller than has been past practice by introducing one — or on larger ships possibly two — additional bulkheads to improve p_i. On gearless bulk carriers or container ships the extra cost of this will not be very great. On ships with cargo handling gear and on refrigerated ships the extra cost of another hold will be considerable and it may pay to look instead at increasing the freeboard to improve s_i — either increasing the depth (but stability must be watched), or reducing the draft or both.

It is clear that these rules require a major rethink about many things that have been accepted practice in ship design.

11.7 PROBABILISTIC RULES FOR PASSENGER SHIPS

This brief treatment is intended to highlight the main differences from the cargo ship rules dealt with in the last section.

The required subdivision index R for passenger ships brings in a factor increasing the standard with the number of passengers and crew carried as compared with cargo ships in which ship length is the sole factor.

$$R = 1 - \frac{1000}{4L_s + N_1 + 2N_2 + 1500}$$

where

N_1 is the number of persons for whom boatage is provided, and
N_2 the remainder of the complement.
The formula for the Attained Subdivision Index is:

$$A = \Sigma a \cdot p \cdot s$$

This looks different from the cargo ship formula:

$$A = \Sigma p_i \cdot s_i$$

but in fact for much of a cargo ship $p_i = p \cdot a$.

On passenger ships, unlike cargo ships, the vertical extent of damage is from the base upwards without limit — although this is qualified in the damaged stability rules by the statement that if a lesser extent of damage is more onerous this must be considered.

In the probability calculations for passenger ships four drafts must be considered as opposed to two for cargo ships.

The formula for p for each compartment appears to have a different basis from that in the cargo ship rules, although this may be more in presentation than in reality. The method of calculating the value for a group of compartments from its components is identical in the two rules.

The formula for *s* is quite different.

Finally the passenger rules contain many more requirements relating to damaged stability, generally on the lines of those in the deterministic rules.

It would be very interesting to know how much these "equivalent" rules have been used since their introduction, and whether designers using them have in general found them to be more or less onerous than the deterministic rules. It may be a case of horses for courses.

11.8 FUTURE RULES

One of the difficulties in writing a book on practical ship design, as opposed to theoretical naval architecture, is the speed with which quite recently written material becomes out of date as new design ideas are brought forward and new rules are set. This section tries to give a brief introduction to some rule changes that have either come into force very recently or are likely to do so in the near future.

The rules governing safety of life at sea have come in for some very significant re-examinations following a number of major catastrophes in recent years whilst the introduction of some radically new ship types has been accompanied by a new approach to safety investigations.

Tom Allan, Director of the British Marine Standards Division, in his 1997 R.I.N.A. paper "The 1995 Solas Diplomatic Conference on Ro-Ro passenger ferries" gives an excellent insight into the many factors which need to be considered when trying to improve the safety of these vessels and summarises the changes in design and operational procedures recommended in the agreements reached.

The factors which it was thought should be considered in arriving at the Solas Conference recommendations were:
- Stability of the ship in intact and damaged conditions.
- The implications of accidents resulting in water on the Bulkhead (Ro-Ro) deck.
- Measures to prevent such accidents and their consequences.
- Construction of the ship, especially of hull doors and closing devices.
- Basic ship design, including the design of the ship to facilitate evacuation.
- The human factor and potential areas of human error.
- Operational factors, including the closure of watertight doors and the lashing of vehicles.
- Interface between passengers and shipboard safety systems
- Crisis management, including access to information needed to manage crises effectively.

- – Training required for personnel to enable them to deal with the special characteristics of, and special requirements on board Ro-Ro ships.
- – Communications, both within the ship and between the ship and the outside world.
- – The adequacy of lifesaving appliances.
- – The adequacy of search and rescue arrangements.
- – Overall safety assessment and risk analysis.
- – Ship/operator relations including the international safety management code.
- – Survey and inspection.

One of the most significant changes from previous rules is the requirement that calculations of survivability after damage be based on the ship being in seas with a wave height appropriate to the sea area in which the ship operates, this figure generally lying between 2 and 4 m.

When Stena Line decided to introduce high speed passenger and car ferries to their Holyhead–Dun Laoghaire service, they wisely decided that the novelty of this type of design demanded the adoption of the safety case approach.

A 1997 R.I.N.A. paper by Kuo, Pryka, Sodahl and Craufurd entitled "A Safety Case for Stena Line's High Speed Ferry HSS1500" gives a useful introduction to safety case methodology.

The questions posed in a safety case examination together with the tasks to be done to provide answers were set out in Table 11.2.

Table 11.2

Question	Tasks to be done to provide answers
1. Hazard Identification	
What aspects of the system can go wrong?	Identify potential hazards systematically.
2. Risk Assessment	
What are the chances and effects of these going wrong?	Assess the risk levels of the identified hazards.
3. Risk Reduction	
How can these chances and effects be reduced?	Reduce risk levels of selected hazards.
4. Emergency Preparedness	
What to do if an accident occurs?	Be prepared to respond to emergencies.
5. Safety Management System	
How can safety be managed?	Manage and control the hazards risk levels.

In the case of the HSS1500, those making the safety case assessment identified 109 potential hazards which for convenience were classified under four headings:

(a) Common potential hazards

Fire in the engine room; collision with another ship; failure of life-saving appliances; fuel leakage from vehicles on board.

(b Less common potential hazards

Bow thruster malfunction; mooring equipment failure; vehicle falling into the water during loading/unloading; collision with submerged object; dangerous goods.

(c) Rare potential hazards

Legionnaire's disease from air conditioning; ballast systems failure; blockage of water jet; landing stage failure; food poisoning; leakage during bunkering operation.

(d) Human-related potential hazards

Navigation error; person overboard; navigation warnings not received.

The likelihood of each type of risk occurring were assessed under five headings:

Scale 1 — Frequent
Scale 2 — Reasonably probable
Scale 3 — Remote
Scale 4 — Extremely remote
Scale 5 — Extremely improbable

The effect of each hazard were classified under four headings:

Scale A — Minor effect
Scale B — Major effect
Scale C — Hazardous effect
Scale D — Catastrophe

The interaction between the probability of an occurrence and the seriousness of the effect produced by it is tabled as a matrix with three regions: intolerable, tolerable and negligible.

Whilst there is clearly a need for good quantitative data if reliance is to be placed on assessments of this sort, even without such data the posing of these questions seems bound to improve decision making and safety.

Chapter 12

Stability and Trim — General

This chapter, in common with the preceding one and the one which follows, deals with the statutory rules governing merchant ships; in this case those concerned with stability and trim. Unlike the other two chapters the subject of this one, stability and trim, is equally applicable to warship design and the treatment has been extended to cover these ships.

12.1 MERCHANT SHIP STABILITY

12.1.1 Merchant ship stability standards

There are a number of different standards which can be subdivided into:
  (i)   standards of intact stability which are applicable to all merchant ships, unless higher standards are required any reason;
  (ii) higher standards of intact stability which are required for the carriage of certain special cargoes;
  (iii) standards of damaged stability which apply to cargo ships permitted to have a reduced freeboard, i.e. less than type B.
  (iv) standards of damaged stability set within the probabilistic subdivision rules for cargo ships and passenger ships respectively, as discussed in the last chapter;
  (v)   standards of damaged stability applicable to passenger ships whose subdivision has been dealt with by the deterministic rules.

12.1.2 Intact stability standards

Intact stability standards applicable to all merchant ships were laid down Internationally in the International Convention on Load lines 1966, and for British ships in The Merchant Shipping (Load Line) Rules 1968. The standards apply to all conditions of loading. The standards are:

(a) The area under the curve of righting levers (*GZ* curve) shall not be less than:
  (i) 0.055 metre-radians (mrad) up to an angle of 30°;
  (ii) 0.09 mrad up to an angle of 40° or the angle at which the lower edges of any openings in the hull, superstructure or deckhouses, being openings which cannot be closed weathertight, become immersed if that angle is less.
  (iii) 0.03 mrad between an angle of heel of 30° and one of 40° or such lesser angle as referred to in (ii)
(b) The righting lever (*GZ*) shall be at least 0.20 m at an angle of heel equal to or greater than 30° (in many cases a reduction in this angle can be accepted subject to the areas under the curve being increased)
(c) The maximum righting lever (*GZ*) shall occur at an angle of heel not less than 30°
(d) The initial transverse metacentric height shall not be less than 0.15 m. In the case of a ship carrying a timber deck cargo which complies with sub-paragraph (a) by taking into account the volume of the timber deck cargo the initial transverse metacentric height shall be not less than 0.05 m.

12.1.3 Influence of these standards on design

These standards are easy to meet in almost all new designs with difficulty occurring only when operational reasons impose unusual dimensional constraints.

The large angle requirements of (a), (b) and (c) can generally be assured by keeping the ratios B/D and T/D within the ranges suggested in Figs. 3.9 and 3.10. The initial stability requirement of (d) should also follow unless there is an unusual amount of top weight. Notwithstanding this, the stability should be checked in detail at as early as possible a stage in the design and should be rechecked when the design nears completion and while any changes that such a check shows to be necessary can still be made without undue expense.

Because the standards are not difficult to meet, it is wise to exceed them quite comfortably, at least in the initial design phase and the author would much prefer to see all ships having a GM in the worst service condition of 0.40 m.

12.1.4 Thixotropic cargoes

In §12.1.1 (ii) it was noted that higher intact stability standards are required for certain special cargoes. Two of these types of cargo which require special stability investigations are grain and dredge spoil. The reason is the same in both cases and is the fact that these cargoes are thixotropic which means that they are liable to move when the ship rolls or heels and act to a greater or lesser degree as a free surface.

Grain cargo stability is dealt with in §12.2, whilst the stability of dredgers is the subject of §12.3.

It is known that some other cargoes such as coal and iron ore may "liquefy" in a similar way and that this may have contributed to the loss of some bulk carriers. Whilst this has not been proven a wise designer should pay attention to minimising the likelihood of this happening and the consequences if it does.

Guidance on the carriage of cargoes which may liquefy is given in the I.M.O. B.C. Code.

12.2 GRAIN STABILITY

12.2.1 Introduction

Grain was formerly carried in general cargo "tramp" ships. In these ships extensive measures had to be taken with shifting boards and feeders to limit any shift of grain. Nowadays grain is almost entirely carried in bulk carriers and no special preparation is required for a grain cargo except for completely filling as many holds as possible and trimming level any partly filled compartments —provided the ship meets the stability requirements given in Chapter VI of the 1974 SOLAS Convention, which has been slightly revised and issued in 1991 by I.M.O. as the International Code for the Safe Carriage of Grain in Bulk.

12.2.2 I.M.O. rule requirements

These rules require heeling curves to be calculated based on an assumed shift of grain. Obviously the greatest heeling moment will be generated in any partly filled compartments, so the first objective is to have as many compartments as possible completely filled and the minimum number, only one if possible, with a complete free surface. It is also helpful if the partially filled compartment is as small as possible, but with the need for a satisfactory trim, it may not be possible to use the same compartment partially filled for all specific gravities of cargo.

Even for full compartments the calculations must take account of settlement of the cargo and the voids which exist between hatch coamings and at the sides of the hatches. As the vessel heels the voids shift from one side to the other thus causing a heeling moment as the grain moves to fill the first void.

After taking into account the heeling moments due to grain shift the ship must meet the following criteria at all times:

   (i)   the angle of heel due to grain shift shall not exceed 12° or a lesser figure if required by an Administration;

   (ii) the residual dynamic stability, as shown in Fig. 12.1, up to 40° or the angle of flooding if this is less, shall not be less than 0.075 mrad;

  (iii) the GM after allowance for liquid free surface shall not be less than 0.30 m.

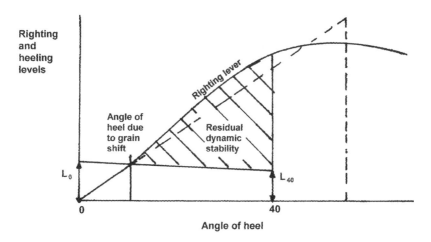

Fig. 12.1. Grain stability. $L_0 = \dfrac{\text{Volumetric heeling moment}}{\text{Displacement} \times \text{stowage factor}}$; $L_{40} = 0.8 \times L_0$

12.2.3 Basis of grain shift calculations

Calculations are based on the following assumptions.

(i)   That in filled compartments there will be a void under all boundary surfaces having an inclination of less than 30° to the horizontal. A formula for the depth of void is given, but in design practice the lesson is that the bottom of top-side wing tanks should be inclined at an angle of at least 30°.

(ii) That within filled and trimmed hatchways the void is to be taken as 150 mm below the lowest point of the hatch or the top of the hatch side coaming whichever is lower. The volume of any open void within the hatch cover to be added. There are other rules which apply if the hatchway is not trimmed.

(iii) The grain shift moment is to be calculated based on the grain surface being at 15° to the horizontal in the void spaces of filled compartments and at 25° to the horizontal in partly filled spaces. The calculated transverse heeling moment is to be multiplied by 1.12 to allow for the effect of the accompanying vertical shift.

(iv) If necessary the partially filled compartment can be overstowed with bagged grain to eliminate the free surface. If this is done the bagged grain must be tightly stowed and extend to a height of 1/16 of the breadth of the grain free surface or 1.2 m, whichever is greater.

It must be emphasised that this is only an outline of the rules. The detailed rules have several other requirements.

12.3 DREDGER STABILITY

12.3.1 Different types of dredger

Before discussing dredger stability, it is worth noting that dredgers are built to undertake two distinctly different roles.

(1) The first of these is the removal of spoil from estuaries and rivers to provide navigational channels. Vessels for this purpose are generally fitted with bottom hopper doors and having loaded up with spoil they sail out to sea to specified dumping areas where the spoil is deposited by the opening of the bottom doors.

(2) The other role is the extraction of sand or gravel from estuaries or from shallow waters further out to sea and the transport of these materials to a quayside where it can be discharged ashore for sale to the building industry. These vessels do not require to have hopper doors and can be built with a conventional double bottom. If intended for use at more than one port these ships are generally fitted with self-discharge facilities, although some ships of the type may be designed for a dedicated service and rely on port facilities.

Although this second type of dredger does not require hopper doors to fulfil its role, these ships are quite often fitted with hopper doors. This may be done so that the other role can be undertaken if it should become financially advantageous to do so, but is more generally done to gain the reduced freeboard and greater cargo deadweight for given dimensions which the hopper type of vessel is permitted to have.

The paragraph on dredger freeboard in §11.2.1 indicated that these ships may be assigned reduced freeboards and outlined the design features giving increased survivability which justifies the reduced freeboards of these ships.

The bottom doors of hopper dredgers must be capable of being operated from the bridge even if the main power fails. They must be capable of being completely opened in not more than four minutes.

Both types of dredger must be designed with spillways so positioned as to limit the hold capacity so that when the ship is full of saturated spoil of the heaviest anticipated density the appropriate load line mark will not be immersed.

Many dredgers are built with a number of different spillways to facilitate conversion to different spoil densities applicable in different operational areas. Non hopper type vessels are required to undertake loading trials which must demonstrate that the spillway in use is appropriate to the specific gravity of the spoil in the operational area. If they move to another operational area with a different spoil a different spillway is brought into use and further trials are required.

12.3.2 Statical stability of dredgers

The stability of dredgers is a very difficult subject, both for the designer who must design the ship so that the stability can be satisfactory and for the Master who must so operate it that the designer's intent is achieved.

The task that the Master of a dredger has in ensuring that the stability of his ship is satisfactory is complicated by the fact that a dredger loads at sea and he can therefore never be completely certain what kind of cargo is going to come on board.

(1) What will be the specific gravity of the spoil?

(2) Will the spoil be essentially a solid, albeit with a partial or complete water free surface or will it be thixotropic with a free surface specific gravity near to the specific gravity of the volume as a whole?

The answers to these questions have a major effect on the ship's stability.

A lightweight cargo will fill the hopper to the topmost weir, resulting in a high VCG, but the surface will be well up in the coaming whose width is usually significantly less than the beam of the ship thereby minimising the free surface.

A heavy density spoil on the other hand will have a lower VCG but the spoil surface will extend from side to side of the ship, as shown in Fig. 12.2, resulting in a big free surface.

The most important question, however, relates to the effective specific gravity of the free surface which may be close to that of sea water at one extreme or at the other near that of the spoil as a whole.

The number of stability conditions required to give guidance on all possible conditions of loading is immense and faced with this problem the author invented "the universal dredge stability diagram" which is presented as Fig. 12.3.

Entering this diagram at the mean draft, the Master can read off the metacentric height for any quantity and specific gravity of spoil, and above all for any specific gravity of the spoil free surface, which need not be identical to that of the spoil as a whole.

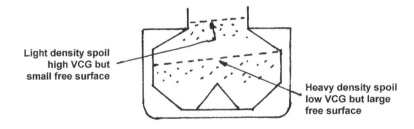

Fig. 12.2. Dredger stability. VCG and free surface of high and low density spoils.

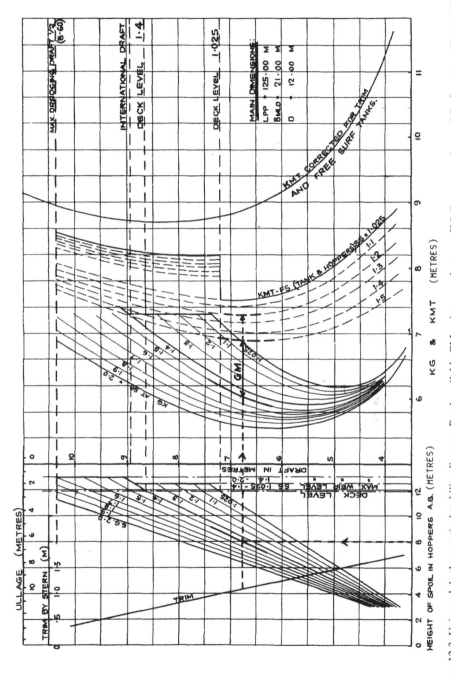

Fig. 12.3. Universal dredger statical stability diagram. Read available GM as intercept between KG line appropriate to SG of spoil and KM–FS line appropriate to type of spoil.

It should be emphasised that the diagram cannot give any guidance on the specific gravity which applies to the free surface, but it does give a warning of how important this can be. The other point which should be made is that the diagram deals only with small angle stability.

The diagram shows clearly that minimum stability will frequently occur when the cargo level is such that the free surface at a small angle of heel will extend beyond the confines of the coaming. Whilst showing that a relatively narrow coaming will improve the stability with a full load of a low specific gravity cargo it gives a clear warning of the danger inherent in relying on this during the loading process.

12.3.3 Large angle stability

For large angle stability it is necessary to make what are known as spill-out calculations. The British Department of Transport (D.Tpt.) require "spill out" type calculations which assume that the spoil surface remains horizontal as the ship heels, with spill out occurring when the spoil level reaches the top of the coamings or weir if appropriate.

Bureau Veritas requires (or did until a few years ago) calculations in which the cargo is assumed to shift not through the full angle of heel (θ) of the ship, but instead through an angle (α) where:

$$\alpha = \theta \frac{(3-U)}{2}$$

where U = specific gravity of spoil.

With this method there is not as much "spill out" as there is with the D.Tpt. method, but on the other hand the wedge causing heel is reduced. The two alternatives are illustrated in Fig. 12.4.

It can be argued that the BV treatment more nearly represents what happens during normal rolling, but on the other hand the D.Tpt. treatment seems to be more correct for the case where for any reason, the ship takes a permanent list. As this is generally the more onerous requirement the D.Tpt. treatment appears to deal with the critical case.

A standard form which can be used for either of these calculations is given as Fig. 12.7. As a preliminary to the use of this form, it is necessary to prepare GoZ (if an assumed CG Go is used) or KN curves for the ship based on two assumptions:

(1)   that there is no flooding of the hopper.
(2)   that the hopper is open to the sea.

Also needed is somewhat similar data giving the volume of spoil in the hopper when the spoil surface is tangential to the coaming top at various angles, together with VCG and transverse CG of this volume (see Fig. 12.5).

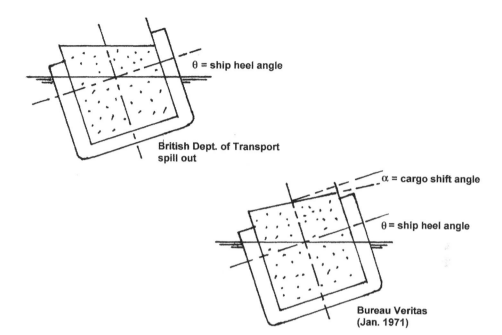

Fig. 12.4. Dredger stability. A comparison of the methods used by the British Department of Transport and Bureau Veritas, respectively.

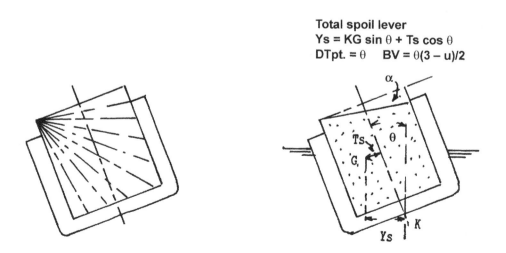

Fig. 12.5. Spoil volumes and levers at various spill out angles.

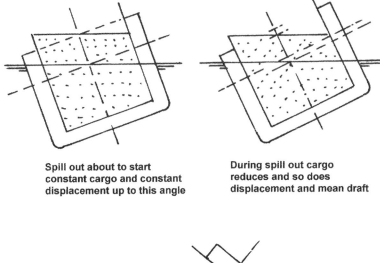

**Spill out about to start
constant cargo and constant
displacement up to this angle**

**During spill out cargo
reduces and so does
displacement and mean draft**

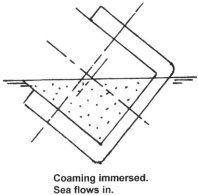

**Coaming immersed.
Sea flows in.**

Fig. 12.6. The different stages during spill out.

The calculation is made in two parts relating respectively to before and after the coaming is immersed. Firstly the spoil volume, which starts to reduce after spill-out commences, as shown in Fig. 12.6, must be calculated. Then using the specific gravity of spoil, and the weight of the ship less spoil from the initial upright case, the weight of spoil and the ship's displacement at each angle are calculated.

The net heeling lever is then obtained by a moment calculation about K in which the weight and moments of the spoil and of the ship less spoil are added.

Up to the point of coaming immersion (see Fig. 12.6) the net righting or heeling lever is the difference between the heeling lever and the KN value for "no hopper flooding" at the appropriate displacement and angle.

After the coaming is immersed the KN value used is changed to that based on the hopper open to the sea and the heeling lever of the spoil must then be reduced by the weight of sea water displaced by the spoil.

Dredgers assigned a 1/2(B-60) freeboard must be capable of surviving after damage to any one compartment; those assigned a freeboard of 1/2(B-100) must survive damage to the engine room and to any two other adjacent compartments.

12.4 DAMAGED STABILITY STANDARDS FOR REDUCED FREEBOARD

Ships with type B-60 or B-100 freeboards must meet damaged stability standards laid down in the freeboard rules, which may be summarised as follows:

12.4.1 Ships with B-60 freeboard

The ship when loaded to the summer load waterline shall remain afloat, after:
- (i)   the flooding of any single compartment other than the machinery space at an assumed permeability of 0.95;
- (ii) and, if the ship's length exceeds 225 m, the flooding of the machinery space at an assumed permeability of 0.85.

12.4.2 Ships with B-100 freeboard

The ship when loaded to the summer load waterline shall remain afloat, after:
- (i)   the flooding of any two compartments adjacent fore and aft, neither of which is a machinery space, at an assumed permeability of 0.95;
- (ii) and, if the ship's length exceeds 225 m, the flooding of the machinery space alone, at an assumed permeability of 0.85.

12.4.3 Condition after damage

In all the above cases, the ship shall be in a condition of equilibrium after damage meeting the following requirements:
- (a)   the final waterline to be below any opening which might permit progressive flooding (abbreviated statement);
- (b)   The angle of heel due to unsymmetrical flooding shall not exceed 15°;
- (c)   The GM shall be at least 50 mm in the upright condition, using the constant displacement method;
- (d)   The ship shall have adequate residual stability.

D	E	F	G	H	I	J	K	L	M	N	O	P	Q
θ Ship heel angle	α Spoil angle	sinθ	cosθ	Spoil vol	Spoil weight	Ship minus spoil	Disp.	GoZ Hopper dry	KGo ×sinθ	KN Hopper dry	Water displ'd by spoil	Disp.	GoZ Hopper open
	α = θ (D.Tp.) or (BV) =(3–u)/2				(H) × u (SG)	L/S + + fuel etc.	(I) + (J)			(M) + (L)	(H) × 1.025	(K) – (O)	
deg	deg			m^3	tonnes	tonnes	tonnes	m	m	m	tonnes	tonnes	m

							Before coaming immersed					After coaming immersed	
0													
5													
10													
15													
20													
30													
40													
50													
60													
70													
80													
90													

CONDITION NO.   FREEBOARD

Displacement   =   tonnes   Mean spoil level =   m

Spoil   =   tonnes   Hopper volume   =   m^3

Lightship + fuel etc =   tonnes   Mean S.G. – u   =

Coaming spill level =   m   KGo   =   m

Spill starts at   deg   KGf   =   m

Coaming immersed at   deg   (allows for free surface of tanks)

Fig. 12.7. Spill-out calculations to British Department of Transport or Bureau Veritas methods (*continued opposite*).

R	S	T	U	V	W	X	Y	Z	AA	AB	AC	AD	AE
KN Hopper open	KGs spoil	Ts spoil	KGs ×sinθ	Ts ×cosθ	Total spoil lever	KGf × sinθ	Sum of moments (not immersed	Lever	Sum of moments (immersed)	Y Lever	GZ (not immersed)	GZ (immersed)	θ Ship heel angle
(M) + (Q)	(based on angle α)	(S) × (F)	(T) × (G)	(U) + (V)			(I)×(W) + (J)×(X)	$\dfrac{(Y)}{(K)}$	(Y) − (O)×(W)	$\dfrac{(AA)}{(P)}$	(N) − (Z)	(R) − (AB)	
m	m	m	m	m	m	m	t×m	m	t×m	m	m	m	deg
		Spoil level Ys					Ship heeling levers				Righting levers		
													0
													5
													10
													15
													20
													30
													40
													50
													60
													70
													80
													90

Fig. 12.7. Spill-out calculations to British Department of Transport or Bureau Veritas methods (*continuation*).

12.5 PASSENGER SHIP DAMAGED STABILITY

Damaged stability standards applicable to passenger ships whose subdivision is based on the deterministic rules are laid down in the 1974 SOLAS Convention as modified by the 1978 protocol and 1981 and 1983 amendments and can be summarised as follows:

The ship shall have sufficient intact stability in all service conditions to enable it to withstand the flooding of the number of main compartments corresponding to the required factor of subdivision.

$F_s > 0.50$	any one compartment
$F_s > 0.33$ and < 0.50	any two adjacent compartments
$F_s < 0.33$	any three adjacent compartments

For damaged stability calculations, volume and surface permeabilities shall be assumed as follows:

Cargo and stores spaces	0.60
Accommodation spaces	0.95
Machinery spaces	0.85
Spaces for liquids either	0 or 0.95 (whichever results in the more severe requirements)

Higher surface permeabilities are to be used for spaces in which there is no substantial cargo, stores, accommodation or machinery in the vicinity of the damaged waterline.

12.5.1 Assumed extent of damage

The assumed damage shall be:

Longitudinal	3.0 m + 3% L or 11 m, whichever is less
Transverse	B/5 inboard from ships side at deepest subdivision waterline
Vertical	From baseline upwards without limit, or any lesser extent which would result in a more severe condition.

12.5.2 Condition after damage

The final condition of the ship after damage, and in the case of unsymmetrical flooding after equalisation measures, shall be as follows:

(a) For symmetrical flooding a minimum positive GM of 50 mm, calculated on the constant displacement method.

(b) For unsymmetrical flooding the total heel shall not exceed 7°. There is a provision for a relaxation of this rule in special cases with 15° as the absolute limit.

(c) The margin line shall not be submerged in the final stage of flooding and if it is submerged in an intermediate stage further investigations and/or arrangements may be required.

Arrangements which may cause unsymmetrical flooding should be kept to a minimum. Cross flooding arrangements should result in an equalisation time not exceeding 15 minutes and their operation should, if possible, be automatic or if controls are needed these should be operable from above the bulkhead deck.

Following the enquiry into the loss of the "Herald of Free Enterprise", the British Department of Transport issued two additional requirements for Ro-Ro passenger ships in June 1989 in notice M 1381:

(1) that ships of this type be inclined every four years to confirm their lightship particulars.

(2) that these ships be provided with "limiting KG (or GM) envelope curves" based on level keel and trims of 0.4% and 0.8% by both bow and stern, over the full operating range of displacements.

The notice defines residual stability criteria on which the envelope curves are to be based.

As this notice shows the requirements for damaged stability of passenger ships are very much under review at the present time and further amendments to SOLAS 74 can be expected shortly.

12.6 CONSTANT DISPLACEMENT AND ADDED WEIGHT

The reference to the constant displacement method in (a) above suggests a brief digression to consider the differences between added weight and lost buoyancy (or constant displacement) methods of dealing with damaged stability. A comparison of the two methods for the simple case of a rectangular box ship floating in fresh water is given in Fig. 12.8.

In this simple case it can be shown that the displacement in the added weight case is the intact displacement increased by the ratio $L/L-l$ whilst the GM in this case is that of the lost buoyancy case reduced by the ratio $L-l/L$ giving, as would be expected, identical righting moments by the two treatments. The formulae for ship shaped vessels are naturally more complicated but are of the same order and the righting moments are again identical.

A rectangular block floating in fresh water is considered for simplicity

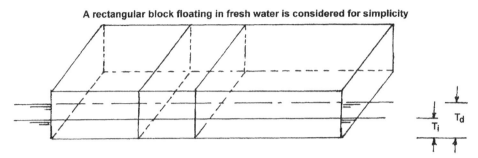

Ship length = L
Breadth = B
Damanged compartment length = l
Intact draft = T_i
Intact displacement = D_i
Damaged draft = T_d
VCG of intact ship = KG_i

	Added weight	Lost buoyancy
Draft	$T_d = T_i \cdot \dfrac{L}{L-l}$	
Displacement	$\Delta_d = L \cdot B \cdot T_d$	$\Delta_i = L \cdot B \cdot T_i$
VCG	$KG_d = KG_i + (T_d / 2 - KG_i) l / L$	KG_i
KM	$= \dfrac{T_d}{2} + \dfrac{1/12 \cdot L \cdot B^3}{L \cdot B \cdot T_d}$	$= \dfrac{T_d}{2} + \dfrac{1/12(L-l)B^3}{(L-l)B \cdot T_d}$
Free surface	$= \dfrac{1/12 \cdot l \cdot B^3}{L \cdot B \cdot T_d}$	Nil
GM	$= \dfrac{T_d}{2} + \dfrac{B^2(L-l)}{12 T_d \cdot L}$ $- KG_i \dfrac{(L-l)}{L} - \dfrac{T_d}{2} \cdot \dfrac{l}{L}$ $= \left\{ \dfrac{L-l)}{L} \right\} \left\{ \dfrac{T_d}{2} + \dfrac{B^2}{12 T_d} - KG_i \right\}$	$= \dfrac{T_d}{2} + \dfrac{B^2}{12 T_d} - KG_i$
Righting moment	$\Delta_d \cdot GM$ (added weight) $= \dfrac{L}{L-l} \cdot \Delta_i \cdot \dfrac{L-l}{L} GM$ (lost buoyancy)	$\Delta_i \cdot GM$ (lost buoyancy)

Fig. 12.8. A comparison between added weight and lost buoyancy (constant displacement) methods. A rectagular block floating in fresh water is considered for simplicity.

12.7 THE STABILITY OF WARSHIPS — GENERAL AND INTACT STABILITY

12.7.1 General

Stability standards for warships are of course a matter for individual navies and their technical staffs. However the 1962 paper "Stability and Buoyancy Criteria for U.S. Naval Surface Ships" presented by Sarchin and Goldberg to S.N.A.M.E. in 1962 has almost become an accepted standard worldwide. It should be read in full by all designers, but a condensed version of it supplemented by some extracts from a more recent comprehensive statement on the stability standards required for British warships and naval auxiliaries laid down in Naval Engineering Standard (NES 109) is given in the following paragraphs.

Although both Sarchin and Goldberg's paper and NES 109 were written with warship stability in mind, both give guidance which is equally applicable to merchant ships — particularly those with unusual specialist tasks.

External hazards which may affect an intact ship are:
(i)   beam winds combined with rolling,
(ii)   lifting of heavy weights — particularly off the ship's centreline,
(iii) crowding of passengers to one side,
(iv)   high speed turning,
(v)   topside icing.

Hazards which may cause internal flooding are identified as:
(i)   stranding,
(ii)   collision,
(iii) enemy action.

The safety of a flooded ship involves considering how the damaged ship will react to:
(i)   beam winds combined with rolling, and whether
(ii)   progressive flooding, may follow.

The criteria suggested for intact stability are primarily dynamic with the following limits set for each of the different cases.

12.7.2 Intact stability — the beam wind and rolling case

As shown in Fig. 12.9, the requirements are:
(i)   a limit on the heeling lever at the angle of steady heel (C) of 0.6 of the max. righting arm;
(ii)   a requirement that the area (A_1) between the righting lever and heeling lever curves should be not less than 140% of area (A_2), where A_2 is defined as the area between the heeling lever and the righting lever from an angle

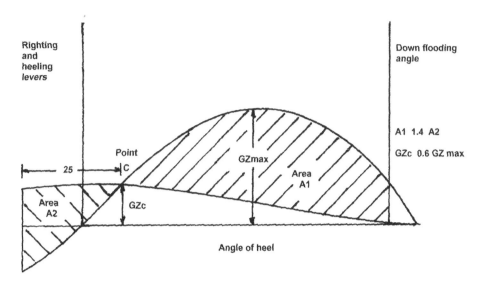

Fig. 12.9. Warship stability versus wind heel.

of 25° through upright to the angle C, i.e., the ship is assumed to have rolled 25° to windward of the angle of steady heel and A_2 is a measure of the kinetic energy as the ship rolls back through C. The margin of 40% of A_2 is intended to take account of gusts, calculation inaccuracies and make sure there is no capsize.

The heeling lever for a beam wind is given by:

$$L = \frac{1}{2} \frac{r \cdot C_{dy} \cdot V^2 \cdot A \cdot l}{g \cdot \Delta} \cos^2 \theta \text{ metres}$$

where

r = density of air = 0.00123 tonnes/m^3
g = 9.82 m/s^2
C_{dy} = lateral drag coefficient
A = projected sail area in square metres
l = distance from centre of sail area to centre of lateral resistance (half draft) in metres
V = wind velocity at centre of sail area in knots
Δ = displacement in tonnes

Sarchin and Goldberg suggested the use of various wind velocities for different types of ships and these seem to have settled down to the use of 90 knots for vessels

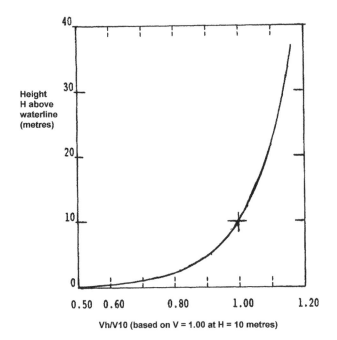

H	Vh/V10	(Vh/V10)^2
37.5	1.173	1.376
32.5	1.154	1.332
27.5	1.132	1.281
22.5	1.106	1.223
17.5	1.073	1.151
12.5	1.029	1.059
7.5	0.963	0.927
2.5	0.820	0.672

Fig. 12.10. Wind velocity ratio (10 min average) versus height above waterline.

which must be expected to weather the centre of tropical disturbances through 70 knots for those which will be expected to avoid these, to 50 knots for vessels which would be recalled to protected waters if winds over force 8 are expected. Velocities 10 knots greater than these figures are, however, often required as a basis for design.

The wind varies with height above the sea surface and the general practice is to use the nominal wind velocity which occurs at 10 m above the water surface as a reference point.

Sarchin and Goldberg's figures for the variation of the wind velocity with height above the water surface has now been replaced by better information, notably as given in R.I.N.A. Maritime Technology Monograph No. 8 by R.W.F. Gould. The modern view is that the wind gradient depends both on the surface roughness of the sea and the averaging time used to obtain the profile.

For practical calculations it is convenient to divide the sail area into 5 m wide horizontal strips, each with its lever about the centre of lateral resistance. As the velocity is squared in the formula it is convenient to multiply the $A \times l$ for each strip by its velocity ratio squared and sum the total to obtain a vertical moment of area corrected to the wind velocity at 10 m. A table for this calculation is shown in Fig. 12.11.

Layer above WL	Layer Description	Dimension s L×D	Area	Wind Gradient Coefficient squared (Vh/V10)	Corrected Area	Vertical Lever from Centre Lateral Resistance	Vertical Moment of Area Corrected
0–5 5–10 etc. etc.							
	Total						

Fig. 12.11. Wind heel moment calculation versus lateral wind area and moment.

A graph of the velocity at different height as a ratio to the velocity at 10 m is given in Fig. 12.10, and annexed is a table of velocities and velocity squared ratios for the centre of the 5 m strips suggested above. The figures given are based on a 10 minute average and it may be noted that the corresponding figures (of velocity squared ratio) for a one hour average would be about 4% more, whilst those for a one minute average would be 5% less.

Caution must be exercised in the choice of value for C_{dy}, as it is known that some wind tunnel operators derive their figures in association with the free stream velocity rather than that at the model scale height of 10 m. The British Ministry of Defence guidance on heeling calculations uses $C_{dy} = 1.16$. Other data suggests a figure of 0.9 as appropriate to a Ro-Ro Ferry.

12.7.3 Intact stability — three other considerations

Case (1) Lifting a weight off centreline
Case (2) Passengers crowding to one side
Case (3) High speed turn

These are all as illustrated in Fig. 12.12. The criteria, which differ slightly between the cases and also between the requirements of Sarchin and Goldberg and those of NES 109, are summarised in Table 12.1. In this table the angle of heel is as indicated by point C. The heeling lever GZ_c at the angle of steady heel not to exceed the fraction of GZ_{max} shown.

The reserve of dynamic stability as denoted by the area A_r lying between the curves of righting and heeling levers up to the down flooding angle is not to be less than the fraction quoted of the total area A_t under the curve of righting levers.

The heeling lever for lifting a weight off the centreline, or for passengers crowding to one side is:

$$L = \frac{W \cdot a \cdot \cos\theta}{\Delta}$$

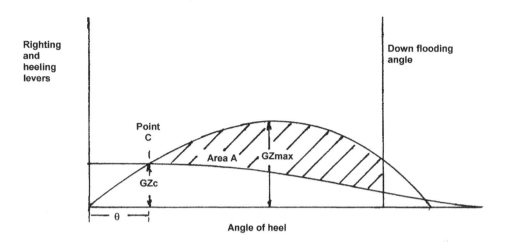

Fig. 12.12. Warship stability.
(a) During high speed turn.
(b) When lifting weights off centreline.
(c) When passengers crowd to one side.

Table 12.1

Case	Angle of heel (°)		GZ_c/GZ_{max}		Reserve of stability area A_r/A_t	
	S&G	NES	S&G	NES	S&G	NES
(1) Lifting off centre	15°		0.6	0.5	0.4	0.5
(2) Passengers crowding to one side	15°		0.6		0.6	0.4
(3) High speed turns	10° (new ships) 15° (in service)	20°	0.6		0.4	

S&G = Sarchin and Goldberg; NES = Naval Engineering Standard.

where
  W = weight being lifted or weight of passengers, in tonnes
  a = distance from centreline in metres to end of derrick or to C.G. of
    passengers
It should be noted that in the case of the weight being lifted the righting lever curve
should allow for the effect this has on the ship's VCG.

For a high speed turn, the heeling lever due to the centrifugal force acting on the ship is:

$$L = \frac{0.264 \cdot V^2 \cdot a \cdot \cos\theta}{g \cdot R}$$

where

V = velocity in knots

a = distance between the ship's VCG and the centre of lateral resistance (half draft) in metres

g = acceleration due to gravity M/sec^2 (0.981)

R = radius of turning circle in metres

θ = angle of inclination, degrees.

12.7.4 Intact stability in general

Sarchin and Goldberg relate all their stability criteria to the hazards mentioned earlier, but intact stability criteria similar to the merchant ship criteria given in §12.1.2 (*q.v.*) but with higher standards are now used by the British and a number of other navies. These require:

(a) Area under the *GZ* curve to be not less than:

      0.08 mrad up to 30°

      0.133 mrad up to 40°

      0.048 mrad between 30 and 40°

(b) GZ_{max} not less than 0.30 m

(c) Angle of GZ_{max} not less than 30°

(d) *GM* fluid not less than 0.30 m

(e) Range of stability to be as large as possible with 70° being the minimum design aim.

12.7.5 Intact stability — topside icing

Although Sarchin and Goldberg deal with topside icing their treatment of this subject is not very detailed and this paragraph draws on other data. Ships designed to operate in waters where icing may occur must meet the rolling and beam sea criteria with the added top weight of whatever thickness of icing is considered appropriate to the operational area. Once ice starts to form it will continue to accumulate unless ice removal measures are taken. The ice thickness the ship must be able to meet therefore depends on what provision is made for ice removal. Although the build up of ice will be greater on horizontal surfaces than on vertical ones it is usual to calculate the weight of ice on the basis of a uniform ice thickness, with 150 mm being the usual value, with the specific gravity of ice taken at 0.95.

The effect on the ship's profile of the added ice is generally ignored. NES 109 allows the wind heeling lever for ice conditions to be based on a wind speed reduced to 70% of the figure applied to intact ships quoted in §12.7.2 and sets the following criteria.

(a) Area under *GZ* curve to be not less than:

   0.051 mrad up to 30°

   0.085 mrad up to 40°

   0.03 mrad between 30 and 40°

(b) GZ_{max} to be not less than 0.24 m

(c) Angle of GZ_{max} to be not less than 30*

(d) *GM* fluid to be not less than 0.15 m

12.8 WARSHIP DAMAGED STABILITY

The damage that a merchant ship must be able to survive which has been described in earlier paragraphs is generally of an accidental nature caused by collision or stranding. After such damage the need for a merchant ship is to remain afloat, and as help can usually be expected within a short time, it is not essential that the ship should continue to be operational.

Whilst damage to a warship may also be accidental from the same causes, it is much more likely to be the result of enemy action. Moreover, continuing enemy action seeking to cause further damage is very likely. In addition there will almost certainly be continuing operational needs.

Whilst after damage the prime need is, as with a merchant ship, to remain afloat, coming close behind this is the need to be able to proceed and manoeuvre under the ship's own power followed by a need to be able to fire weapons to counter enemy action as the best way of avoiding further damage.

From an operational point of view, quick recovery from a large angle of heel is of great importance.

Sarchin and Goldberg (and NES 109) divide warships into three groups based on length and suggest that:

(i) those up to 30 m should be able to meet damage to any single compartment,

(ii) those between 30 and 92 m should meet damage to any two adjacent compartments,

(iii) those of over 92 m should meet a damage of 15% *L* or 21 m, whichever is greater.

This last requirement generally equates to a three-compartment standard for frigate sized ships, but may mean four or even five compartments in a big ship such as an aircraft carrier.

In general the most severe condition will arise from damage to a number of compartments all forward or all aft of amidships as shown in Fig. 12.13.

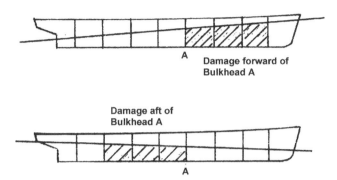

Fig. 12.13. Most severe damage conditions. Cross-hatched compartments are flooded.

The damage, unlike that in the deterministic rules for merchant ships, is assumed to extend laterally without limit, unless a more severe condition results from a reduced penetration. Similarly, it is assumed to be of unlimited vertical extent, unless intact buoyancy (say in a double bottom) results in a more severe condition.

The permeabilities to be used in calculations are generally as those given in §12.5.

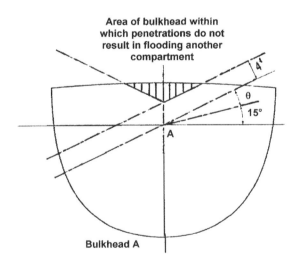

Fig. 12.14. Effect of static heel of 15°, an angle of roll of θ and a 4-ft (1.22 m) wave height (above still water) on flooding. Point A corresponds to the deeper of the two trimmed waterlines as shown in Fig. 12.13.

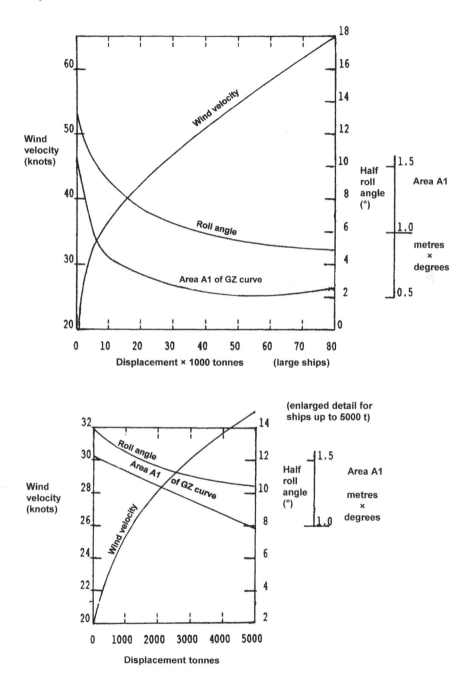

Fig. 12.15. Warship stability. Roll angle and wind velocity assumptions required area A_1 under GZ curve for a given displacement.

Sarchin and Goldberg give three criteria setting conditions which determine whether a ship can reasonably be expected to survive damage and some degree of wind and waves. These are:

(i)   ship is assumed to have a static heel of 15° due to asymmetrical flooding;

(ii) ship is assumed to be rolling to an angle θ; values of θ are given plotted against displacement (these represent roll amplitudes of various sizes of ship in 4 ft waves and are reproduced as Fig. 12.14);

(iii) a rise in the water level at any critical flooding point of 4 ft corresponding to waves of this height.

The combination of these three conditions gives rise to flooding of the form shown in Fig. 12.14. One interesting thing that this diagram shows is the possibility of there being an area near the centreline free of flooding where penetrations in bulkheads may be permitted to facilitate the running of the many systems which must be run fore and aft on a warship. It is somewhat surprising to find warships being permitted penetrations in watertight bulkheads that would not be allowed by merchant ship rules, but it appears to be an eminently practical idea.

The wind velocity which it is considered a damaged ship should be able to withstand is set somewhat lower than that required for intact ships and is also shown in Fig. 12.15.

The damaged stability criteria set in NES 109 are as follows, all as illustrated in Fig. 12.16.

(i)   angle of list or loll < 20°,

(ii) *GZ* at point *C* < 60% GZ_{max} (to down flooding angle or 45° if less),

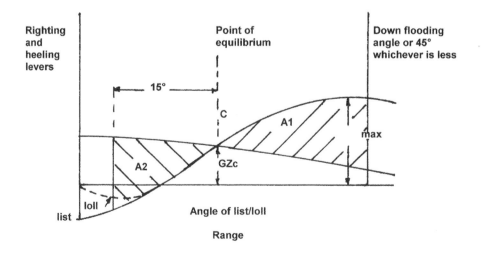

Fig. 12.16. Warship damaged stability.

(iii) area A_1 > value given in Fig. 12.14,
(iv) area A_1 > 1.4 × area A_2,
(v) longitudinal trim must not cause down flooding,
(vi) longitudinal GM > 0.

12.9 TRIM

12.9.1 General discussion

The other subjects dealt with in this chapter have all been governed by statutory rules. There are no such rules in relation to trim, but it seems appropriate to deal with this subject in close conjunction with stability as the one set of calculations usually deals with both subjects.

12.9.2 Trim and stability booklets

Two alternative methods are in general use in the presentation of trim and stability conditions. These may be presented either as extreme conditions with departure conditions having maximum fuel, fresh water and stores and arrival conditions having minimum quantities of these consumables or alternatively the conditions may be presented on a round voyage basis simulating the ship's operation as closely as possible.For liner type operations the latter presentation is probably the better but for other operational patterns the former is usually adopted.

12.9.3 Trim fully loaded

Most ships are designed with the intent that they will float on a level keel when fully loaded. One advantage of this is that it keeps the draft to a minimum; another is that the power required for the designed speed is usually lowest if there is no trim. There are, however, exceptions to both these statements.

The first exception brings in the concept of the change in trim which is sometimes brought about by the ship's speed. Full lined ships appear to trim by the head when under way, and therefore from the point of view both of minimum draft and minimum power it may be better for the ship to have some trim by the stern in the static condition. Possibly more important than either of these considerations is the fact that a ship with trim by the head tends to be directionally unstable.

Small ships are frequently designed to have a trim in the loaded condition or a raked keel. This is generally arranged to enable a larger diameter propeller to be fitted than would be possible with a level keel and as it is part of the designer's intention from the start the waterlines are arranged to suit.

The factors which determine the ease or otherwise with which a desired trim can be met in the fully loaded condition with homogeneous cargo are the LCB position at the load draft; the lightship weight and LCG; the cargo deadweight and its LCG and the weights and LCGs of oil fuel, fresh water and stores.

The LCB position will generally be arranged to suit minimum powering; and whilst both the lightship weight and the cargo deadweight and their respective centres can be altered in the early stages of the design, they do tend to be fixed by other considerations and it is altering the disposition of the oil fuel and the fresh water that provides the easiest method of making adjustments in the design to suit trim should this be necessary.

This is particularly easy in ships like bulk carriers and tankers in which it is usual to have one large fuel bunker forward and one large bunker abaft the cargo space. If an excess of fuel capacity of about 50% is provided this should give an ability to trim that will meet most eventualities. Care must be taken, however that the cargo disposition is such that the arrival trim with minimum oil and water is also satisfactory although this can usually be achieved without too much difficulty with the use of water ballast in the trimming tanks.

12.9.4 Ballast trim

Two main factors govern trim in the ballast condition. The first of these is the desirability of having a forward draft which is sufficiently great to avoid, or at all events to minimise, slamming.

For tankers and bulk carriers which tend to undertake frequent ballast voyages and have a full hull form making the avoidance of slamming something that required careful consideration, the aim used to be the provision of a draft forward of about $0.035\ L$ — this slightly generous figure being adopted because of the ease with which a large ballast capacity could be provided on these ships.

More recently, with the advent of the requirement for segregated ballast tanks in tankers, it is no longer reasonable to aim for as deep a draft as this and the MARPOL requirement is that the segregated ballast capacity should be such as will give a mean draft of $2 + 0.02\ L$ with a trim not exceeding 1.5% L.

For other ship types with finer lines, less frequent ballast voyages, and on which the provision of ballast capacity is less easy a forward draft of $0.025\ L$ may be adopted.

The second consideration is that of propeller immersion. In the past designers used to aim to provide about 0.3 m over the propeller tip, or at very least to immerse the propeller tip. This is still quite a general aim on smaller ships, but can result in a considerable trim and a high ballast displacement on larger ships.

Buxton and Logan in an excellent 1986 R.I.N.A. paper "The ballast perform-ance of ships with particular reference to bulk carriers" demonstrated very clearly that the reduction in EHP obtained as a result of reducing the displacement more than offsets the reduction in propeller efficiency provided about 90% of the propeller is immersed.

The point is made that additional immersion can be expected from the stern wave when under-way. Although recommending a light draft for fair weather ballast voyages, these authors wisely suggest incorporating in the design an ability to increase the ballast displacement to about 55% of the load displacement in heavy weather. As in practice the heavy weather deep ballast condition is usually obtained by flooding a cargo hold this deep ballast condition need not involve any reduction in cargo capacity.

Chapter 13

Other Statutory Rules

In addition to the rules discussed in previous chapters, there are a number of other statutory rules with which some ships must comply. Although all of these impact in some way upon the outfit, the arrangement and/or the specification, only a few of these have a significant effect on ship design meriting treatment in this chapter.

13.1 FIRE PROTECTION

13.1.1 Zones

Ships carrying more than 36 passengers are required to have the hull, superstructure and deckhouses divided into main vertical fire zones by "A" class divisions. Steps and recesses in these divisions are to be kept to a minimum and constructed as "A" class divisions. The mean length of a zone on any deck is not to exceed 40 metres. The rules require that fire zone divisions above the bulkhead deck should, as far as practicable, be in line with watertight subdivision bulkheads below — a rule that designers are happy to follow as it minimises weight and cost.

13.1.2. A class divisions.

"A" class divisions are to be constructed of steel or equivalent, suitably stiffened and insulated so as to be able to meet a fire test of one hour duration. "A" class divisions are divided into A0, A15, A30 and A60 according to the fire risk within the adjacent compartments. A0 is uninsulated steel, with each of the others having different amounts of fire resistant insulation. By arranging spaces such as lavatories which have a low fire risk and therefore do not require insulation adjacent to the "A" class divisions the cost and weight of these can be minimised.

The boundaries of certain other spaces which either contain items giving rise to a particular fire risk or whose function is specially important to the operation and safety of the ship are also required to be constructed as "A" class divisions. In the first category are machinery bulkheads and casings and casings surrounding galleys. In the second category are control stations such as the wheelhouse, chartroom, radio room and stairways with their important role in escape. Other divisions within the zones are required to be "B" or "C" class.

13.1.3 Means of escape

At least two separate means of escape are to be provided from each watertight compartment below the bulkhead deck and from each main fire zone above the bulkhead deck. At least one of the escapes below the bulkhead deck shall be independent of watertight doors and at least one of the escapes above the bulkhead deck shall give access to a stairway forming a vertical escape.

13.1.4 Other fire rules

Although the foregoing are probably the most important rules from a design point of view, there are many other detailed requirements with which a passenger ship naval architect must make himself familiar.

The rules for cargo ships are broadly similar.

As well as requirements for fire protection the rules lay down requirements for fire detection and extinguishing with some special provisions applying to tankers and ships carrying dangerous cargoes.

13.2 LIFESAVING APPLIANCES

13.2.1 Passenger ship requirements

Passenger ships on International voyages are required to carry lifeboats on each side with a total capacity of not less than 37.5% of the complement plus life-rafts for 50% of the complement, with these being served by at least one launching device on each side of the ship.

For passenger ships on short international voyages which are designed to comply with a special standard of subdivision, the boatage required may be reduced to 30% of the complement on each side plus life-rafts for the remainder plus 25% extra.

13.2.2 Cargo ship requirements

Cargo ships are required to carry lifeboats on each side capable of accommodating everyone on board plus life-rafts, capable of being launched on either side also able to accommodate the complement. Alternatively, they may be fitted with one or more lifeboats capable of being free-fall launched over the stern and able to accommodate the complement plus life-rafts on each side of the ship again capable of accommodating the total number on board.

The rules set standards for the lifesaving appliances, their stowage, launching and embarkation.

13.3 MARPOL — MARINE POLLUTION RULES

13.3.1 General discussion

The International Conference on Marine Pollution 1973 set a number of rules that apply to all ships, limiting the discharge of harmful substances. For ships other than oil tankers the rules have limited design implications, the main requirements being the provision of oily water separators, sewage disposal plants and waste incinerators.

13.3.2 Rules for oil tankers

For oil tankers the Marpol rules were however the start of major design changes. These were extended in the 1978 International Conference on Tanker Safety and Pollution Prevention, and modified by United States Government legislation, whilst at the date at which this is being written, further changes are under consideration by IMO.

The 1973 and 1978 Conferences laid down rules both for existing ships and for new designs, with the requirements for the latter being more severe. The rules are intended to reduce the pollution which the ship might cause during its normal operation and that which might result from an accident to the ship such as stranding or collision.

13.3.3 Two types of water ballast

Two types of water ballast were defined:
  - "clean ballast" (CBT) is ballast carried in a cargo tank, which since it last carried oil had been so cleaned that ballast discharged from it would be practically oil free (15 parts per million),
  - "segregated ballast" (SBT) is ballast carried in tanks completely separated from the cargo oil and oil fuel system and permanently allocated to the carriage of ballast or non-noxious substances.

"Slop tanks" are tanks specifically designated for the collection of tank washings and other oily mixtures.

13.3.4 Crude oil washing

"Crude oil washing" (COW) is the method used to clean cargo oil tanks whilst the cargo is being discharged to make tanks suitable for loading water ballast. As the name implies, it uses the cargo oil itself jetting this onto the structure through high pressure rotating jets positioned to provide effective washing of all areas.

As crude oil washing brings with it a risk of explosion a further associated requirement is the fitting of an inert gas system.

Existing ships were generally converted to CBT and COW.

13.3.5 Segregated ballast

New designs of crude oil tankers of 20,000 dwt and above and new product carriers of 30,000 dwt and above are required to be provided with segregated ballast tanks (SBT) of a capacity that enables the ship to comply with the following require-ments — based on lightweight plus ballast only.

(i) $T = 2.0 + 0.02\ L$ metres (moulded, amidships)
(ii) trim $< 0.015\ L$
(iii) propeller fully immersed.

Although the intention is that ballast is only carried in cargo tanks on rare voyages when weather conditions are so severe that additional ballast is necessary for safety, the cargo tanks of crude oil tankers are required to have a COW system fitted.

To minimise the outflow of oil in the event of stranding or collision, segregated ballast tanks are to be arranged so that the area of side and bottom shell which contain segregated ballast is to be a proportion (J) of the area of the side and bottom shell of the cargo tank length.

$$[PA_c + PA_s] > J[L_t(B + 2D)]$$

where

PA_c = area in m^2 of side shell in ballast spaces
PA_s = area in m^2 of bottom shell in ballast spaces
L_t = cargo tank length
J = 0.45 for 20,000 dwt
  = 0.30 for 200,000 dwt;

interpolate for deadweights between 20000 and 200000. For ships over 200.000 dwt, J is to be taken as either 0.20 or the figure given by a formula which takes into account the likely outflow if this is greater.

13.3.6 Double bottom and double skin

Following the Exon Valdez disaster, the United States government introduced a requirement that all ships trading in their waters should have complete double skin protection.

This requirement poses problems in inspection and maintenance and there are indeed also worries about whether it is the complete answer. A double bottom must of necessity be of limited depth and a double skin must be of limited width. Provided they are not breached this combination will effectively stop pollution, but a stranding is quite likely to penetrate the double bottom whilst a collision may well penetrate the double skin.

Considerations like these have led to the development of other designs such as the mid-deck tanker. This concept involves the provision of a deck which divides the cargo tanks horizontally at a level which ensures that if the lower tank is breached the hydrostatic pressure of the sea outside exceeds that of the oil in it.

In theory, water should therefore enter the tank and no oil should escape; but in practice the ship's motions may well modify this and cause oil leakage to occur.

The position of the deck is, however, such that the chance of its being breached is remote, whilst the fact that for the same ballast capacity the side tanks can be wider than those in a tanker with a double bottom reduces the risk of the these being breached.

Protagonists of both systems are vociferous in favour of their own ideas and it is too early to be sure what will be permitted by new rules in the making and will become general practice in the future.

13.4 TONNAGE RULES

13.4.1 General discussion

The current tonnage rules are set out in The Merchant Shipping Tonnage Regulations 1982. As well as giving the new tonnage rules this booklet also gives the old rules which can still be applied to certain ships. The following sections deal with the new rules although there are some comments on the relationships that these have with the old ones.

13.4.2 Gross tonnage

The formula for this is:

$$GT = K_1 V$$

where V = total volume of all enclosed spaces in cubic metres
$K_1 = 0.2 + 0.02 \log_{10} V$

The first thing to note about this is that V for steel ships is to be measured to the inner surface of the shell; the second thing to note is that there are no exclusions.

These contrast with the old rules in which the tonnage was measured to the face of sparring and in which a number of spaces were exempt (chartroom and wheelhouse, chain locker, washing and sanitary spaces for Master and crew, galley, water ballast tanks etc.)

A modern ship with a total capacity of 100,000 m³ would have a K_1 value of 0.3 and a gross tonnage of 30,000.

The total tonnage volume of such a ship under the old rules might have been multiplied by about 0.88 to allow for measurement inside sparring whilst a further 5% or so might have been exempt giving a capacity in cubic feet of 100,000 × 35.316 × 0.88 × 0.95 = 2,952,000 and a corresponding gross tonnage of 29,520.

It may be noted that the formula for K_1 changes with ship size from 0.22 for a small ship of 10 m³ capacity to 0.32 for one of 1,000,000 m³. This seems reasonable when the considerable impact which exclusions under the old rules had on small ships and the relatively much smaller impact on large ships is considered.

The simplification is of course greatly to be welcomed.

13.4.3 Net tonnage

The formula for net tonnage is:

$$NT = K_2 \, V_c (4d/3D)^2 + K_3(N_1 + N_2/10)$$

where
   V_c = total volume of cargo spaces in cubic metres
   $K_2 = 0.2 + 0.02 \log_{10} V_c$
   $K_3 = 1.25 \, (GT + 10000)/10000$
   N_1 = number of passengers in cabins with not more than 8 berths
   N_2 = number of other passengers
   d = draft
   D = depth
There are a number of provisos:
   (a) that $(4d/3D)^2$ shall not be greater than unity,
   (b) that the whole of the first term shall not be taken as less than 0.25 GT,
   (c) that N_1 and N_2 shall be taken as zero when $N_1 + N_2$ is less than 13,
   (d) that NT shall not be taken as less than 0.30 GT.
The factor involving the ratio of draft/ depth is interesting as is the proviso that in effect means that the ratio shall not be taken as more than 0.75. The factor gives a tonnage reduction to ships with a relatively low draft and therefore deadweight — a reasonable recognition of their probable lesser earning ability. The proviso avoids penalising ships with A or B-60 freeboards.

Whilst the first term takes account of the cargo carrying ability, the second adds in an allowance for the earning power of passenger spaces.

The biggest change from the old rules is the disappearance of the allowance for the propelling machinery space. The original function of this, introduced in the early days of steamships, was the encouragement of the fitting of machinery of adequate power in engine rooms with reasonable space. In the new rules it was recognised that the need for such an incentive had long since ceased.

13.4.4 Segregated ballast in oil tankers

An entry may be made on International Tonnage Certificate indicating the total tonnage of these tanks. The formula is:

$$ST = K_1 V_b$$

where V_b = total volume of segregated ballast tanks in m^3, and K_1 is as already defined.

13.4.5 Deck cargo

Where required for the payment of dues any space so used shall be measured and the additional tonnage computed as follows:

$$DT = 0.353 \times \text{mean length} \times \text{mean breadth} \times \text{mean depth (all in metres)}$$

13.4.6 Tonnage as a factor in design

As mentioned in Chapter 2, tonnage used to be quite an important factor in ship design, particularly in the case of small ships, but today it is of much reduced significance.

Chapter 14

Special Factors Influencing Warship Design

14.1 WARSHIP ROLES

14.1.1 General discussion

Section 2.6 discusses some of the problems involved in setting the staff requirements for a warship, but some further discussion on the wide variety of roles that a warship may have to undertake seems appropriate as a lead in to consideration of the special factors influencing warship design.

Whereas most merchant ships have one, or at the most two or three roles, a warship may be asked to undertake a whole series of different tasks ranging from completely peaceful ones to those occurring in full hostilities. The following list of roles is abstracted from a paper by John Sadden and S. McComas entitled "Modern corvette design and production" presented to IMEC 92 — a paper which is well worth studying for its insight into many aspects of warship design.

- courtesy visits,
- disaster relief and search and rescue,
- anti-smuggling operations,
- support to the Civil Power,
- protection and policing of an economic zone,
- anti-piracy operations,
- observation of hostile or potentially hostile forces,
- defence of merchant shipping,
- blockade of hostile ports,
- support to forces ashore by naval gunfire,
- destruction of surface ships,
- destruction of submarines,
- destruction of aircraft.

14.1.2 Capabilities required

To carry out all these roles requires a very comprehensively equipped ship with capabilities including:
  – surveillance both above and below surface,
  – a comprehensive command system matched to the likely threats,
  – anti-surface ship armament,
  – anti- submarine armament,
  – a gun for naval gunfire support,
  – effective defences against aircraft and missiles,
  – flexible communications,
  – low signatures,
  – good survivability,
  – seakeeping characteristics and endurance matched to the intended area of operations,
  – sufficient speed to ensure that the ship can be in the right place at the right time.

A lack of ability in any of these characteristics may mean a serious reduction in a ship's value at a time of crisis. On the other hand providing a high degree of capability in all of them is expensive. For larger navies specialised vessels can be considered for some specialist roles, but smaller navies will generally need to go for versatility in most, if not all of their ships.

The importance of the various roles will vary with the perceived threat. Every country has aircraft and missiles so self defence against these must be emphasised. Whilst the major navies see an anti-submarine capability as of major importance, not all nations have a submarine service so this capability may not be seen as important for one of the smaller navies, although this may well change if sales of second-hand submarines continue.

14.2 THROUGH-LIFE COSTING

14.2.1 Economic criteria for warships

Whilst for merchant ships the profitability of ownership can be established by the difference between the income from the freight carried and the operational costs incurred there is no such simple answer to the economics of warship ownership as it is quite impossible to put a value on the "income" side.

The expenditure side can, however, be estimated and this is generally done on a "through-life" basis, with a life of twenty years being the usual assumption.

Through-life costs start with the capital cost of the original design and construction (see Chapter 18) to which are added each year's operational costs (generally calculated as described in Chapter 19, although this was written for merchant ships) together with a number of other major costs which arise from the shore based facilities which are essential if the ship is to operate efficiently.

In a through-life costing estimate all these costs are discounted to a common time base to give a net present value.

14.2.2 Other through-life costs

The "other" through-life costs typically include:
1. A contribution to the cost of dockyards required for maintenance and in particular to the cost of providing any new facilities such as a new dry dock or syncrolift which may be required if the existing facilities are inadequate, for any reason, for a new design.
2. The cost of providing crew training and especially any training required for new equipment being introduced into service.
3. The cost of purchasing and storing spares and special tools required for maintenance and particularly those items needed for equipment not already in service.
4. The estimated cost of refits during the service life and in particular that incurred in the mid life up-date if this is included in the life cycle plan.
5. A *pro rata* contribution to all naval research, training, administrative and other overheads.

The first of these costs clearly provides a major incentive to keep a new design within the scope of existing dockyard facilities unless the improvement in performance obtained by going outside these is so significant as to justify this expenditure — as for example in the change from Polaris to Trident submarines, where a step change in capability justified very major expenditure on new base facilities.

The third, and to some extent the second, provides an incentive to choose equipment already in use, unless the improvement in performance clearly outweighs the cost benefit of standardisation.

The fourth provides an incentive towards including a degree of spaciousness in a design to ease refit work and maybe particularly towards the adoption of modularity. Both of these will involve immediate extra costs but may result in significant savings in the future.

Through-life cost estimates can be used to assist in high-level decisions on the number of ships that a navy can run within a financial constraint: to evaluate the relative costs of different ship types, to determine which of a number of alternatives will be most advantageous when there is a complex interaction between capital, operational, and other costs.

14.3 COMBAT SYSTEMS

14.3.1 General discussion

As a warship's capability resides in the combat systems with which it is fitted together with the effectiveness with which these can be used, the inclusion of a section on these is plainly necessary. The subject is, however, a highly specialised one in which the author cannot claim any great expertise, although working closely with weapons engineers on a number of warship designs has given him a reasonable knowledge about these systems and in particular about the interfaces that these have with ship design. As the literature on the subject is limited by security constraints, this section is included in the hope that readers lacking other guidance will find it helpful.

The term "combat systems" includes weapons, sensors, command and control systems, electronic warfare systems, internal and external communication systems and ancillary support systems.

14.3.2 Combat system costs

In a typical modern frigate the cost of the combat systems may amount to between 35 and 40% of the total unit production cost and, in addition, very large sums will have been spent on the development of weapon and sensor prototypes. The development time for these will often be greater than that required to design and build the ship, which raises an important question for a navy wanting an up-to date ship. Should the design incorporate weapons still under development with all the uncertainties that this will entail about performance, weight, space, services required, completion date, etc., or should the best available proven units be accepted to reduce shipbuilding time even though their performance will be below that which might be attainable from new developments?

14.3.3 Single role or multi-purpose

The difficulty facing naval staff when setting the design requirements for a warship in deciding whether to concentrate on single role such as anti-submarine or anti-aircraft or try to build a multi-purpose ship has already been mentioned. The single purpose ship will be cheaper and may be better at its primary task but whether the exigencies of war will permit it to be used only for this task can be a matter for keen debate.

The variety of targets that a warship may wish to attack together with the variety of opponents against which it must defend itself have led to the development of a wide range of weapons and sensors.

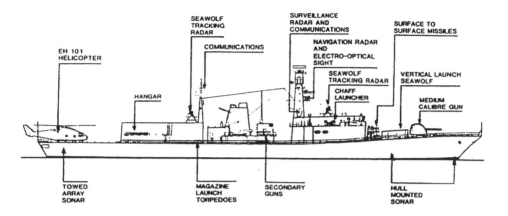

Fig. 14.1. Weapons fit on the Type 23 frigate.

In the following brief description of the weapons, some or all of which will be comprised in a frigate's weapon fit some outline performance data is given to colour the picture — this data being abstracted from a recent *Jane's*.

Figure 14.1 abstracted from a 1991 R.I.N.A. paper "The Type 23 Duke Class Frigate" by T.R. Thomas and M.S. Easton, shows the weapons fit on a very recent and particularly comprehensively armed frigate and provides a suitable background to a discussion of combat systems.

14.3.4 Anti-submarine weapons

Today the principal anti-submarine weapon is the helicopter. In the case of Type 23, the ship is designed to take a helicopter still under development — the new EH101 Merlin — but until this is ready a Lynx is being carried. This can carry 4 Stingray torpedoes or a number of depth charges and has electronic and sonar detection and tracking gear. It has a speed of 125 knots and a range of 320 nautical miles, enabling it to attack targets whilst they are still at a considerable distance from their parent ship.

The Merlin is very much larger and will have greatly improved performance in all respects. Because of its size a specially equipped landing deck is required and it and the hanger between them take up a major portion of the ship.

Submarines in closer proximity can be attacked by a number of underwater weapons such as torpedoes or depth charges which can be launched from the ship in a variety of ways. Type 23 ships have two twin torpedo tubes for launching Stingray anti-submarine torpedoes. These have a range of 5.9 nautical miles and travel at 45 knots.

14.3.5 Surface ship attack weapons

The primary weapon against other surface ships are surface-to-surface missiles (SSM), Exocet and Harpoon being examples of these. Those fitted on Type 23 are Harpoon with two quadruple launchers fitted just forward of the bridge. Harpoon has a range of 70 nautical miles.

A secondary weapon against surface ships, which can also play a part in anti aircraft defence, is a modern automatic medium calibre gun. This weapon has the advantage that it can be used in the minor engagements of peacetime such as anti-piracy, anti-terrorist or fishery protection duties, etc., where a missile would be inappropriate and indeed all that is required is the traditional "shot across the bow".

These guns range from 76 mm (3 in) to 126 mm (5 in), with firing rates of from about 25 up to 120 rounds per minute for the smaller calibre. Type 23 has a Vickers 4.5-in calibre gun with a range against surface targets of 11.9 nautical miles and in an anti-aircraft role of 3.3 nautical miles.

14.3.6 Anti-aircraft and anti-missile weapons

Surface-to-air missiles (SAM) can be divided into two categories: area defence missiles such as Seadart and point defence missile systems (PDMS), such as Seawolf. The salient features of these two systems are shown in Fig. 14.2. Between them, these systems are the main defence against both attacking aircraft and incoming missiles. Until fairly recently, these defence missiles were fired from trainable launchers which had to have large clear overhead arcs, necessitating very careful positioning of the launchers. Vertical launch missile systems now in use reduce the required clear arc to an overhead cone and have the additional great advantage that all the missiles in their storage position are "ready to fire" with no reloading required.

Type 23 is equipped with a silo of 32 vertical launch Seawolf missiles, with a range of 6 nautical miles.

14.3.7 Close-in weapon systems (C.I.W.S.)

This is a "last ditch" defence against incoming missiles or aircraft which have evaded or survived hits by the SAM systems. It consists of a number of self contained automatic Gatling type guns with a high fire rate and large built-in magazine capacity. Examples of this are PHALANX and GOALKEEPER. Other machine guns are sometimes fitted and Type 23 has two 30 mm Oerlikons capable of firing 650 rounds per minute with a range of 5.4 nautical miles in an anti-surface role and 1.6 miles in an anti-aircraft role.

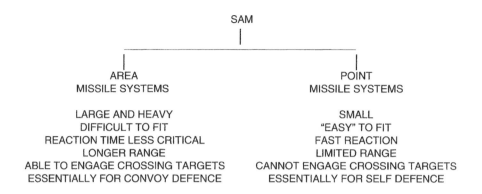

Fig. 14.2. The two different types of surface-to-air missiles.

14.3.8 Decoys

A number of systems are carried which do not actually attack the enemy but are designed to confuse him.

Above water these weapons launch material designed to confuse or jam the guidance systems of incoming missiles whether these are of radar or heat seeking type. Type 23 has two four-barrelled launchers for this duty.

Underwater decoys can either be launched from or towed by a ship.

14.3.9 Sensors

The information required to bring all these weapons on target are provided above water by radar/optical systems and below water by sonars.

Above water sensors must be positioned to have clear fields of view and must be kept out of the path of projectiles from the ship's own weapons.

As the close positioning shown on Fig. 14.1 demonstrates this is, of itself, a strong argument for the length of the "long thin" ship.

Sonars provide the information on submarines. There are two types — active and passive. The former has the disadvantage of letting a submarine know that a search for it is in progress, but was the more effective search tool and hull fitted sonars are commonly of this type. The recent development of the towed array, which consists of a very long chain of hydrophones towed some distance astern of anti-submarine frigates has made possible detection at long distances, but with the major snag that operation of the array requires very low ship speeds.

Type 23 has both a bow-mounted active sonar and a passive towed array, the latter deployed by a large winch.

Not shown on the plan of the Type 23 is the command and control centre or centres (because this function is so vital it is one there is a strong case for duplicating) nor the extensive network of communications bringing information from the sensors and carrying instructions to the weapons.

14.4 SIGNATURES

14.4.1 General discussion

The effectiveness of a warship is greatly increased if it can avoid being detected by enemy vessels, both surface ships and submarines or, if complete avoidance is impracticable, it can give the enemy a misleading impression of the vessel type and size.

Ships can be detected by sight, by noise, by infra-red emission, by radar and magnetic signatures. Collectively, all these indications of a ship's presence are called its signature and some are so distinctive as to identify the ships type or even class.

The importance of minimising signatures is possibly at its greatest in anti-submarine warfare, but it is also important in avoiding exploding mines and attack from the air. Reducing signatures not only makes ships more difficult to detect, but also improves the effectiveness of decoys.

14.4.2 Sight

The measures to minimise sight signature are the familiar ones of keeping the profile as low as other design constraints permit, the use of naval grey paint, of camouflage including fake bow and stern indications.

14.4.3 Noise

Noise originates mainly from machinery and propellers and travels both through air and water, although it is the latter transmission path which is the more important. The main measures to reduce machinery noise consist of flexible mounts for all machinery which may be noisy and/or a vibration source. Where noise reduction is particularly important, the main noise sources may be raft mounted and isolated from the hull with a double mounting system. When this is done particular attention must be paid to avoiding "shorts" through pipes or other attachments.

Minimising the noise from propellers is of particular importance in anti-submarine vessels and significant reductions can be achieved by careful attention to the propeller design with the changes required fortunately being generally in a direction which also improves propeller efficiency.

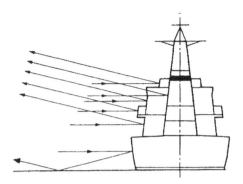

Fig. 14.3. Methods of reducing radar signature.

Underwater noise reduction in anti-submarine vessels not only decreases the submarine's ability to detect the ship but also increases the performance of the anti-submarine vessel's SONAR. Indeed in many cases the emphasis on noise reduction has this as its main motive.

14.4.4 Infra-red

Reduction in infra-red signature requires cooling the exhaust gases before they are emitted from the funnel, keeping hot machinery uptakes away from the funnel structure and insulating this. Insulation fitted on the ship's side in accommodation spaces also serves this function as well as improving habitability.

14.4.5 Radar

Reduction in radar signature is achieved by minimising the superstructure, sloping the ship's side and other surfaces away from the vertical, avoiding re-entrant corners and curved surfaces and by the application of radar absorbing materials. The way in which incoming radar beams are deflected away from the sender by sloping surfaces is shown in Fig. 14.3.

14.4.6 Magnetic

Measures to reduce magnetic signature by degaussing are taken on all warships, but more comprehensive measures avoiding the use of magnetic materials are usually only taken on mine-hunting vessels; on which ships reducing the magnetic signature is so important as to virtually determine the whole design and specification.

14.4.7 Active measures

All the measures mentioned so far are passive, but the active measures to decoy attacking weapons already described in §14.3.8 are also included in the term "susceptibility". Successful deception requires a combination of both the active and passive decoy measures (see also §9.1.12).

14.5 VULNERABILITY

14.5.1 General discussion

Measures to reduce a ship's vulnerability to damage are an important feature in warship design. Apart from the damage to which all ships may be subject through collisions, stranding and accidental fires and explosions, warships, in time of war, are subjected to deliberate attempts to damage them by the enemy, using the wide range of conventional weapons shown in Fig. 14.4, which is abstracted from the 1988 R.I.N.A. paper "The naval Architecture of Surface Warships" by D.K. Brown and E.C. Tupper, which should be read by everyone involved in warship design. In addition a nuclear attack may be possible, but obviously there is no protection against a hit or near miss from such weapons.

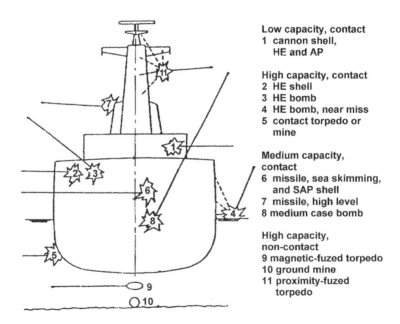

Fig. 14.4. The attacks to which a warship may be subjected.

There are, however, a number of steps that can be taken to provide protection against nuclear weapons detonated in the general area. Such measures include the strengthening of the superstructure to resist blast loads, the protection of electrical circuits against the large pulses of electromagnetic radiation associated with nuclear weapons and the sealing of the ship to prevent the ingress of fallout, a protective measure that also provides the best form of defence against chemical and biological warfare. Such sealing of the ship is usually further supplemented by a wash-down system designed to wash away fall-out and other contaminants.

In the measures to reduce vulnerability, designers aim to enable the ship to continue to "Float, Move and Fight", generally in that order of importance. Clearly, if the ship is going to sink an ability to move and fight is of little value; if it can no longer move it will be so easily attacked that any continuing ability to fight is unlikely to last for very long.

The measures which contribute to the achievement of these aims include efficient structural design, watertight subdivision and good damaged stability, fire divisions and minimisation/avoidance of the use of inflammable material, redundancy and separation of vital machinery and equipment, zoning and containment.

Structural design has been discussed in Chapter 10, watertight subdivision in Chapter 11, damaged stability in Chapter 12, and fire protection measures in Chapter 13.

14.5.2 Redundancy and separation

Vulnerability can be reduced by duplicating essential machinery, equipment and services and separating them widely. The machinery of a twin-screw ship can be so arranged that all the essential propulsion machinery for each shaft are located in different compartments, if possible with one or two intermediate compartments. Wide separation of the electric generators can be taken even further with some of these being located well away from the machinery spaces and even above the waterline and of course switchboards should be similarly separated. The alternative routes for electrical mains should be also be widely separated with one route possibly low down and near the centreline whilst another is well above the waterline.

14.5.3 Zoning and containment

The concept of zoning started with the need to limit the damage caused by nuclear, chemical and biological attacks by dividing the ship into a number of zones which in an alert situation could be effectively isolated from one another by boundaries which were fire-proof and blast- and splinter-resistant. To do this necessitates self contained electrical supply, ventilation and all vital services required both for personnel and for any weapons situated in the zone. Taking the concept further

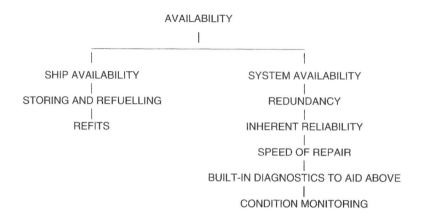

Fig 14.5. Contributors to ship and system availability.

brings the thought that all crew members should have their quarters in the zone in which they will be when in action — with an end to the traditional segregation of officers and ratings, but there is some doubt whether this will ever be found practicable.

14.6 AVAILABILITY

14.6.1 General discussion

High availability is a most desirable characteristic to build into a warship. It may mean that four ships can perform the role that would in the past have required five or six ships with a lower availability. As measures to improve availability are usually fairly inexpensive this argument can make them financially attractive.

Availability can be looked at in at least two ways — the achievement of ship availability in which the ship is able to be in the right place at the right time, and system availability in which the achievement is insuring that the ship can then do the task it is set. Each of these availability criterion raises different requirements as shown in Fig. 14.5.

Some of the main ways of increasing ship availability are given in the following sections.

14.6.2 By reducing the time spent storing and fuelling

Whilst the actual time spent in storing and fuelling may not in any case be very long, the need to store or refuel may require the ship divert to a port. Availability can therefore be considerably increased by either increasing the amount of stores and fuel carried or by equipping the ship with an ability to restore and refuel at sea — with, of course, the corollary of the need to have the necessary replenishment vessels.

14.6.3 By reducing the time spent in refits

This can be attacked in at least three ways:

(i) By the use of repair by replacement in which defective machinery or equipment is removed as a unit and sent for repair in a factory with its immediate replacement by a stock unit. If this procedure is to be successful it must have been planned into the ship design, so that there are easy removal routes. These will require more space but this is a small price to pay. Not all items justify repair by replacement and it is equally important to allow the good access to any equipment which is to be repaired *in situ*.

(ii) By designing parts of the ship to accept modules which can contain a complete system, enabling the ship to be comprehensively updated in a fraction of the time such a refit would normally take. This system was taken to its logical conclusion by the Royal Danish navy in their Standard-flex 300 class of vessel which can be converted from a missile patrol boat role to a mine hunter role in a matter of 24 hours or so. A typical module for this scheme comprises an Oto-Melara 75 mm gun complete with its barbette, requiring only a minimum of service connections to be ready to fire.

(iii) By the use of long lasting materials and of paints and systems such as cathodic protection that will improve the resistance to corrosion of the hull steel. Modern anti-fouling paints have so reduced marine fouling even in the tropics that the period between dry-docking can now be three years. Epoxy and similar paints ensure freedom from corrosion, with only a need for touch-up where there has been mechanical abrasion.

Chapter 15

The General Arrangement

15.1 INTRODUCTION

Although the general arrangement plan should be the embodiment of everything the naval architect wants to achieve in the design, a surprising amount of detail is often left to be decided by the draughtsman who draws the general arrangement plan. A draughtsman's skill and knowledge — usually learnt on the job — are not to be decried, and indeed their importance is one of the themes of this book; but the naval architect responsible for the design should take a detailed interest in everything that is shown on the general arrangement plan and the skills to enable him to do this should be taught at universities and technical colleges to all young naval architects as every bit as important as the more mathematical aspects of ship design on which academic courses tend to concentrate.

15.2 THE SCALE OF THE GENERAL ARRANGEMENT PLAN

The choice of the scale on which the general arrangement plan is drawn may seem a minor matter but is, in fact, quite important. The author strongly advocates the use of small scales, seeing the following advantages in their use.

1. With a small scale it becomes possible to draw several decks on one sheet of paper. This greatly eases the work of lining up items such as stairs and hatches that appear on two or more decks and helps to ensure that no inconsistencies develop between decks and that structural continuity is kept in mind throughout the design work.
2. A small-scale plan does not demand detail in the way that the big empty spaces of a larger scale plan seem to do. This saves time in drawing and avoids decisions being taken at an unnecessarily early stage in the design.

Where detail needs to be investigated this should be done on a small separate plan — probably using a larger scale than could be considered for the general arrangement.

3. Designs are nearly always drawn against a tight deadline and improvements requiring changes which involve rubbing out a lot of work can be very unpopular, so unnecessary detail tends to act against the introduction of improvements.

Whilst designing with the use of C.A.D. reduces the argument in favour of the use of small scales, the author thinks there are still cogent reasons for working on a small scale.

Another factor that should enter into the choice of scale is that of familiarity. It greatly eases the designer's task if the scale used is one in which he is used to working and in which he has good guidance drawings, preferably "as fitteds" of appropriate ships, readily available.

For ships of from about 100 m to about 200 m in length the best scale to use is undoubtedly 1:200. Below 100 m it is probably better to adopt a larger scale. For ships between 30 and 100 m the best scale is 1:100, and for ships of less than about 30 m a 1:50 scale becomes desirable. Above 200 m the choice is not so easy, there being fairly well balanced arguments between sticking to 1:200 — accepting that the number of decks on a sheet will be limited and moving to a smaller scale to preserve the advantages mentioned above even if this makes it difficult to convey the desirable level of detail.

15.3 FACTORS INFLUENCING THE GENERAL ARRANGEMENT

The feature which should have the greatest influence on the general arrangement is, of course, the main purpose of the ship and the designer's ideas on how this purpose can best be achieved.

For cargo ships the arrangement should be such that the type of cargo for which the ship is designed can be carried as cheaply as possible in stowage arrangements that ensure it is delivered in good condition and with methods of loading and discharge that are speedy and economical.

For passenger ships the arrangement should be such that the cabins, public rooms and the services provided to passengers will result in them so enjoying their voyage that they will want to travel again with the same company and will recommend the service to their friends.

For service ships the arrangement should be such that the ship is able to perform its service functions efficiently.

For warships the main driving force in the arrangement should be the positioning of each of the combat systems so that all its components will function at near to their optimum capability and, if possible, will continue to do so after enemy attack.

Features which are particular to individual ship types will be considered in the next chapter. The rest of this chapter examines features common to most ships.

It is worth commenting that, whilst the special features should drive the general arrangement, all the factors discussed in earlier chapters of this book must play a part in determining the optimum arrangement.

As a starting point, the general arrangement must be accommodated within and be built upon a lines plan developed to meet the deadweight, capacity, and speed requirements and which takes into consideration the needs of stability, trim, seakeeping and manoeuvrability. As well as the part they have played in the development of the lines plan, these factors must be given continuing attention throughout the development of the general arrangement.

The weight equation will have required some initial consideration of the structural design and the drawing of an outline midship section should usually be progressed in parallel with the general arrangement plan. A decision on frame spacing should be taken on an amalgam of structural and arrangement reasons. Whilst from a structural/weight point of view there will be sensible upper and lower limits, arrangement details may enter into the final decision with the spacing being adjusted so that bulkheads, hatches, etc. can be positioned on frame stations whilst also being located to suit arrangement aspects.

The ends of deckhouses should be positioned above bulkheads; the sides of deckhouses should be positioned to line up with deck longitudinals.

Ways of minimising building and operational costs (subjects that will be discussed in later chapters) must be considered.

Keeping all these factors continually in mind during the development of the general arrangement is not an easy task. In most designs conflicts will develop between the various factors and the success of the design depends on the skill with which these are resolved.

15.4 THE AESTHETICS OF SHIP DESIGN

A design consideration almost entirely related to the general arrangement and not dealt with elsewhere in this book is that of external aesthetics — the appearance of the ship.

It has to be admitted that an appreciation or otherwise of a ship's appearance is to some extent a matter of personal taste and with some exceptions it is also true that the fact that a ship has a good appearance makes no contribution to the "bottom line" so beloved of accountants. Obvious exceptions to this where good aesthetics have at least a small and possibly quite a large pay-off are cruise liners and passenger ferries.

Warships, with a secondary role as peacetime representatives of their nations, also merit having some attention paid to their looks.

Unfortunately, however, aesthetics seem nowadays to come very much at the end of the priorities for most merchant ships —something that was not always the case as shipbuilders and shipowners used to vie with their competitors in respectively producing and operating handsome ships.

15.4.1 Aesthetics in the past

Some of the features which at one time helped to make a ship look well had adverse effects on its economics either by increasing building cost or by reducing operating income. A parabolic sheer line, for example, is a lovely thing but increases fabrication costs and produces odd-shaped spaces that are not stowage friendly. A large midships accommodation house looks well, but the advantages of having the accommodation aft necessarily override aesthetic appeal.

The curved deckhouse fronts and tiered arrangement of successive decks were particularly beautiful features of passenger liners in the past, but in the interest of cramming in as much accommodation as possible modern cruise liners tend to have vertical "wall" fronts, although there are some pleasing exceptions such as the new P&O cruise liner "Oriana".

The positioning of masts and funnels and the shape and rake of these used to be given a lot of thought — with the rake of the foremast being increased on each successive funnel and the mainmast. Curiously enough, this gave the impression that these were parallel whereas if all were given the same rake they tended to look as though they were "falling together".

On warships curved bridge fronts and indeed funnels with curved plating have disappeared for the very good reason that these were found to be very good radar reflectors. New designs with flat sides to minimise radar reflection unfortunately do not have the same aesthetic appeal.

15.4.2 Aesthetics today

There are still a number of features which can, if carefully designed, help to give a ship a handsome appearance at almost no additional cost. Amongst these are:
 – the choice of colours for the hull and superstructure;
 – careful setting of the line dividing these colours; the use of contrasting colour bands;
 – the profile of the bow and stern, and how these match the lines and to some extent one another;
 – bulwark sweeps of attractive appearance and kept all to the same general pattern throughout the ship;
 – and even today, the positioning and shape of the funnel.

15.5 LOCATION OF THE PRINCIPAL FEATURES

Taking a decision on the locations to be given to various important features provides the starting point for most arrangement work. Probably the most important location decision on a merchant ship is that involved in fixing the relative positions of the cargo space and machinery with, as will be seen later, the crew accommodation generally being located above the machinery space.

15.5.1 Historical space location — cargo and machinery spaces

When the author started work in a shipyard in the 1940s, the engine room of most general cargo ships (and, for that matter, of most passenger ships) was located at or near amidships with two or three cargo holds forward and two aft. Tankers and small coasters were then exceptional in having their machinery aft and all the cargo space forward. The reason for having the machinery amidships at that date may have been partly tradition, but there was also the fact that in the size of ships then being built the space required for the large engines of the day could only be provided satisfactorily in an amidships location.

The big crews then carried required large deckhouses and these were arranged round the engine casings providing these with protection and giving the crew ready access to the engine room and locating the accommodation in a position subject to the least motion in a seaway.

The division of the cargo spaces by the engine room meant that satisfactory trim was easily obtained in almost any condition of loading.

15.5.2 Space location today — cargo and machinery spaces

Cargo ships today either have the machinery right aft with all cargo forward, or may have one, usually small, hold abaft the machinery space. The former arrangement is seen as the ideal arrangement when it is practicable and is universally used for bulk carriers and slower speed ships, although the space requirements of high powered machinery sometimes dictates the use of the latter arrangement for fast, fine lined ships such as container ships and refrigerated cargo ships.

The change from machinery amidships to aft came about partly through a realisation of the advantages to be gained and partly because the increasing size of ships and associated increase in the fullness of the lines, together with a reduction in size of main engines made such a change practicable.

The advantages sought consisted of a gain in cargo capacity and an improvement in cargo stowage. These were obtained partly by gaining as cargo space the best and most rectilinear space in the ship and partly by eliminating the need for a shaft tunnel which ceased to be necessary with no holds abaft the engine room. In addition, there was a reduction in construction cost with the reduction in length of shafting and the elimination of shaft tunnel structure.

The reduction in crew numbers occurring in the same time frame made the reduced accommodation area available round engine casings in an aft position acceptable.

One disadvantage of having the machinery aft is the greater difficulty there is in trimming the ship satisfactorily in all conditions of loading, but this can generally be overcome by a judicious arrangement of the tankage.

The development of lifeboats that can be launched from slipways in association with a change in the rules for lifesaving appliances which permits ships to be equipped with one such lifeboat launching over the stern in lieu of the former requirement for lifeboats to be fitted port and starboard and positioned clear of the propeller(s) has recently enabled the accommodation block to be moved even further aft. This can result in the main engine(s) being positioned completely forward of the accommodation block with a shipping and removal route directly overhead through hatch(es) forward of the accommodation with the corollary that the role of the engine casing can be confined to accommodating the uptakes, ventilation ducting and personnel access. The fact that container ships using this type of arrangement can carry additional containers is illustrated in Fig. 15.1.

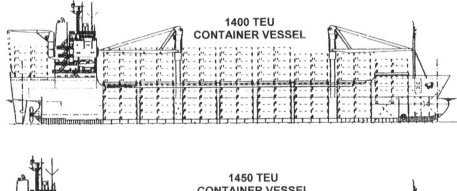

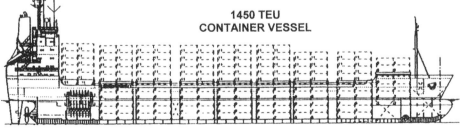

Fig. 15.1. Shifting accommodation right aft can increase container numbers.

Passenger ship designers, seeing the advantages that having the machinery aft gained in cargo ships, followed the same trend some years later, although the fact that these ships are almost all at least twin-screw meant that the machinery space has to be kept some distance forward so that the tank top at the aft end of the engine room is sufficiently wide to accommodate the main engines.

15.5.3 Location of the galley

Possibly the next most important decision concerns the location of the galley because of the influence this has on the location of associated spaces which either serve the galley or are served from the galley. Although a decision on the location of these spaces is not of the same order of importance as the decision on the cargo/machinery location which has just been discussed, it is a relationship which plays quite an important part in ship design, particularly on passenger ships.

On a passenger ship a decision on the position of the galley leads fairly directly to decisions on the position of the dining saloon or saloons. These, in turn, determine positions for the main access stairs and lifts which then dictate the arrangement of bureaux, shops and other public rooms on higher decks. To ensure that the supplies needed in the galley reach it with ease, the position of dry and refrigerated stores become a further corollary of the galley position and these, in turn, dictate the position of stores loading hatches, side doors and passages. Whilst a selected galley position does not automatically choose the locations of these other spaces, it is most desirable to have all these corollaries in mind when trying to decide on the galley position.

This example of the importance of "associated" spaces can be repeated in other areas of design. In specialist ships and on warships the desirability of proximity and good provision for ready access between associated spaces is an important consideration which should be in the designer's mind when the general arrangement is being laid out.

15.6 ARRANGING ACCOMMODATION

The main objectives which a designer should have when setting out the arrangement of accommodation are:
1. the comfort and well-being of the occupants;
2. the ease with which the accommodation can be maintained;
3. keeping the construction cost as low as possible and certainly to an acceptable budget.

These objectives apply both to passenger and crew accommodation, whilst in addition crew accommodation should have convenient, generally covered, access to the places in which the crew work.

Although much that is covered by the words 'comfort' and 'well-being' may be thought to depend more on the standard of specification of furniture and fittings than on the general arrangement the latter can also have a significant influence.

Rules relating to the design and specification of crew accommodation on British merchant ships are given in the Merchant Shipping (Crew Accommodation) Regulations 1978. These rules give some very good practical advice although the standard set is a minimum which is comfortably exceeded in most ships now being built.

15.6.1 Number of tiers of decks, and allocation to tiers

When designing an accommodation block for a cargo ship, the first decision needed is the number of decks on which this is to be located. The basis of this decision is commonly the provision of adequate visibility from the wheelhouse over the forecastle and/or over the maximum obstruction caused by containers or other deck cargo the ship is intended to carry. The considerable lengths to which this principle can be taken is illustrated in Fig 15.2 which shows a nine-tier superstructure, inclusive of a poop, fitted on a comparatively small container ship to provide visibility over six tiers of containers on the hatch covers.

Whilst it is not absolutely necessary for there to be decks at each deck level from the upper deck to the required wheelhouse height, this is generally the best arrangement. The alternative of building the wheelhouse on a framework of pillars with only an access stairway is possible but will usually be more expensive and may result in a structure liable to vibration.

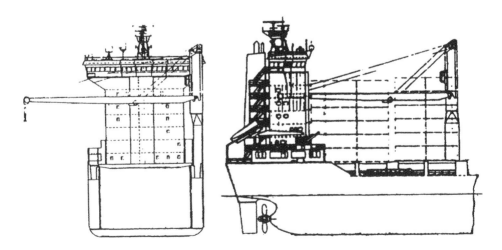

Fig. 15.2. A nine-tier superstructure provides visibility over six tiers of containers on the hatch covers.

15.6.2 Accommodation area calculation sheet

The next step is to complete a sheet such as that given as Fig. 15.3 (taken from the Watson and Gilfillan 1976 R.I.N.A. paper) filling in the areas required for all the accommodation spaces. Once all the areas are available in the third column, these can be allocated to individual decks. In this allocation the aims should be:

(1) to keep related rooms on the same deck if possible and if not on as near a deck as can be arranged;

(2) to arrange accommodation for senior personnel on higher decks than junior personnel — possibly more for tradition than for any environmental advantage!

(3) to keep the total area on each deck either the same or less than the area of the tier below it; this helps to eliminate overhangs and makes for good structural design.

15.6.3 Space and shape of cabins

With today's small crews the pressure which there used to be to minimise the space allotted to each type of officer and crew cabin has diminished except on small ships. This is partly because of the reduction in the numbers of rooms required and partly because it is recognised that small crews need higher standards of accommodation and that this means more space in their cabins. Having decided on the approximate area which is to be given to each type of room together with the furniture to be fitted, the designer should then make a sample cabin layout to try to establish the shape of room which will accommodate the furniture in a pleasing way and fit conveniently into the potential deckhouse space.

An interesting decision concerns the orientation and position of beds. In theory these can lie either fore and aft or athwartships and the former can be positioned either adjoining a passage or at the deckhouse or ship's side. Those sleeping in fore and aft beds are probably less subject to ship's motions with rolling being a more severe motion than pitching — unless the ship is stabilised in which case the pitching motions will be worse and athwartship beds are better. Beds adjacent to passages may suffer some disturbance from those using the passages, whilst the need for the cabin width to be such as will accommodate both the length of a bed and the width of a door may not be acceptable. Beds adjacent to the ship's side are in the most dangerous possible location in the event of a collision and this position is consequently not now used. The same argument, but to a much diminished degree, applies to beds at a deckhouse side with a more important disadvantage of this position being drips from windows or sidelights.

If the room is to have a couch as well as a bed this may be the next item to be positioned with a preference for it being located at right angles to the bed. Thereafter it is a matter of finding wall space for wardrobe, chest of drawers, desk,

		No. of units	Unit area (m²)	Gross area (m²)	Area Allocation				
					Deck	No. 1	No. 2	No. 3	No. 4
1	Owner's suite								
2	Senior officers' suites								
3	Deck officers' cabins with toilets								
4	Engineer officers' cabins with toilets								
5	Chief steward's cabin with toilet								
6	Pilot's cabin with toilet								
7	Cadets' cabins								
8	Passengers' cabins with toilets								
9	P.O.s' cabins								
10	Deck ratings' cabins								
11	Engine ratings' cabins								
12	Catering staff cabins								
	Total of 1 to 12								
13	Wheelhouse/chartroom								
14	Radio room								
15	Engs' change room/officers' toilet								
16	Officers' laundry & drying rooms								
17	Offices								
18	Officers' dining room & duty mess								
19	Officers' lounge								
20									
	Total of 13 to 20								
21	P.O.s' & crew's messes								
22	Crew's recreation rooms								
23	P.O.s' & crew's toilets								
24	Crew's laundry & drying rooms								
25									
	Total of 21 to 25								

Fig. 15.3. Calculation sheet for accommodation areas (*continued opposite*).

		No. of units	Unit area (m²)	Gross area (m²)	Deck	Area Allocation			
						No. 1	No. 2	No. 3	No. 4
26	Galley & pantries								
27	Hospital, bath & dispensary								
28	Hobbies room								
29	Fan rooms								
30	Emergency generator/battery room								
31	Cold rooms								
32	Dry provision store-room								
33	Bonded & other store-rooms								
34	Deck store-rooms								
35	Lockers								
36	Refrigeration machinery								
37	Deck machinery equipment spaces								
38	Swimming pool								
39	Engine casings								
40									
			Total of 26 to 40						
41	Passages/stairs	≡	%Σ(1 to 12)						
42	Outside deck area	≡	%Σ(1 to 41)						
			TOTAL AREA						

Fig. 15.3. Calculation sheet for accommodation areas (*continuation*)

etc. With most cabins now having their own private toilets, the positioning of these becomes an important factor. This has to be considered both in relation to the cabin arrangement and generally in relation to the toilet of the next cabin. A good arrangement can help to reduce piping costs quite considerably.

One extreme of cabin shape is a long narrow room in which most of the furniture is arranged along one of the long sides with not much more than an access passage. Such a room can accommodate all the required furniture and may even have quite a lot of space in it, but will be an aesthetic disaster and not contribute to the well-being of its occupant. Another extreme is a square room, but unless the

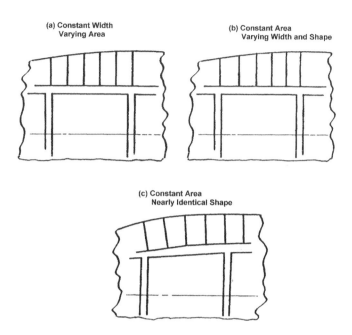

Fig. 15.4. Alternative cabin shapes.

specification calls for some island furniture such as a coffee table and/or easy chairs this shape of room will have too much centre space relative to the wall space needed for doors, windows and furniture which is usually wall supported. The ideal shape is probably a rectangle with proportions not dissimilar to those of A4 paper.

Meek and Ward in one of the very few technical papers devoted to accommodation design — "Accommodation in ships" (R.I.N.A. 1973) —make a useful suggestion for the design of cabins in a block in which the form of the ship is a factor. A modified version of this is reproduced as Fig. 15.4. Arrangement (a) in this has passages parallel to the centre line and cabins of constant fore and aft length but with widths and areas changing with the ship's form. As the smallest cabin must accommodate all the required furniture and provide at least the rule area, it is obvious there is waste space in the larger rooms. A more serious objection is the possibility that the occupant of one of the smaller cabins may feel unfairly treated. Arrangement (b) retains passages parallel to the centreline but the fore and aft length of each cabin is arranged in association with the mean athwartships width to give the same area. The snag to this arrangement is that each cabin is of a different shape and the arrangement of furniture may have to be different. In arrangement (c), the passages are no longer parallel to the centreline but all the rooms are to all intents and purposes identical in both shape and area.

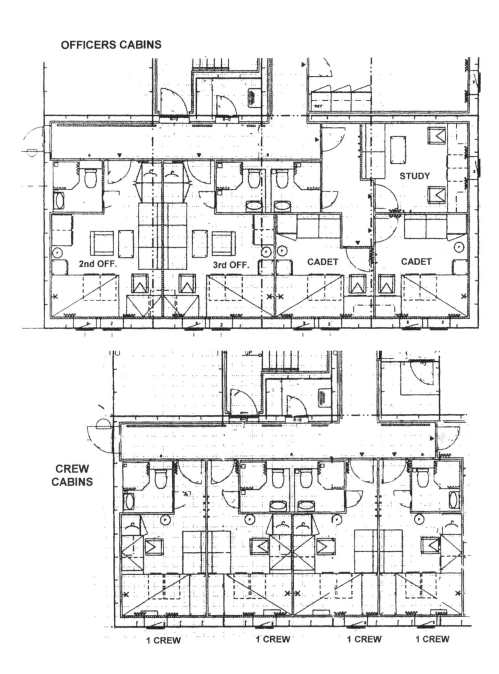

Fig. 15.5. Accommodation on a modern cargo liner. Scale 1:100.

Figure 15.5, photocopied at a 1:100 scale from the plans of a modern cargo liner, shows the high standard of accommodation now given in these ships. It will be noted that single berth cabins with private toilets are provided for crew as well as for officers, with the officers status being acknowledged by bigger settees and wardrobes and the addition of a coffee table and easy chair. The space allocations (cf. §5.2) are officer's cabins 13.4 m^2; crew cabins 9.6 m^2 in both cases excluding toilets which are 2.6 m^2.

15.6.4 Passages

Passages are essential for access, but the available space will generally be put to better use if it can be added to a room and the space devoted to passages should be reduced as much as possible.

It is, of course, the length of passages which should be reduced as the width must not be skimped but should be tailored to suit the traffic for which the passage is intended; working alleyways in particular should have generous width. On the other hand, whilst wide accommodation passages may look well on a ship in port, they can pose problems to those using them in a seaway and there is much to be said for limiting the width to not much more than a metre so those using them can "bump" from side to side without falling over as the ship rolls.

In passenger ships in which long fore and aft passages are required, the line of these should be altered at intervals to reduce the vista which can otherwise be quite frightening when the ship is pitching.

15.6.5 Stairs

If possible stairs serving several decks should be run in a vertical stair well. This makes finding one's way about the ship as easy as possible — vital in an emergency. It also minimises the structural discontinuity which is a corollary of deck openings and facilitates the enclosure of these for fire protection. Again for structural reasons stairs should, if possible, be arranged within the "line of openings".

Stairs running fore and aft are probably safer to use in a seaway with rolling motions being more severe than pitching; this used to be standard practice on passenger ships until the fitting of stabilisers reversed this, leaving pitching as the more severe motion.

Figure 15.6 shows a number of cargo ship stair arrangements, of the two fore and aft arrangements: (i) uses the least space and makes the least structural discontinuity, but (ii) is more convenient for the users. The athwartship arrangement shown in (iii) with the stair arranged at the forward end of the engine casing usually is usually a convenient and economical arrangement.

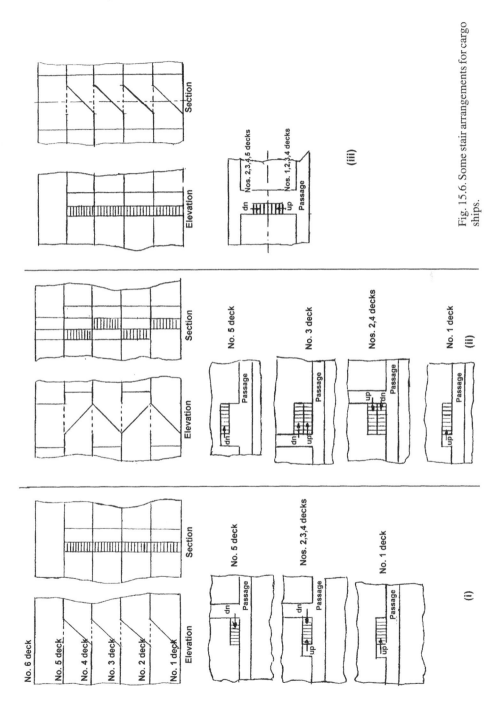

Fig. 15.6. Some stair arrangements for cargo ships.

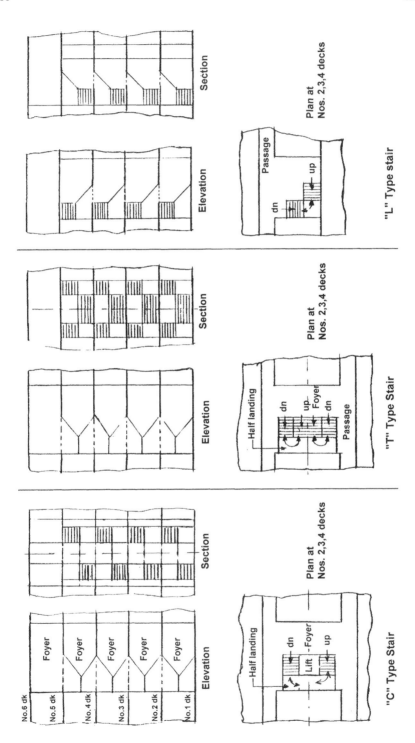

Fig. 15.7. Some stair arrangements for passenger ships.

Stairs should be made an architectural feature on passenger ships, usually in association with lift wells. Half landings can make them easier for passengers and contribute to the appearance. Figure 15.7 shows some possible arrangements.

15.6.6 Lavatory spaces

Although with private toilets becoming more general, the number of these is diminishing, it may still be worth giving some ground rules for their layout. A check should first be made to find out what is on the deck below the projected lavatory space as discharge pipes should not be run through refrigerated spaces or over switchboards. If possible, the room should be arranged so that WCs are grouped side by side in a block so that the discharges can all be directed into one pipe running athwartships. A similar arrangement should be adopted with showers.

15.6.7 Galley

The range, which is the main feature of the galley, is usually an island fitting. It should be arranged with its length athwartships to minimise the danger of the ship's rolling motion throwing someone working at it onto a hot plate. The remainder of the fittings are generally arranged round the boundary bulkheads of the room, with the arrangement carefully thought out to minimise movement during food preparation, cooking, serving and washing-up. Space must be provided for the opening of ovens doors, dish washers, refrigerator doors, etc. A typical galley arrangement with adjoining mess and lounge is shown on 1:100 scale in Fig. 15.8 photocopied from the plans of a bulk carrier.

15.6.8 Storerooms

The shelves of storerooms should generally be about 600 mm wide and access passages should be of about the same width. The best use of space is obtained by having shelves on both sides of the passage making 1.8 m a suitable clear width — or for larger storerooms, multiples of this width with a connecting passage..

15.7 ANCHORING, TOWING AND MOORING ARRANGEMENT DESIGN

Although the frequency with which these functions are carried out and the conditions for which this equipment must be designed vary widely with ship type and service, anchoring, towing and mooring are features about which some general guidance can be given.

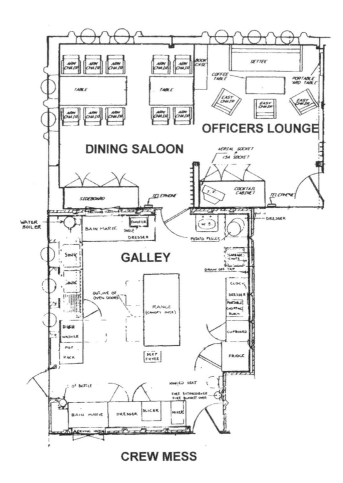

Fig. 15.8. Galley, dining saloon and officers' lounge on a bulk carrier. Scale 1:100.

15.7.1 Anchoring arrangements

The weight of anchors and the length and diameter of cable to be carried are stipulated by Classification societies based on an "equipment number" common to all the societies which provides an approximate measure of the forces that act on a ship when at anchor. High holding power anchors of reduced mass are permitted and the diameter and weight of the cable can be reduced if the cable is made of special quality steel.

Anchors and cables are commonly handled by windlasses, although capstans are sometimes used. The positioning of windlasses is partly dictated by the position of the hawse pipes and partly by the lead of the chain pipes to the cable locker — with, in both cases, the reverse also being the case. The hawse pipe must be so positioned that the anchor will drop free and the chain when at anchor will not foul the bow (particularly important for bulbous bows). The pipe must be long enough to accommodate the shank of the anchor and the shell casting must provide a secure resting place for the flukes.

If a windlass with barrels and/or warping ends for mooring purposes is being considered, the positions of these must be checked to ensure they suit the mooring arrangement. The difficulty of meeting all these requirements with a conventional twin gypsy windlass which used to be fairly standard, led to the development of single gypsy windlasses. With two of these fitted, arrangement problems are greatly eased and this has now become almost standard practice on all but small ships.

15.7.2 Towage

With many vessels now fitted with bow thrusters or bow and stern thrusters the use of tugs has greatly diminished, but provision must still be made for taking towlines both forward and aft. This requires the provision of suitably sized bollards (towage wires are heavier than mooring ropes) positioned to suit leads from the warping ends of the winches to the fairleads used. Even if the operation is infrequent, securing and letting go towlines can be labour intensive and a special towmaster unit with a small capstan both fore and aft can help towards minimum manning.

15.7.3 Mooring

The mooring arrangements for most ships consist of bow and stern lines plus bow and stern springs and bow and stern breast lines.

The bow lines and bow springs are generally led from the fore deck or forecastle, if fitted; the stern lines and the stern springs from the aft deck; the forward breast line used generally to be led from the upper deck just abaft the forecastle whilst the aft breast line was led just forward of the accommodation house.

These breast line positions had several disadvantages and modern mooring arrangements try to keep all machinery in two blocks one forward and one aft, as shown in Fig. 15.9, which is abstracted from "Advanced Mooring Design", a study carried out on behalf of the British Dept of Transport. The particular ship used as the basis for this study was a 21000 tonne deadweight products carrier.

Another development shown on this figure is a reduction in the number of lines. Both bow and stern lines remain doubled but springs and breast lines are now single line giving 8 lines in all as opposed to 12 lines when doubling of springs and breasts was standard practice.

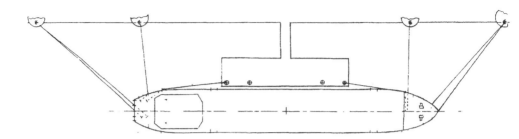

Fig. 15.9. Outline mooring arrangements. Scale 1:2000.

The advent of self-tensioning winches some thirty years ago eliminated what used to be a major task requiring a large harbour watch, namely the adjusting of mooring ropes as the vertical position of the ship altered relative to the quay with the rise and fall of the tide and/or a change in the ship's draft as cargo was loaded or unloaded. The task of handling all these ropes when mooring or unmooring remains a major task and if this is to be carried out by today's small crews these must have the backing of a carefully designed system of winches well positioned in relation to the fairleads. An arrangement, suggested in the study, which reduces the manning to two persons at each end of the ship, is shown in Fig. 15.10.

An alternative arrangement in which the number of winches has been reduced by combining the foredeck winches with the windlasses and fitting double-barrelled winches aft, all appropriately angled to meet their multiple duties, is shown on Fig. 15.11.

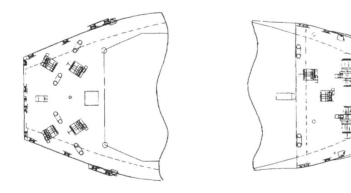

Fig. 15.10. Mooring arrangement details. Scale 1:500.

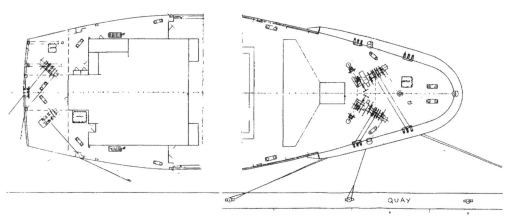

Fig. 15.11. Mooring arrangement details with reduced number of winches. Scale 1:500.

Chapter 16

General Arrangement of Some Ship Types

16.1 INTRODUCTION

Although this chapter has been given the name "General arrangement of some ship types" it goes beyond the general arrangement to explain the reasoning which led to the main features of the designs discussed.

It was originally intended that the chapter would include a general arrangement plan of each of the main ship types with a short accompanying analysis of the design. Now that *Significant Ships* is published annually, providing plans of about fifty of the more important ships built worldwide each year, the inclusion of so many plans in this book has become unnecessary.

Unfortunately the descriptions of the ships in *Significant Ships* —useful though they are — do not extend to giving an explanation of the reasoning behind the designs. Believing that an understanding of the "whys" behind a design is of particular value to designers, led to a decision to limit the number of ships featured but to deal more fully with the reasoning behind their designs.

Four types of ship types were chosen for this treatment, each of them, ships in whose design the author had been closely involved and each having a number of unusual requirements and constraints. These are given in §§16.1 to 16.5.

The original intention was to present all the plans included in this chapter at the same 1:1000 scale (courtesy of modern photocopying) so that their features could be easily compared and approximate measurements made if required. In the event, the size of the book's pages has limited the use of the 1:1000 scale to ships not exceeding 150 m in length with reduced scales for larger ships.

At a late stage, it was decided to revert to some extent to the original idea by adding some general guidance on the design of other ship types, noting the major factors involved in each case and summarising references to these which appear in the text, and these are given in §§16.6 to 16.8.

Section 16.9 gives a series of tables covering the main ship types aimed at directing readers to the best "models" for their design for which information is given in the 1990 to 1995 volumes of *Significant Ships*.

16.2 MULTI-PURPOSE CARGO SHIPS — TWO QUITE SIMILAR DESIGNS

Two designs are used to illustrate the design of this type of ship. Figure 16.1 shows the "Clyde" class designed by the author as a standard ship for U.C.S. in 1969. Figure 16.2 shows the M.V. "Serenity", a ship of the "Neptun" class built by Schiffswerft Neptun and completed in 1990.

It is interesting to see on the one hand the great similarity there is between these designs, in spite of the lapse of time, and on the other to note the many detail improvements in the later ship, some of which may reflect the fact that it was probably aimed at a somewhat higher market.

The dimensions are remarkably similar as are the main features of the two general arrangements: four holds forward and engines aft; three long holds and one short hold; and twin hatches at the three main holds designed to give access to maximum numbers of containers.

The main differences stem from the higher speed provided by the "Neptun" class with the corollary of finer lines and lower deadweight and cubic capacity. Other differences will be commented on later when discussing the improvements made in the more recent ships. A comparison of the main particulars of the two designs is given in Table 16.1.

Table 16.1

Comparison of the main particulars of the "Neptun" and the "Clyde"

	Neptun	Clyde
Length B.P. (m)	146.95	138.41
Breadth mld (m)	23.05	22.86
Depth to upper deck (m)	13.40	13.72
Draft (m)	10.09	10.01
Deadweight (tonnes)	17175	18800
Displacement (tonnes)	24668	23880
Lightship (tonnes)	7493	5080
Grain capacity (m^3)	22284	26616
Containers T.E.U.	1033	450
Speed service (knots)	16.2	14.9

16.2.1 The "Clyde" class

Although it is now more than twenty-five years since the design of the Clyde class of multi-purpose ships, the fact that the design process which led to its development was distinctly unorthodox means that it may still be of interest today.

It had been decided by the shipyard that there was a market for a ship capable of carrying a wide range of cargoes — general break bulk cargoes, bulk cargoes such as grain or ore, timber and containers.

It was realised that although general cargo and the smaller bulk cargoes would be those most frequently carried, the cargo which should be given most consideration in the design should be the only one which always came in the same large unit size, namely containers.

To maximise container numbers, it was decided to fit twin hatches and to make these capable of taking three containers in their width. This necessitated a beam of about 22 m and it was decided to increase this to 22.86 m, the St Lawrence Seaway limit. With this beam it was possible to provide a depth of 13.72 m enabling five tiers of containers to be accommodated below decks and two tiers of containers to be carried on deck. The lengths of each of the main hatchways were then arranged to take three containers in the length.

It was decided to have three main holds and a short No. 1 hold; the space required for cargo gear and hatch cover stowage then determined the length of the cargo spaces.

Possibly the most striking fact was that up to this stage in the design no firm decisions had been taken on the deadweight, the cubic capacity or the speed of the ship. Various alternatives for these were then examined against market research indications before the final specification was fixed.

Figure 16.1 shows the stowage of the various cargo alternatives for which this class of ship was designed.

16.2.2 The "Neptun" class

Probably the most striking advantage of the "Neptun" class is the much larger number of containers that this class can carry and it is interesting to examine how this has been achieved.

(i)   The use of cranes in lieu of derricks reducing the deck space required and enabling the length of the hatches to be increased and giving two extra container bays in the length.

(ii)   The adoption of two different widths of hatch covers and moving the "centreline" bulkhead a little off centreline so that the starboard hatches can accommodate four containers in the width. (It may be remarked that symmetry in arrangement of a ship is normal and usually helpful, but designers should not hesitate to depart from symmetry if to do so is advantageous.)

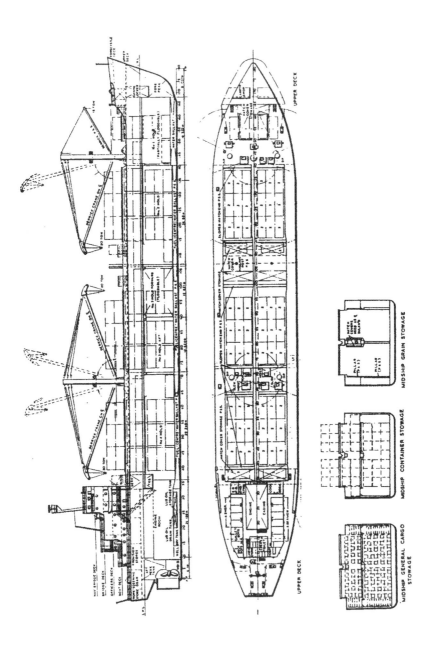

Fig. 16.1. Multi-purpose cargo ships. "Clyde" class designed by the author in 1969 built by Upper Clyde Shipbuilders, U.K.

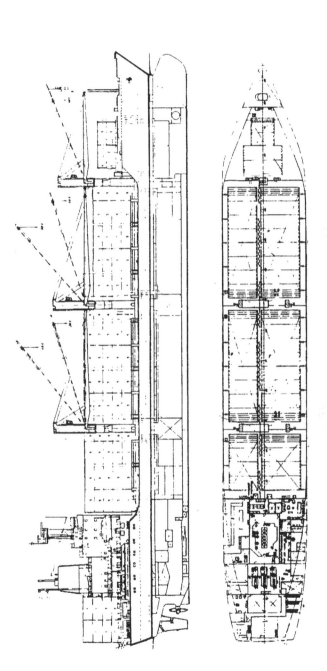

Fig. 16.2. Multi-purpose cargo ships. M.V. "Serenity" built in 1990 by Siffswerft Neptun, Germany.

(iii) The removal further aft of the accommodation giving another bank of containers on deck in the length whilst the wider stern and the provision of supports gives two more banks abaft the accommodation.

(iv) It is in deck carried containers that the biggest gain has been made, with these being stowed on supports right to the ship's side giving nine in the width rather than six as on the "Clyde" class.

(v) The wheelhouse has been raised so that it is a seventh tier on the "Neptun" as against the fifth tier that it occupies on the "Clyde" class. This enables deck carried containers to be stowed five high as against the modest two provided in the "Clyde" class. (It may be remarked that the stability of the "Clyde" class would have permitted many more containers on deck, but at the time of its design the ships were of more interest as general cargo carriers and this was not pursued.)

Other differences between the two designs are the long forecastle and the bulbous bow. The desirability of having a long forecastle on the "Neptun" class stems partly from the increased speed of this class, but it also helps to increase the number of containers that can be safely carried. A bulbous bow, although not shown on the plan of the "Clyde" class, was in fact an option available on that class, as was a higher powered engine increasing the speed to 15.5 knots.

16.3 BULK CARRIERS

Four ships are presented to illustrate different features which can determine the arrangement of these ships.

16.3.1 Panamax bulk carrier "China Pride"

The "China Pride", shown in Fig. 16.3, was built in 1990 by Jiangnam shipyard in China. This "Panamax" ship has what has come to be accepted as the most standard configuration for a bulk carrier, with the only unusual feature being the sloping hatch coamings at both sides and ends. It is probably fair to say that one of the main objects of the designer of this type of ship is the achievement of the utmost structural efficiency and through this the minimisation of steelweight and cost.

Points to note in the design are the identical hold and hatch lengths of Nos. 2 to 6 holds inclusive; the side sliding hatch covers; and the alignment of the house front with the bulkhead under.

Homogeneous cargo can be carried in all holds; heavier ore cargo can be carried in Nos. 1, 3, 5 and 7 holds with the other holds empty; grain can be carried in all holds with one hold slack without shifting boards. The ballast capacity provided in double bottom wings and hopper tanks can be augmented by the use of No. 4 cargo hold, giving a total ballast capacity of about 32000 m^3.

Table 16.2

Principal particulars of the "China Pride" and the "Solidarnosc"

	China Pride	Solidarnosc
Length B.P. (m)	215.00	224.60
Breadth (m)	32.20	32.24
Depth (m)	18.00	19.00
Draft scantling (m)	13.11	14.10
Draft design (m)	12.50	12.50
Deadweight scantling (tonnes)	65665	74000
Deadweight design (tonnes)	61687	63000
Displacement (tonnes)	77288	
Lightship (tonnes)	11623	
Cubic capacity grain (m^3)	78067	85525
Ballast capacity (m^3)	32000	28745
Speed service (knots)	14.90	13.80
Power MCR (kW)	9451	8134

The principal particulars of the "China Pride" and the "Solidarnosc", which will be discussed in the next section, are given in Table 16.2.

16.3.2 "Solidarnosc"

Figure 16.4 shows the arrangement of a new bulk carrier design by Burmeister and Wain, the first of which — the "Solidarnosc" — was completed in 1991. This is another "Panamax" ship, but there are a number of striking departures from the standard practice shown in the ship described in the previous section. The side shell has double-skin construction and the transverse bulkheads are also double-plated, giving a smooth surface within the holds. This not only reduces time and cost of cleaning and the susceptibility of these surfaces to corrosion (see comments in Chapter 10) but also makes the ship suitable for completion as an OBO or a product tanker. It is also possible that this structural design may with some ingenuity be made particularly construction friendly.

In spite of the loss of cargo cubic entailed by double skin aspects of the design, the design is claimed to provide increased cubic in comparison with similar sized vessels. This seems to have been achieved partly by increasing the depth, partly by the use of a continuous trunk deck 2.3 m high into which the topside tanks are incorporated and partly by reducing the length of the engine room and arranging the accommodation further aft. Another interesting feature is the fact that the side rolling covers are fitted directly to the trunk deck eliminating hatch coamings.

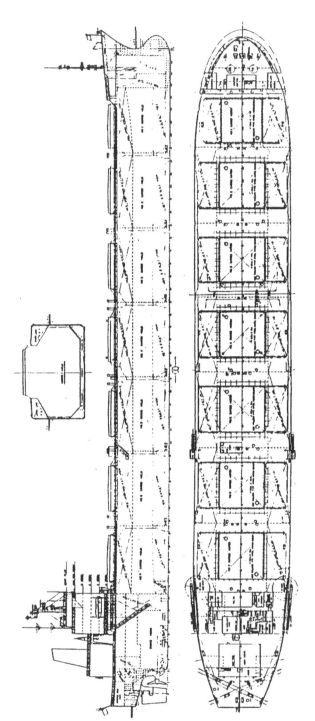

Fig. 16.3. Bulk carriers. M.V. "China Pride" built in 1990 by Jiangam Shipyard, China.

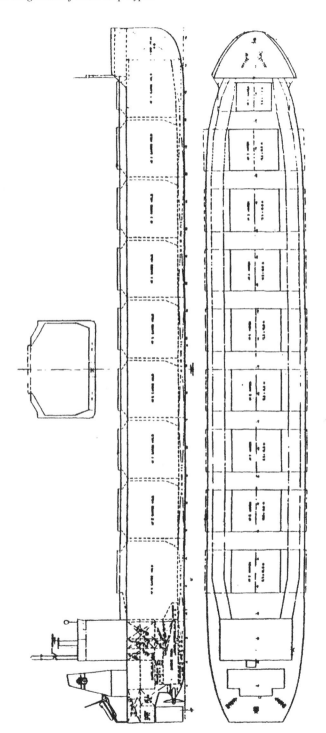

Fig. 16.4. Bulk carriers. M.V. "Solidarnosc" built in 1991 by Burmeister and Wain, Denmark.

As Fig. 16.4 shows, the arrangement provides nine holds, with the ship having the DNV HC/EA notation which allows heavy cargoes to be carried at full draft with specified combinations of empty holds. The maximising of the proportion of the length given over to cargo space should also reduce the still water bending moment and therefore the scantlings. As a general comment applicable to both ships, it may be worth noting that bulk carriers generally have an odd number of holds with the odd numbers being used for ore cargoes with the even numbers empty.

The slightly slower speed coupled with the longer length gives a reduced Froude number and advantage has been taken of this to permit the use of a fuller block coefficient and a simpler rounded bow in lieu of the bulbous bow which is more usual on this type of ship.

16.3.3 "Sir Charles Parsons"

Figure 16.5 shows the arrangement of the bulk coal carrier "Sir Charles Parsons", one of three vessels built by Govan shipbuilders in 1985 for the British C.E.G.B. The design for these vessels was prepared by YARD and the author was closely involved. Because of this and because the design had to meet a number of unusual requirements it seems to merit inclusion here.

The aim of the design was to greatly reduce the cost of shipping coal from ports on the British North East coast to power stations on the Thames and Medway and this was done by:

1. Making the deadweight as large as possible within the operating constraints using maximum dimensions and minimising weight.
   (i) *Length*: the provision of powerful bow and stern thrusters enabled the ship's length to be increased to just a few metres less than the channel width in which the ship has to turn.
   (ii) Breadth: the breadth was arranged to use the full outreach of the mechanical unloaders.
   (iii) *Depth*: the depth was arranged to provide capacity for a full deadweight of the lightest coal.
   (iv) *Draft*: whilst the specified deadweight can be carried on the limited draft which may be imposed by the tidal water depth, increased deadweight can be carried by loading to full statutory draft when conditions permit. A heavy-duty ballast system enables trim to be controlled during rapid cargo discharge.

2. By fuel saving economies
   (i) *Speed*: an economical speed was set to meet the desired voyage schedule with suitable margins for tidal and weather constraints.
   (ii) *Lines*: the use of low resistance lines developed by tank testing.

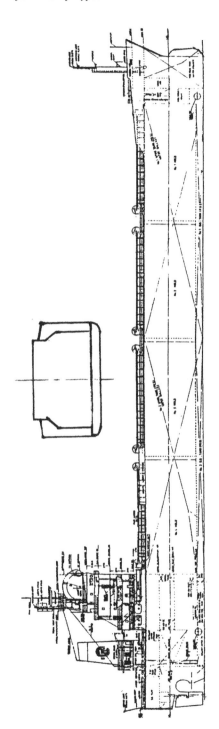

Fig. 16.5. Bulk carriers. M.V. "Sir Charles Parsons" designed by YARD (the author's firm), built by Govan Shipbuilders, U.K. Scale 1:1000.

Table 16.3

Main particulars of the "Sir Charles Parsons" class

Length B.P. (m)	148.0
Breadth (m)	24.5
Depth (m)	13.5
Draft max./designed (m)	9.02/8.02
Deadweight (tonnes)	22530/19187
Cubic capacity coal (m^3)	26147
Speed (knots)	12.5
Power (kW)	4355 kw 600 rpm geared to 110 rpm at the C.P. propeller Complement

   (iii) *Propeller*: the use of a slow running high efficiency propeller.

   (iv) *Fuel efficiency*: the choice of fuel efficient main propulsion machinery; the use of a shaft driven alternator to carry the sea load.

3. By operational economies

   (i) *Manoeuvring*: bow and stern thrusters eliminate the need for tugs in normal berthing.

   (ii) *Rapid loading*: clear holds arranged with double skin and unloading and large hatches; push button operation of hatch covers.

   (iii) *Labour saving*: heavy duty shore power connection to enable a complete shut down of ship's generators when alongside.

   (iv) *Ballast system arranged for rapid ballasting and deballasting*: to speed turn round.

   (v) *extra accommodation*: to enable rapid change-over of crews, again to speed turn round.

None of the features outlined above is in any way unusual but bringing them all together halved the cost that C.E.G.B. had been paying for carrying coal.

The main particulars of the "Sir Charles Parsons" class are given in Table 16.3.

16.3.4 *"Western Bridge"*

Featured in Fig. 16.6 is the self-unloading bulk carrier "Western Bridge" built in 1991 by Hashihama Shipbuilding, Japan for British Steel.

It was a matter of some regret to the author when designing the "Sir Charles Parsons" that the C.E.G.B. had already ordered new land-based unloading equipment for the terminals to which that class of ship was to operate as he thought that the case for self-unloading equipment carried on the ships merited examination and might have made for an even more efficient service.

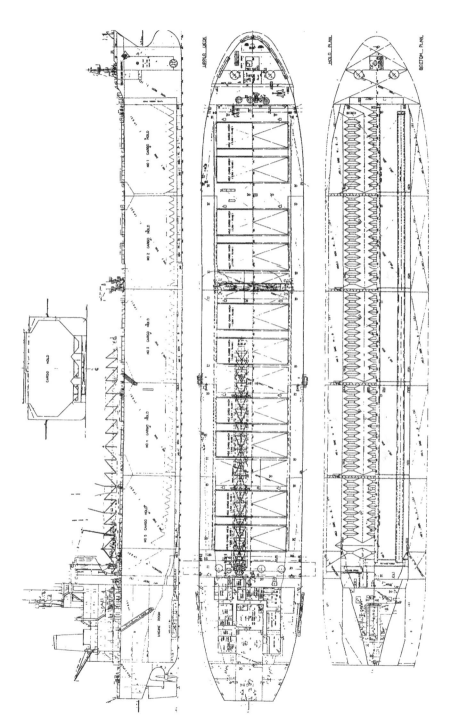

Fig. 16.6. Bulk carriers — self-unloading. M.V. "Western Bridge" built in 1991 by Hashima Shipbuilding, Japan.

In recent years the use of self-unloading equipment has greatly increased and this trend seems likely to continue. The argument between whether it is better to spend money on a shore-based unloading system or put the investment into a ship-based system depends on the relative numbers of ships and terminal ports. If there are a few ships serving a relatively large number of ports, the argument favours ship-based equipment; if there are a lot of ships serving a small number of ports, the argument favours shore-based plant.

16.4 FRIGATES AND CORVETTES

16.4.1 General considerations

The fact that the design of frigates and corvettes is essentially volume based has been stated in §5.4; the importance of length in minimising the power requirement of these very fast but relatively small ships has been referred to in §6.8; the fact that seakeeping ability is largely a function of hull size is stated in §8.7.

In §10.3.3 it is suggested that from a structural point of view it is desirable to arrange that as much of the required space is provided in the hull and the superstructure is kept to a minimum; in §14.4 the desirability of minimising super-structure to reduce both sight and radar signatures is mentioned.

In §3.4.5 it is noted that the relationship of depth to beam of virtually all ship types should be kept within a narrow band if the design's stability is to be satisfactory. Use of this relationship presents some difficulty for a warship as the depth is not a continuous variable but must be made up by the sum of sensible values of double bottom height, the height from the tank top to the lowest deck and one or more tween deck heights.

16.4.2 Different configurations of corvettes and frigates

In the discussion in §5.3.1 on how to fix dimensions when using the volume design method, it was suggested that the total volume required should be divided into hull and superstructure volumes. This is all very well as a principle but in fact the proportions of hull and superstructure that are well suited to one size of ship may cease to be suitable as the ship size is increased or reduced.

As the volume required in a design increases there must be an increase in some or all of the principal dimensions. Length and beam can be increased progressively but, although a minor change can be made in depth by, for example, increasing the double bottom height, at some point it becomes necessary to make the step change of adding another tween deck.

The process of progressive change in ship configuration as the hull volume increases is shown in Fig. 16.7.

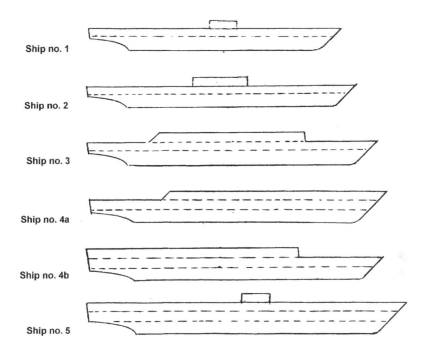

Fig. 16.7. Different superstructure configurations with increasing size on a warship of corvette and frigate type.

Ship 1: Two deck ship with minimum superstructure; optimum L/B and B/D ratios; "sensible" value of depth D.

Ship 2: L and B increased, but D unchanged as it is already a "sensible" value so extra volume provided by increasing the superstructure — initially as a deckhouse.

Ship 3: L and B increased again; superstructure becomes full width and contributes to longitudinal strength and to stability at large angles.

Ship 4: Further increases in L and B; superstructure now extends to either bow (a) or stern (b) depending on which gives best arrangement; seakeeping is obviously greatly improved by the first alternative.

Ship 5: At the next increase in volume, L and B again increase and the depth is increased by a tween deck height and the ship reverts to a minimum superstructure configuration.

16.4.2 The "Fifty Man" frigate

The example of a frigate design used to illustrate this section and shown in Fig. 16.8 is unlike all the other designs featured in this chapter in that it features a design that has never been built and is unlikely ever to be built. The design has, however, several merits which justify its choice as an illustration of the factors involved in frigate and corvette design, notably the fact that it is not shrouded in security and that it is undoubtedly at the forefront of much modern warship design thinking.

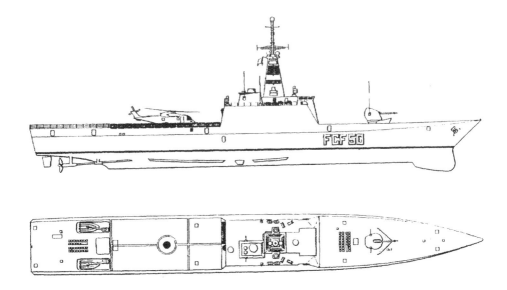

Fig. 16.8. The "Fifty Man" frigate. An imaginative new design (1988) from the author's firm Y.A.R.D.

It was prepared in 1987, by Y.A.R.D. (now part of B.Ae Systems) for their stand at the Royal Naval Engineering Exhibition of that year. The object was to provide a vehicle to demonstrate their expertise in naval architecture, marine engineering, weapons and controls technology, etc. To do this, the very interesting theme of "low manning" was chosen and the ship, a model of which was featured at the exhibition, was christened the "Fifty Man" frigate or FCF 50.

In pursuit of the theme there were two distinct stages. In the first of these all the ways in which the crew might be reduced were explored in considerable detail. When this had shown that a reduction to 50 from the 170 to 250 manning of recent designs was practicable, the second stage then looked at the changes in ship design which the reduction in manning either permitted or required.

16.4.3 Reducing crew numbers

It was suggested that there were four main ways of reducing crew numbers:
  (i)   reductions in on board maintenance (this involves the use of materials and machinery with either low maintenance requirements or whose maintenance can be deferred so that it can be carried out in port);
  (ii) rationalisation in manpower usage: a patrol cycle of 60 days with crews rotating at the end of each patrol; all crew members trained for a number of tasks;

(iii) automation of shipboard systems; expert and knowledge based systems; paperless communication.

(iv) integrated logistic support; transferring to shore as many functions as possible.

The study confirmed that crews of the order of fifty were a practical proposition with the technology already available or in course of development.

16.4.4 Changes in ship design stemming from reduction in crew

With the reduced crew permitting a major reduction in accommodation area, a significant reduction in ship size with a corresponding reduction in cost became possible. This was, however, rejected on the grounds that good seakeeping, quite largely a function of size, remained an essential feature or possibly became even more important if a reduced crew was to cope with all their duties in adverse sea conditions.

Although a reduction in hull size was rejected, the superstructure was happily pruned to a minimum, with much of that remaining being used as the helicopter hangar with the machinery casing offset to one side. Because of its small size, the superstructure is located near amidships enabling the landing platform to be much nearer the pitch centre than in any existing design and giving the helicopter an exceptional operational window.

Even with the reduced superstructure, the ship still has spare space by conventional warship standards. This is made use of in a number of ways all of which contribute in some measure to improving the ship:

(i)   increased weapons fit — increased capability;

(ii) improved accommodation — helps efficiency of reduced crew;

(iii) space for modularity — reduces refit time;

(iv) better access to machinery — reduces maintenance time;

(v)   better removal routes;

(iv) space below decks for anchoring and mooring fittings and machinery — reduces maintenance. reduces radar signature'

(vii) reduction in the complexity of structure and outfitting — reduces building cost.

16.4.5 Main particulars of the design

The main particulars of the design are:

LOA 136.0 m

Beam 16.5 m

Draft 4.3 m

Displacement 4200 tonnes
Speed 28 knots
Machinery CODAG: 1 gas turbine, 2 diesels
Range 11000 miles at 16 knots.

16.4.6 Weather deck and superstructure layout

Although the arrangement adopted in FCF 50 is shown quite clearly in Fig. 16.8, it may be useful to give some more general comments on the factors which helped to decide this part of the arrangement.

The most important of these is the disposition of the weapons and their trackers in positions where these have clear arcs in both azimuth and elevation. The weapon positions in turn dictate the position of magazines below deck and the implications of this must be examined.

Probably next in priority is the allotment of space for the replenishment-at-sea (RAS) equipment which requires sheltered but unobstructed deck space and easy routes to stores.

Other items for which suitable space must be found include boats and anchoring and mooring machinery and equipment.

16.4.7 Internal layout

Because of its large space requirement, the positioning of the main machinery block must take precedence over most other decisions. The fine hull form means that the large items such as diesel engines, gas turbines and gearing cannot be sited very far aft nor very far forward and must be located in the vicinity of amidships.

An advantage of locating the machinery as far aft as possible is a reduction in the length of shafting and in the number of compartments whose arrangement is spoiled by having to accommodate shafting.

An advantage of locating the machinery as far forward as possible is the corresponding movement forward of the machinery casings and the possibility this provides of bringing the helicopter landing area (if required) nearer amidships.

The number of machinery compartments provided plays an important part in determining the ship's vulnerability. Where the main propulsion machinery comprises two shaft sets it is most desirable that these should be unitised with each set in a different W.T. compartment and, if at all possible, with further separation between these compartments by a non-propulsion machinery compartment.

Much the same should apply to the generators and other important auxiliaries which should be divided between at least two compartments and, once again, it is better if there is an intervening compartment, possibly one of the propulsion machinery compartments.

The uptakes, downtakes and removal routes for machinery can have a considerable influence on the arrangement of the space above the machinery compartments and should be carefully planned to suit the machinery and the accommodation space requirements and minimise any discontinuity in the intended structural design.

Much of the space below the lower deck forward and aft of the machinery space will be taken up with storage tanks for fuel and fresh water together with water ballast tanks to maintain stability as these are used. The pros and cons of alternative arrangements of tanks depend on assessments of what will happen in the event of damage.

When arranging the space above the lower decks, it is wise to start by deciding on an access and routing policy covering both longitudinal and vertical access for personnel in the course of their duties, for escape, for supply of stores and ammunition, and for the routing of services.

In principle, longitudinal access can consist of a single passage on or near the centreline or two passages at or near the ship's side. The former is better protected and gives better access to vertical access stairs. The latter provides a degree of redundancy, but at the expense of using double or nearly double the amount of space plus the need for extra transverse interconnecting passages and passages providing access to stair wells.

The selection of the position for the operations room should have the next priority. This can be a difficult decision to make against conflicting wishes to have this in a location where it has protection against above-water damage from missiles, and to have it in a location where it avoids vulnerability to under-water damage and has good access.

The accommodation tends to have to be fitted into the remaining space, but if possible it should be kept away from the ends of the ship to avoid extreme motions. Historically, officers' accommodation was located in one block whilst ratings were grouped by rank and departments and preferably located reasonably close to their working areas. More recently, attention has been given to the concept of zoning under which accommodation for officers and crew are located together in the zones in which they will be under action conditions and the zones are largely equipped with self contained services.

16.5 A SPECIALIST SHIP — R.D.N. FISHERY INSPECTION SHIP "THETIS"

There are so many different types of specialist ship, each with its own particular problems, that it is impossible to give any general guidance to the best design approach to adopt as this can differ widely. An account of how the author's team tackled the design of one particular specialist ship may however be helpful.

The ship in question is a fishery inspection ship which was designed by YARD and Dwinger Marineconsult for the Royal Danish Navy. Three vessels were built to the design by Svendborg Vaerft with the first of these — the "Thetis" — entering service in 1991.

YARD carried out the feasibility phase, which Dwinger thereafter developed to design definition with YARD as a major subcontractor.

The design process is described in some detail in a 1991 R.I.N.A paper "A new Danish fishery inspection ship type" by D.G.M. Watson and A.M. Friis from which the following extract, which concentrates on the less frequently described feasibility study aspects, is abstracted.

16.5.1 Staff requirements

The principal staff requirements set by the Danish Navy's Naval Material Command (NMC) were as follows:

- Ship to be ice-strengthened to DNV ICE 1A and to be able to break 80 cm of solid ice. All exposed structures to be designed to minimise ice accretion.
- Ship to be able to stand up to wind gusts of 150 knots in light ice conditions.
- Maximum continuous speed: in calm seas not less than 22 knots, and in 4 m waves not less than 20 knots.
- Slow speed: to be capable of operating in all sea conditions at speeds of 4/5 knots.
- To have a bow thruster capable of holding the bow against an athwartship wind of 28 knots.
- Endurance: provisions for 4 months; fresh water 100 tonnes storage plus the output of two freshwater generators of 15 tonnes per day; fuel for 24 days and 8300 nautical miles at a matrix of speeds.
- Dimensional limitations: overall length not to exceed 100 m; draft max not to exceed 4.5 m.
- Armament: to include a 76 mm gun; a Lynx helicopter with hangar weapons control system including sonar and radar; an outfit of depth charges.
- Deck equipment to include a motor barge suitable for Arctic waters.
- Accommodation for 64 officers and crew.
- Propulsion machinery: to be suitable for ice-breaking and towing duties; to be divided between at least two watertight compartments; to be reliable, economical and easy to maintain.

Another set of requirements for a fishery inspection ship for Faroese waters was to be considered in parallel and if sensible combined as one design.

In addition, if it was possible, without incurring significant extra cost, to pre-plan the ships so that they were fairly easy to convert to a frigate or mine-laying role, this would be favourably regarded. A further alternative would be to so configure the design that ships for these roles could be built using much of a common design.

16.5.2 Feasibility study

The first thing the consultants did was to question some of the staff requirements and get as much additional information as possible, both of which are always well worth doing.

The first question sought to obtain RDN experience with twin and single-screw propulsion in the ice conditions specified, and elicited the answer that on such a comparatively small ship only a propeller on the centreline would be adequately protected from the ice in which the ship had to operate.

This immediately focused attention on the limited draft given as one of the requirements. With a draft of 4.5 m the power which could be absorbed by a single propeller was clearly limited even if relatively high RPM were used accepting the resultant low efficiency and the consultant reported that the desired speed of 22 knots seemed unlikely to be attainable. They then asked the reason for the draft limitation and whether it could be changed. The answer was that this draft had been specified so that the ship could enter some Baltic ports but that this was not essential to the main purpose of the ship and the maximum draft could be increased to 6.0 m.

Even with this increased draft, powering appeared to present a problem with the overall length limited to 100 m corresponding to a length B.P. of about 89 m giving an F_n of 0.38 for the required speed of 22 knots. Even at this stage it seemed probable that a longer ship would have other advantages so the reason for the length limitation was queried. This turned out to be the fact that the ship must be able to turn in some very narrow Greenland fjords so it was suggested that increased length could be made acceptable if the ship was provided with a device or devices to improve her turning ability, and this was accepted provided later calculations confirmed that an increased length was desirable.

The severity of the winds in which the ship was required to operate had been a surprise and provoked a question as to their frequency, which elicited the reply that these were Katabatic winds and were almost a weekly event in the Greenland fjords. Needless to say, very careful attention was thereafter given both to reducing windage and to providing good stability.

16.5.3 Fixing dimensions

16.5.3.1 Length

In spite of the consultant's view that the ship's length would need to be increased from the original staff requirement, it was decided to prepare a design with an overall length of 99.5 m and a length B.P. of 89 m.

A second design was progressed on the basis of an LOA of 115.0 m and an LBP of 105.0 m. This reduced the Froude number for 22 knots to 0.35 and was expected to permit a considerable reduction in power.

Later, at the suggestion of NMC, a third design was progressed based on an length BP of 120.0 m (overall length 130.5 m) to see if this gave a worthwhile further reduction in power.

16.5.3.2 Depth

It was decided that the depth of the ship should provide for three tween decks heights plus a double bottom, which in view of the high rise of floor and the rake of keel contemplated had to be about 1.9 m amidships, the resultant depth being 9.2 m.

16.5.3.3 Breadth

The beam clearly had to be adequate in relation to the depth to provide very good stability in view of the heavy icing and severe winds in which the ship has to operate. After careful consideration, a B/D ratio of 1.5 was proposed giving a breath of 14 m. As this B/D ratio assumed that the ship would have minimum superstructure, it was accepted that the beam might have to be increased if considerable superstructure was found necessary as the design developed.

16.5.3.4 Draft and block coefficient

The following factors were considered in relation to draft:
- the desirability of good freeboard,
- the desirability of having a deep draft for good seakeeping in rough seas,
- the need for as deep a draft as possible to accommodate a large diameter propeller, the lower tip of which should not project below the keel so that it had the best possible protection from ice.

From a powering point of view it was concluded that there was little advantage in reducing the block coefficient below about 0.53. From a first estimate of the weight equation it seemed that the draft with a suitable block coefficient need only be about 4.6 m. If a draft of 6.0 m was to be achieved, the lines would therefore need to be extremely fine which would bring attendant arrangement problems. To avoid these problems whilst still achieving a draft aft of 6.0 m to suit a large diameter propeller, the following steps were taken.

Firstly, the ship was fitted with a box keel. This both increased the draft amidships and provided a most desirable substitute for bilge keels which had been found by the Royal Danish Navy (RDN) to have a very short life in ice conditions. The box keel proposed was 0.6 m deep and 1.0 m wide. Secondly, the ship was given a rake of keel of 1.6 m in the length BP with a resulting draft aft of 6.0 m.

16.5.4 Weights, displacement and hull form

These were calculated using the approximate methods given elsewhere in this book, supplemented by the use of some additional data relevant to the ship type and some direct calculation. The steel weight calculation paid attention to the fact that the ship would have double hull construction and scantlings appropriate to a fairly severe ice class. The machinery weight which assumed the use of medium speed diesels, was calculated in detail and refined as the accuracy of the power estimates improved.

The powering calculations were based on the following displacements:

89 m LBP ship: 2410 tonnes

105 m LBP ship: 2730 tonnes

120 m LBP ship: 3130 tonnes

The basis hull form was that of a high-speed trawler of known good tank test results and service performance. This was modified to incorporate an ice-breaking bow, the box keel and rake of keel already mentioned.

16.5.5 Propulsion, powering and machinery

The decision to opt for a single centreline propeller was subjected to further examination with the advantages and disadvantages of water jet propulsion, twin screws, Voith Schneider propellers and rotatable Z-drive thrusters all being considered. All of these were eventually rejected on the grounds of being too vulnerable to ice damage.

As a single propeller has the disadvantage that it provides no redundancy in the propulsion system, it was decided that a ship operating in the very severe environment envisaged should have an independent thruster to give it a get-you-home capability. A C.P. propeller was selected to enable the multi engine fit to operate at optimum RPM over a wide range of loads and avoid the need of a reversing gearbox.

The powers required by the three sizes of ship at the two specified speeds are given in Table 16.4.

Table 16.4

Powers required by the three sizes of ship at two specified speeds

Ship size (m)	Power (kW × 1000)		Speed 22 knots
	Speed 20 knots		
	Calm seas	4 m seas	Calm seas
89	7.0	11.8	11.3
105	6.0	9.4	9.3
120	5.2	7.8	8.0

At the main endurance speed of 16 knots the 89 m ship required a power of 2,320 kW and would use about 465 tonnes of fuel for the specified range, whilst both the larger ships were estimated to require 2,130 kW and use 415 tonnes of fuel. It was concluded that the 105 m hull would give the required performance.

16.5.6 General arrangement

The profile of the 89 m LBP design, that of the 105 m LBP design and that of the "Thetis" as built are given in Fig. 16.9. The 89 m ship arrangement followed the pattern of the ships which the new class were to replace, featuring a long bridge. The ship was sufficiently big to accommodate within this arrangement all the specified equipment and crew, but there was insufficient space to provide space for modules to permit rapid changes of role.

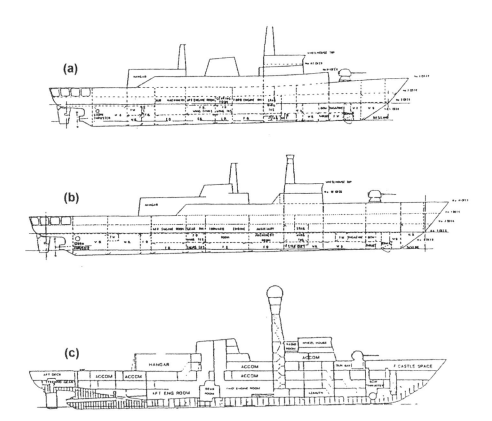

Fig. 16.9. Royal Danish Navy Fishery inspection ship designed by Y.A.R.D. and Dwinger Marinconsult. (a) The 89 m LBP outline design. (b) The 105 m LBP outline design. (c) The 99.5 m LBP design to which the R.D.N.S. "Thetis" and sister ships were built.

The most obvious change between the 89 m and the 105 m ships is the extension of the bridge forward as a combined bridge and forecastle, and the elimination of sheer forward made possible by the resulting increased freeboard at the bow. This change was the designer's reaction to seeing a film taken on the existing ships in which the speed and extent of ice accumulation was made dramatically clear and the desirability of having completely clear decks with all anchoring and mooring equipment below decks was seen as probably the only satisfactory answer. As a by-product, the elimination of sheer significantly eased the fitting of the additional modules which the increased space made possible.

Another advantage of the increased length was the ability this gave to shift the helicopter landing platform and hangar nearer amidships improving the operational window.

The definitive design of the "Thetis" as built closely followed the 105 m LBP feasibility study design with the main change being a small reduction in length to 99.75 m LBP.

The virtue of having spare space in a ship of this sort became apparent even before Thetis entered service as a need arose for the conduct of an oil exploration seismographic survey in the waters off Greenland and no existing ship could be found with the necessary capabilities. The conversion of Thetis to this role proved relatively easy and she carried out this most unusual and valuable service in her first year of operation.

16.6 OIL TANKERS

As stated in §2.9.2, the dimensions of a modern tanker with the large segregated ballast capacity required by Marpol regulations (see §13.3.5) are now controlled by the hull volume rather than by the displacement.

The length can be derived from eq. (3.9):

$$L = \left[\frac{V_h (L/B)^2 (B/D)}{C_{bd}} \right]^{1/3}$$

The hull volume V_h can, in turn, be derived from the specified cargo volume by dividing this by the constant K_c derived from Fig. 3.5.

Taking as an example a ship with a specified cargo capacity of 56,000 m^3 and a service speed of 14.5 knots, particulars quoted in *Significant Ships* 1994 for a standard design of products tanker the "Hadra".

From Fig. 3.5, $K_c = 0.66$
From Fig. 3.8, $L/B = 5.5$, $B/D = 1.91$
and substituting these in the equation

$$L = 4.43 \left[\frac{V_r}{C_{bd}} \right]^{1/3}$$

To solve this it is necessary to estimate C_{bd}. Guessing a first approximation to L of 200 m gives $F_n = 0.168$ and $C_b = 0.824$ from Fig. 3.12 or eq. (3.14) and $C_{bd} = 0.85$ from Fig. 3.8.

This gives $L = 178$; $B = 32.5$; $D = 17$ and $T = 11.4$.

For comparison the figures for "Hadra" are:

$L = 174$; $B = 32.2$; $D = 18$ and $T = 11.0$.

Having fixed the main dimensions, the next decision relates to the tank config-uration, a subject discussed in §13.3.6. Although the arguments for the mid deck tanker still appeal to the author, most companies feel obliged to meet the require-ments for trading in U.S. waters and build twin-hull ships.

16.7 CRUISE LINERS

The main requirements against which a cruise liner design is to developed will usually have been studied in an economic transportation study before the designers are given their brief, which will generally include statements covering at least the following matters:
- the passenger market for which the ship is intended to cater — de luxe, high class, middle market or package tours.
- the passenger numbers to be accommodated, divided into cabin types — outboard and inboard cabins; single berth, twin berth, four berth; with private toilet, with private bathroom etc.
- the maximum service speed and the intended normal service speeds on various legs of cruises.
- the sea areas and times of year in which the ship is to undertake cruises.
- the maximum duration of cruises between re-storing and re-fuelling ports.

The required crew may be stated, but if no guidance on this is given a reasonable estimate can be made by referring to §5.5.1.

With this data in front of him, a designer is well advised to look through the data available in *Significant Ships* and the technical press and collect photocopies of the plans, dimensions and other data on about six recently built ships which appear to have characteristics close to those required in the new design. This is not to suggest that the new design should slavishly copy an existing ship but so that the good features of these guidance ships (or some of them) can be adopted and the bad features avoided.

Whilst the advice given in the previous paragraph is applicable to some extent to the design of all types of ship, it is mentioned here because the assistance that a study of this sort can give can be particularly helpful in the complex task of designing a passenger ship.

Although many of the tasks involved in the design process are interactive and many of the decisions taken during the design need to be amended frequently as the design develops, it is possible to suggest an order of attack which will speed the design and minimise the need for alterations.

1. Establish the total volume required within the hull and superstructure that will enable all the requirements to be met, using the routine set out in §5.2.1. The accuracy of the volume calculation can be improved by basing the required cabin volume on specially drawn typical cabin plans. Two decisions which can have a significant effect on the required total volume are whether the ship should have an atrium and where the lifeboats are to be stowed.

   Atriums have become a feature of many recent large cruise liners and do appear to introduce a touch of luxury to these, but on the other hand they take up a lot of space.

   Lifeboats can either be stowed "up top" or under overhanging decks. An examination of the guidance drawings should help to decide which of these alternatives is more suitable to the size of ship that is being designed. The "up top" position is clearly essential on smaller ships, whilst the "under deck" position appears to have advantages on larger ships.

   The work involved in calculating the required total volume is quite considerable. Although this approach is the most methodical one that can be adopted and is essential if the ship requirements differ radically from those of any guidance ship identified, an alternative approach in which preliminary dimensions are estimated directly from those of one or more of the "similar" ships can be adopted.

   In using this approach it may be worth noting that the dimension that it is most important to get "right" is the breadth.

2. Assuming the methodical volume approach has been adopted, the next step is to establish the proportions of the total volume which will be provided within the main hull and superstructure respectively and from this to determine the main dimensions of $L \times B \times D$, using the method detailed in §5.3.

3. Having established preliminary main dimensions, a small scale (say 1:1000) sketch profile can then be drawn and used to establish the broad outlines of the design.

4. Although volume is the main controlling factor in the design of a cruise liner, it is best at this stage to make preliminary weight calculations using

the methods given in Chapter 4 and establish a suitable block coefficient and load draft.

5. This brings in another interactive stage as estimating the machinery weight, and for that matter the machinery space volume needed in item 1 above, requires a power estimate and a decision on the type of machinery to be fitted, with diesel, diesel–electric and gas turbine–electric being the main contenders.

6. A machinery power and machinery type decision is also needed to enable calculations to be made of the fuel space and weight which are also needed for the volume and weight estimates.

7. With the draft fixed, a depth to the bulkhead deck that gives a freeboard ratio likely to meet subdivision requirements can be established and this deck together with other decks above and below can be drawn in with tween deck heights arranged to suit the probable uses to which the decks will be put.

8. Decide on the spacing and number of watertight bulkheads which are likely to be required to meet subdivision and damaged stability requirements.

9. Decide on the position of the engine room and add the fore and aft peak bulkheads and the engine room bulkheads to the profile, followed by the other bulkheads — equally spaced at this stage, although their ultimate positioning will depend on frame spacing, accommodation module dimensions and of course subdivision.

10. To minimise steps in fire-resisting bulkheads it is best to position these in line with selected watertight bulkheads and this can with advantage be done at this stage (see §13.1.1).

11. Indicate lifeboat stowage as decided when making volume calculations also the extent of the atrium if one of these is to be fitted.

12. Decide on how the public rooms are to be positioned and allocate deck space to these. Based on the dining saloon position(s) decide where to locate the galley and use this in turn to position storerooms (see §15.5.3).

13. Position and allocate space to entrances, main stairs and lifts.

14. Allocate decks and space to the various types of passenger cabins and to the accommodation for officers and crew.

15. Carry out a preliminary stability assessment using the "profile method" shown in Fig. 4.21.

Now, and only now, is the time to start on a proper plan which is probably best developed on a 1:200 scale. Many details of the sketch plan will almost certainly be quickly left behind but the overall intent of this should persist.

Amongst things worth remembering as the plan develops is the desirability of positioning air conditioning/fan rooms so that vents from these can be conveniently led to the area they are intended to serve. The positioning of bathrooms and toilet spaces in tiers to minimise piping.

16.8 RO-RO FERRIES

Much of the guidance given on cruise liner design in the previous section applies also to Ro-Ro ferries and this section will concentrate on items which are specific to ferries. The following additional requirements will usually be set for the design of a Ro-Ro ferry:

1. The number of vehicles to be carried, generally stated in terms of a mixture of cars, buses and heavy goods vehicles. If the ship may have to serve a number of different routes or if the traffic is likely to vary at different seasons, it may be desirable for the ship to be designed to be able to accommodate more than one mixture of cars and bigger vehicles — with dimensions and weights for each of these categories.

2. Details of the loading/unloading ramps at each of the ports the ship is intended to use. Width and length of the ramp; distance to the centre of the ramp from the side of the quay against which the ship will be berthed; the height of the fixed end of the ramp at LWOST; the tidal range at springs and neaps.

As described in §5.6. and shown in Fig. 5.3, the dimensions of a Ro-Ro ferry are largely determined by the vehicle deck(s) layout. Obviously the largest vehicles to be carried exercise the greatest influence on the design. These are generally carried in the lanes nearest to the centre line, usually occupying at least two, but if a large number are to be carried three, four or more lanes may have to be arranged to suit these vehicles. If an even number of lanes is required for heavy vehicles, the engine casing and access stairs etc should be arranged in a centrally positioned casing, if however an odd number of lanes seems more suitable then the casings can be arranged asymmetrically, generally with one side on the centreline. There are at least two advantages in having the heavy vehicles near the centreline, one being that this gives these less manoeuvrable vehicles a straight or near straight run to the bow and stern doors, another being that the lever of any weight difference between the lanes is minimised and, thirdly, the fact that hoistable car decks can be more easily arranged in this position. By the end of this process a fairly clear idea of the ship's length should have emerged.

With the ship's length known, the fact that the breadth of the ship outboard of the heavy vehicle centre lanes should be a multiple of car widths plus access space fairly quickly determines the beam of the ship. With this known, the next step is to see how many cars can be carried on the main car deck and on the upper car deck that is generally fitted in at the ship sides to utilise the tween deck height required by the heavy vehicles. If the number of cars to be carried along with the maximum number of heavy vehicles cannot be met in this way the possibility of running cars either down ramps to a lower deck or up to a higher one should be examined.

At this stage it is worth examining whether the berthing arrangements suit the ship design as it is, and if not, whether the berthing arrangements should be modified or if the ship must conform to these, which of the ship design requirements can be modified.

With the preliminary length and breadth figures settled, it is now time to carry out weight calculations and determine a suitable block coefficient and draft.

The next decision is one the author feels strongly about. The tragedies of the "Estonia", the "Herald of Free Enterprise" and other Ro-Ro passenger ferries has focused a lot of attention on the need to improve the safety of this type of vessel and many ideas have been put forward. Some, like partial height car deck bulkheads and inflatable air bags at the ship side, provide solutions which can be applied to both new and existing ships. In the author's view, these proposals provide useful answers to the problem of existing ships but for a new design another and better solution is to increase the freeboard to the car deck considerably above the figure required by current subdivision rules. This will limit the amount of water that can get onto the car deck as the result of any type of accident and provide a head that will ensure that the water drains away through scuppers. Just what increase in freeboard compared to present practice is required needs investigation but an additional metre of freeboard would undoubtedly effect a very considerable improvement in the safety of most existing ships.

Raising the car deck by a figure of this order would raise the centre of gravity by a similar amount and to keep the metacentric height unchanged would require either a reduction in the superstructure to restore the KG value or a similar increase in KM.

Reducing the superstructure would reduce earning capacity and be unpopular but fortunately, as discussed in §8.6.2, the KM value of most ships could be increased by adopting high stability lines without any need to increase the breadth.

With the depth to the car deck fixed the design can then be developed along much the same lines as those discussed under cruise liners.

16.9 GUIDANCE ON DESIGNS FROM SIGNIFICANT SHIPS

The plans and data in the six volumes of *Significant Ships* published between 1990 and 1995 are an excellent source of design information to ship designers. Unfortunately the index in each of the volumes is not really arranged to take a designer quickly to the ships which are most likely to assist him and with a new volume each year this will be an increasing problem.

The following pages try to fill this need, with Table 16.5 giving an indication of the numbers of designs given in the 1990 to 1995 volumes for each of the main categories of ship.

Table 16.5

Overall guide to *Significant Ships* 1990–1995

Ship type	1995	1994	1993	1992	1991	1990	Total
Container ships	11	20	5	13	6	9	64
Bulk carriers	7	7	4	6	7	3	34
Tankers	10	11	22	11	18	13	85
Specialist cargo ships	3	3	2	5	2	5	20
Ro-Ro ferries	9	2	12	5	9	7	44
Passenger ships	4	1	2	5	4	6	22
Specialist service ships	5	2	2	3	2	2	16
Totals	49	46	49	48	48	45	285

Each of the six main categories of ships in Table 16.5 includes a number of more specialised ship types:
- container ships (Tables 16.6A and B) include multi-purpose container/bulk
- bulk carriers (Table 16.7) include OBOs, self-unloaders and bulk cement
- tankers (Tables 16.8A, B and C) include crude oil carriers, product carriers, shuttle and FPSO, bitumen/oil, palm oil, chemicals, sulphur, fruit juice, LNG tankers, and LPG tankers
- specialist cargo ships (Table 16.9) include refrigerated cargo ships, livestock carriers, car carriers, steel coil carriers, and coasters
- Ro-Ro ferries (Table 16.10) include passenger and car ferries, freight ferries, and train ferries
- passenger ships (Table 16.11) include cruise liners, sail cruise ships, passenger/cargo ships
- specialist service ships (Table 16.12) include dredgers, cable layers, factory trawlers, heavy lift ships, research vessels, icebreakers, and offshore supply ships.

In Tables 16.6–16.12 the data and plans given in *Significant Ships* are indexed by year, ship type and main particulars. The deadweight quoted is usually the design one if this is available. The dimensions given are LBP, breadth and draft (matching the deadweight). The capacity given is that appropriate to the ship type: grain, bale, liquid, refrig, container TEUs, car numbers or lane length for Ro-Ro ferries, and passenger numbers for cruise liners and ferries.

The order in which ships are given is one of descending size — in terms of deadweight, one of the capacity measurements or the length BP, whichever seems more appropriate to the ship type and the available.

Some special features are mentioned in the remarks column, but limited space has confined this to one or two items so please consult the full data in the reference.

Table 16.6A

Container ships in order of container TEUs

Year	Name	Dwt	Main dimensions			Speed	Capacity TEU	Remarks
			L	× B	× T			
95	OOCL California	50037	262	40	12	24.6	4960	Largest container ship to 1995
95	APL China	49253	262	40	12	24.5	4832	
95	APL Korea	49350	262	40	12	24.6	4826	
95	NYK Procyon	47300	283	37.1	11.2	23.5	4743	
92	Hyundai Admiral	52233	263	37.1	12.5	25.1	4411	
91	Hannover Express	55590	281.6	32.25	12.5	23.8	4407	
95	Neptune Sardonyx	52320	281.6	32.25	12	24.5	4388	Highest capacity Panamax
94	Ever Racer	56100	281	32.22	12.5	22.7	4229	
94	Nedlloyd Hong Kong	51151	265	37.75	12.53	24	4112	Largest hatchcoverless ship to 1994
92	Hanjin Osaka	54622	277	32.2	12	24	4024	
94	Zhonge	44037	264.2	32.2	11.5	24	3764	
91	Nedlloyd Europa	36400	253	32.24	11	21.5	3568	Hatchcoverless
94	Ville de Vela	37128	225.2	32.2	10.8	22.5	3538	
95	Chesapeake Bay	37500	232	32.2	11.3	23.5	3467	
93	Tokyo Senator	35734	206.16	32.2	11	20.5	3017	
94	Norasia Hong Kong	35380	229.5	32.24	11	22.5	2780	Hatchcoverless
91	Vladivostok	40250	225.25	32.2	11	22	2668	
94	Trade Sol	31470	191.96	30.6	9.95	19.5	2480	
92	Zim Hong Kong	37865	224.5	32.2	10.5	21	2402	
94	Nuevo Leon	29256	191	32.2	11	20	2396	
92	R J Pfeiffer	21500	203.15	32.2	10.51	22.5	2292	
95	Canmar Fortune	33800	204	32.2	10.78	19.9	2268	
92	Betelgeuse	23000	179.8	32.2	10	19	2232	Ro-Ro/Lo-Lo; 3 cranes
90	Cap Polonio	22263	188.19	32.2	10	18.5	1960	Two cranes
90	CGM Provence	26288	166.96	27.5	10.52	18.6	1799	
92	Muscat Bay	23805	172	28.4	10.1	18.7	1742	Four cranes
93	Contship Pacific	23276	153.7	27.5	10.66	19.4	1684	Three cranes
92	Atlantic Lady	16160	160	18.8	8	19.25	1646	Hatchcoverless
94	Westerdeich	17600	156	26.7	9.7	20	1572	Three cranes
93	San Lorenzo	17205	156	27.4	8.75	19.5	1512	Three cranes
94	Cecilie Maersk	19350	180.15	27.8	8.25	19	1501	Travelling gantry crane
94	Nordlake	22450	167.26	25.3	9.9	18.95	1496	Three cranes
94	Nedlloyd River Plate	19762	158	27.2	8.75	19.36	1444	Two cranes
94	Marwan	18985	158.71	25	9.5	18.6	1400	Three cranes

(continued)

Table 16.6A (*continuation*)

Year	Name	Dwt	Main dimensions			Speed	Capacity TEU	Remarks
			L	× B	× T			
91	Bunga Kenari	21571	165	27.3	9.5	15.5	1201	
91	Kota Wijaya	20755	174	27.6	9.02	19.1	1138	
90	Nordlight	11420	145.2	22.86	7.65	17	1050	Two cranes
92	Uni-Crown	17374	141	25.6	9.5	16	1038	
92	Kairo	12580	140.14	22.3	8.25	18.5	1012	Two cranes
90	Katherine Sif	9766	1207	22.7	7.6	17.2	976	Two cranes
92	Cape Bonavista	10410	126.4	22.7	7.5	16.55	930	Two cranes
92	Cape Hatteras	12855	134	23.5	8.65	18.1	925	Two cranes
94	Sea Arctica	8500	118.5	24	7.65	17	780	Two cranes; for Greenland trade
95	Bunga Mas Satu	10500	124.55	20.8	7.50	17	668	Bridge and accommodation forward
92	Secil Angola	8371	115.45	20.8	6.5	15	650	Two side cranes
91	Hanjin Bangkok	8075	114	20	6.61	14	414	
90	Bell Pioneer	3900	106	16.92	5.2	14.5	301	First hatchcoverless ship

Table 16.6B

Multi-purpose container/bulk in order of container TEUs

Year	Name	Dwt	Main dimensions			Speed	Capacity TEU	Remarks
			L	× B	× T			
90	Serenity	17175	146.85	23.05	10.09	16.2	1033	Three cranes
90	Admiralengracht	12100	121.24	18.9	8.61	–	678	Three side cranes
94	Frotabelem	10640	124.4	22.8	7.8	17.18	666	Two twin cranes
94	Germania	8790	116.95	19	8	16	645	Two side cranes; Ro-ro ramp
94	Arktis Fighter	5212	93.6	18.8	6	15.7	444	Two side cranes
95	Sloman Challenger	5665	94.73	17.8	6.4	14.5	400	Two side cranes
93	Tropic Tide	3441	110	22	6.2	15.7	400	Ro-Lo; lane length 200 m
94	Cari Sea	4766	93	16.5	5.5	14	390	No gear; suitable grab discharge
95	Irena Arctica	5238	99.2	21.5	6.5	14	378	For Greenland trade
94	Arcadian Faith	5273	96.7	16.4	6.55	16	373	Two side cranes
92	Celtic Crusader	5750	84.5	17	6.4	13.5	300	

(*continued*)

Table 16.6B (*continuation*)

Year	Name	Dwt	Main dimensions			Speed	Capacity TEU	Remarks
			L ×	B ×	T			
90	Sea Bird	4250	86.5	15.5	5.6	13.5	276	Two side cranes
90	Roberta Jull	3100	83.96	15.1	5	13.18	270	Two side cranes
94	Leknes	4226	84.8	13.8	5.6	12.5	232	
95	Bermuda Islander	2800	84.65	13.75	4.3	13.5	205	
93	Morgenstond I	4292	84.86	13.17	5.7	11.8	190	
94	Fischland	3540	81	12.8	5.45	11.4	168	Two side cranes

Table 16.7

Bulk carriers in order of deadweight

Year	Name	Dwt	Main dimensions			Speed	Capacity grain	Remarks
			L ×	B ×	T			
Bulk carriers								
92	Bergeland	272132	327	55	20	15.2	174324	Second largest ore carrier to 1992
90	Hanjin Gladstone	207000	300	50	18	13		Ore/coal
94	Erradale	152000	276.73	44.4	16.75	15.5	181000	
95	Merchant Prestige	149674	270	45	16.5	16.5	175753	Cape-size
93	Erridge	114012	256	40.5	14.52	14.11	136042	
94	Corona Ace	77447	220	36	12.79	13.8	91045	Design dwt 69940 for coal cargo
95	Brazilian Venture	70728	215.4	32.26	13.7	14	81315	Panamax
90	China Pride	65655	215	32.2	13.11	14.91	78067	
91	Solidarnosc	63000	224.6	32.24	12.5	13.8	85525	
94	Romandie	62600	221	32.24	12.5	14.75	85200	
94	Thalassini Tyhi	62158	216	32.5	12.2	14.6	85600	
91	Dixie Monarch	44679	194	32.2	10.7	14.3	99704	Three cranes; deep hull; wood chip cargo
94	Angel Wing	44950	176	32	10.72	14.3	56297	Four cranes
92	Pacific Endeavour	40750	176.8	30.5	10.7	14.3	53860	Four cranes; deck cargo fittings
94	Saga Spray	37543	190	30.5	10	15	51946	Side tanks; protected cargo handling

(continued)

Table 16.7 (*continuation*)

Year	Name	Dwt	Main dimensions			Speed	Capacity grain	Remarks
			L ×	B ×	T			
92	Alam Selaras	33710	171	30.5	9.75	14.5	46112	Four cranes (Freedom class)
95	Atlantic Bulker	27492	169.4	26	9.32	14	38239	Four cranes; log carrier
94	Erna Oldendorff	18355	136	22.8	9.15	14	23212	Four cranes
90	Igor Ilinsky	7365	122	19.86	6.87	15.2	–	Two twin cranes; timber carrier
95	Arklow Brook	7182	95	17	6.75	11.75	8892	Mini-bulker
95	Baumwall	3873	92.05	15.3	5.46	14	5950	Side loading newsprint carrier

OBOs

91	Front Driver	152001	275	45	17	14.73	175289	Double hull
92	Scanobo Trust	76694	234	38	12.2	14.5	111192	Double hull
93	Sibohelle	66175	242.88	32.24	12.5	14	89431	Double hull + 9 oil tanks
92	Futura	61355	224.6	32.24	12.5	13.48	83336	Double hull + 9 oil tanks

Self unloaders

91	Western Bridge	96725	239	38	15.02	15	89897	Iron ore; conveyors
91	Yeoman Burn	77500	235	32.2	14	14.61	72104	Aggregates; conveyors
91	Pearl Venus	48495	217	32.2	10	13.5	114470	Three cranes and conveyor for wood chips
95	Hai Wang Xing	37944	178.8	29	9.8	14.5	39564	Collier
93	Malmnes	9891	120.95	15.8	7.45	13.4	11767	Cargo scrapers and elevators

Bulk cement

93	Goliath	15539	134	23.5	8.3	15.3	13729	Pneumatic loading and discharge
95	Koralia	8500	113	19	7	14.5	6646	Self discharge system
91	Halla No 2	8050	106	17.8	7.03	13	7216	Fluidising and conveyor discharge
92	Kanyo Maru	7535	108	17.5	7.07	12.6	6021	Fluidising and conveyor discharge

Table 16.8A

Tankers in order of deadweight

Year	Name	Dwt	Main dimensions			Speed	Capacity liquid	Remarks	
			L ×	B ×	T			Hull type	Tanks abreast

Tankers — crude oil

Year	Name	Dwt	L ×	B ×	T	Speed	Capacity liquid	Hull type	Tanks abreast
93	Berge Sigval	306430	317	58	22.37	16	350344	Double	3
93	Siam	302377	320	58	22	15.86	346717	Double	3
95	Crown Unity	300000	314	58	22.2	14.7	345096	Double	3
93	Arosa	291381	315	58	21	15	332700	Double	3
90	Argo Electra	285000	315	57.2	20.8	14	330000	Single	3
92	Golar Stirling	282030	320	58	20.8	15.5	351670	Double	2
93	New Vitality	279865	319	56	21.5	15.4	332835	Single	3
93	New Wisdom	279863	317	59	20.95	15.3	357753	Double	3
95	Murex	277800	320	58	20.8	15.5	345000	Double	3
95	Yukong Navigator	277798	315	51.2	20.45	15	330647	Single	
93	Eleo Maersk	269480	327	56.4	19.8	14.5	340800	Double	3
90	Sea Duke	261604	313	56	19.6	14	318544	Single	3
93	Okinoshima Maru	259552	319	60	19	15.7	316074	Single	3
93	Cosmo Delphinus	258095	311	58	19.49	15.25	318147	Single	3
91	Prosperity	258080	321	58	18.48	16	322815	Single	3
92	Nisyros	143932	263.3	44.5	16.9	14.3	167169	Double	2
90	Jahre Traveller	142000	260	44.5	15.6	14	170000	Single	3
91	Landsort	141844	264	48	15.2	14.7	172850	Double	1
94	Ankleshwar	139115	264	46	15.9	14	170315	Double	2
92	Chevron Atlantic	137678	258	46	15.77	15.1	166986	Double	2
93	George Schultz	136055	245.4	48.3	16.76	15.7	162300	Double Alt 1 and 2	
92	New Fortune	135830	267	44.4	15.6	14.5	169110	Single	
93	Knock Clune	135287	267	44.4	15.9	14.4	163279	Double	1
91	Knock Allan	135000	267	44.4	15.6	14.4	167600	Single	3
92	Wilomi Tanana	134003	265	43.2	16	14	163578	Double	1
94	Eco Africa	134000	264	45.1	15.75	14	166405	Double	1
94	Hanne Knudsen	120000	257	42.5	15	15	140800	Double	Diesel elect
92	Mayon Spirit	98507	234	41.2	14.42	14.8	120043	Double	1
91	Olympic Serenity	96733	221.12	42	14.2	13.9	114580	Double	1
90	Onozo Spirit	89315	234	41.2	13.1	14.5	123698	Single	3
90	Dicto Knutsen	87067	233	42.5	12.25	14	125488	Single	
93	Eos	85914	235	45.64	11.58	16	109970	Double	1
94	Yuhsei Maru	84100	233	41.8	12.2	14.8	109800	Double	1
93	Glenross	82474	236	41.6	12.2	14	108030	Double	1

(continued)

Table 16.8A (*continuation*)

Year	Name	Dwt	Main dimensions			Speed	Capacity liquid	Remarks	
			L	× B	× T			Hull type	Tanks abreast
Tankers — Products									
92	Futura	91000	231.2	40	14	14.4	105000	Double	1
90	Zafra	54000	218.7	32.24	11.58	14.4	90500	Double	2
94	Hadra	40546	174	32.2	10.97	14.5	56000	Double	1
91	Salamina	40260	174	32.2	11	14	56407	Double	1
94	Kandilousa	40068	174	32.2	11	14	52750	Double	1
91	Arbat	39700	174	32.2	10.97	14.5	56000	Double	1
90	BP Admiral	33000	168	30.8	10	14	48277	Single	3 Side tanks WB; centre oil
91	Stolt Markland	29999	167.2	29.5	9.79	15.5	38070	Single+DB	4 parcel tanker
95	Lista	26400	163	24.3	10.1	14	32970	Double	2 Twin screw
95	Jian She 51	13144	135.7	19.6	8.4	14.75	13900	Double	2
93	Anchorman	6200	95.64	17.5	6.85	12.66	8686	Double	2
90	Agility	2680	74.9	14.5	5.1	11.5	3328	Single+DB	2 Coastal tanker

Table 16.8B

Specialist tankers in order of deadweight

Year	Name	Dwt	Main dimensions			Speed	Capacity liquid	Remarks	
			L	× B	× T			Hull type	Tanks abreast
Shuttle and FPSO									
95	Heidrun	123000	254	46	15	15	140800	Double	Riser loading
93	Tordis Knutsen	116596	251.5	42.5	14.5	14.7	138715	Double	Shuttle
93	Griffin Venture	97962	230.6	41.8	14.17	8.5	130602		FPSO oil and gas
Bitumen/Oil									
91	Theodora	5200	103	17	6.1	14.5	5245	Molten bitumen up to 250°C	

(*continued*)

Table 16.8B (*continuation*)

Year	Name	Dwt	Main dimensions			Speed	Capacity liquid	Remarks	
			L ×	B ×	T			Hull type	Tanks abreast
95	Tasco II	4592	99.9	15.8	5.71	14	4350	Molten bitumen up to 240°C	
Palm Oil									
91	Bunga Siantan	16924	133	22.4	9.1	13.5	19733	Double	3
Chemical Tankers									
91	Fandango	46087	173	32.2	12.25	14.5	52437	Double	1
93	Jo Selje	37300	176.1	32	10.73	15.5	39260	Double	3
91	Tirulami	31045	165	31.3	9.12	15	19235	Double	Phosphoric acid centre tanks
91	Conger	23400	160	24.6	10.33	14.5	27740	Double	3
95	Brage Atlantic	16094	132.4	22.8	9.2	14.7	19587	Double	Products/chemicals
91	Jo Alder	12600	126.4	21.25	8.06	14.5	14300	Double	3 and 2 at ends
92	Weserstern	8795	103.6	17.7	8.4	12.5	10000	Double	2
93	Nathalie Sif	8603	110	19	7.51	14.1	10940	Double	2
92	Marinor	7930	105	18	7.5	14.4	8505	Double	2
91	Trans Arctic	7000	108.4	17.5	7.71	15	7553	Double	1 and 2; Stainless steel tanks
91	Katarina	6000	95.52	17.5	6.1	10.2	7440	Single + DB	2
Sulphur									
94	Sulphur Enterprise	25838	151.64	27.43	10.06	14	–	Insulated tanks for molten sulphur	
93	Janana	8850	119.6	19	7.6	14	7600	4700 molten sulphur; 2900 oil.	
Fruit Juice									
93	Ouro do Brasil	18600	160.6	26	9.5	20.5	12455	Refrig stainless steel tanks	

Table 16.8C

LNG and PLG tankers in order of deadweight

Year	Name	Dwt	Main dimensions			Speed	Capacity liquid	Remarks
			L ×	B ×	T			Tank nos. & type
LNG Tankers								
95	Ghasha	68351	280	45.75	10.97	20.76	137709	5 spherical tanks
90	Northwest Snipe	66695	259	47.2	11.37	18.5	125670	4 spherical tanks
94	Hyundai Utopia	63621	260	47.2	10.97	18.5	127088	4 spherical tanks
94	Puteri Intan	62265	260.8	43.3	10.86	19.9	130300	4 membrane tanks
95	Hanjin Pyeong Taek	61436	257	43	11	19.02	130637	4 membrane tanks
93	Polar Eagle	42031	226	40	10.1	18.5	89880	4 prismatic tanks
LPG Tankers								
90	Pacific Harmony	49701	212	36	11.01	16	75208	4 prismatic tanks
92	Berge Clipper	49082	212	36	11.25	16.75	78549	8 prismatic tanks
92	Baltic Flame	48572	210	34.2	11.7	16.55	76664	4 prismatic tanks
93	Gas Al-Gurain	48495	219	36.6	10.6	16.7	78474	4 prismatic tanks
91	Helice	35600	193.6	32.2	10.5	16	57000	4 prismatic tanks
91	Jakob Maersk	22982	173.4	27.4	9.4	17.3	35559	8 prismatic tanks
91	Annapurna	17562	153.5	25.9	8.3	15	22937	3 prismatic tanks
90	Norgas Christian	9500	122.02	17.8	8.62	16.25	8237	3 bi-lobe tanks
93	Vallesina	6350	96	16.8	7.8	14.5	6000	3 bi-lobe tanks
94	Tarquin Ranger	5771	95.6	15.7	6.9	15.5	5600	2 cylindrical tanks
94	Pointe Clairette	5278	99.9	14.75	5.5	12.36	6067	+640 m^3 liquified Butane
90	Gitta Kosan	2900	90	16.2	5.3	14.75	4300	3 bi-lobe tanks
90	Maria Cristina Giralt	1165	72.8	12.95	4.1	13.5	1600	2 dome end cylindrical

Table 16.9

Specialist cargo ships in order of deadweight

Year	Name	Dwt	Main dimensions			Speed	Capacity	Remarks
			L ×	B ×	T			
Reefers								Refrig + container TEUs
90	Ditlev Lauritzen	16600	150.6	24	10	20.2	21684	186 TEU
92	Courtney L	13620	192	27.2	7.8	21.5		868 TEU; banana carriage
91	Chiquita Deutschland	13500	145	24.4	10	22.25	18280	299 TEU
94	Carmen Dolores H	11004	125.72	20.5	8.26	18.8		754 TEU; 468 @ +4°C; 42 @ –25°C
93	Barrington Island	10358	165	25.2	8	22.2	17631	322 TEU
92	Justinian	6585	138.3	22.6	7.4	21.8	14360	178 TEU
92	Hudson Rex	6321	140	20.6	7.02	19.2	14620	38 TEU; Derricks
90	Del Monte Pride	6300	147.5	23.5	6.7	20	16332	145 TEU
94	Dole America	6263	138.5	22.6	7.4	21.8	14686	264 TEU
90	Hornbay	5900	141.5	13	7.3	20	13169	Multi-purpose; Ro-Ro; cranes
92	Crystal Pride	4500	121	19.6	7.25	20.4	9900	68 TEU; banana cargo
90	Ice Star	3187	84.27	15.1	5.3	13.3	5240	42 TEU
Livestock Carriers							Livestock nos.	
95	Bison Express	3173	93.19	15.85	5.64	16	1750	
Car Carriers							Car nos.	Decks/Loading
92	Otello	18424	190	32.26	9.5	20.2	6151	12/side and quarter
95	Hual Trooper	16319	190	32.26	9	20.34	6480	12/side and quarter
94	Titus	15199	190.5	32.26	9.5	20.3	6134	13/side quarter
93	Fides	12130	164	26.8	7.6	19	2589	9/quarter
Steel Coil Transporter							Bale	
91	Hakuryu Maru	2510	110	18	5	15	3968	Self loading/unloading
Coasters								
90	Union Jupiter	3220	96	12.5	4.27	11	4477	Suitable Rhine waterways
95	Sea Rhone	2046	77.4	11.3	3.73	10.8	2835	Low air draft

Table 16.10

Ro-Ro ferries in order of length BP

Year	Name	Dwt	Main dimensions			Speed	Capacity and Remarks				
			L ×	B ×	T		Cars	Trail-ers	Lane (m)	Pass-engers	Cabins
Ro-Ro Ferries: Passengers and cars											
91	Ferry Lavender	2689	181	29.4	6.75	21.8			2880	796	86
90	Silja Serenade	3500	180.7	31.5	6.8	20	464			2500	–
91	Ishikari	6146	175	27	6.7	21.5	151	76		854	71
95	Isle of Innisfree	5285	172.3	23.4	5.6	22	605			1650	34
93	Silja Europa	5380	171.6	32	6.8	21.5	400			3013	1194
94	Finnhansa	9005	171.3	28.7	6.8	21.74			3380	90	32 future trains
90	Sabrina	5770	171	24.7	6.75	23.2	140	170		694	–
91	European Seaway	7550	170	27.8	6	21		124		200	81
93	Pride of Burgundy	6000	170	27.8	6.27	21.5	600			1320	79
95	Aretousa	–	166.4	27.0	6.3	23.8	634			1500	125
95	Robin Hood	6600	166	27.2	6	18.5			2400	300	
93	Majestic	7150	163.05	26.8	6.7	23			2139	1500	339
92	Pacific Express	5113	158	25	6.5	26.2	90	100		660	62
95	Superfast 1	4592	158	24	6.25	26.8			2245	1400	200
93	Norbank	6170	157.65	23.4	5.8	22			2000	114	57
93	Spirit of Brit.Colum.	3109	156	26.6	5	19.5	470			2100 day passengers	
94	Paglia Orba	6325	152.6	29	6.63	19			2330	267	95 Freight 9795 dwt
93	Kalliste	6600	150	29	6.5	20			2375 +	202	98
91	Prins Filip	3899	150	27	6.2	21	600			1200	121
92	Frans Suell	2962	149.8	27.6	6.25	21.5			2200	2172	44
92	Normandie	4225	146.4	26	5.65	20.5	680			2120	217
92	Barfleur	4130	146.35	23.3	5.4	19.5			2510	1212	78
90	Olau Britannia	5118	144	29	6.5	21.3	575			1600	–
91	Stena Challenger	4598	142	24	5.5	18			1800	120	60 (drivers)
91	Wakanatsu Okinawa	4336	138	23	6.4	21	180	42		150	24
93	Juan J Sister	5585	133.3	26	6	18			1680 +	550	139
95	Manuel Azana	4871	126.8	18.4	3.68	20	450			1184 seating only	

(continued)

Table 16.10 (*continuation*)

Year	Name	Dwt	Main dimensions			Speed	Capacity and Remarks				
			L ×	B ×	T		Cars	Trail-ers	Lane (m)	Pass-engers	Cabins
95	Bang Chui Dao	3873	125	23.4	5.4	20	226			938	169
93	Kong Harald	902	103.8	19.2	4.7	18	50			490	230
93	Ibn Battouta 2	1492	103	18.9	5	17.85			1304	1300 seating only	
93	Las Palmas de Gr.Can.	2700	101.83	20.7	5.3	16			993	378	39
93	Caledonian Isles	600	85.2	15.8	3.15	15	52			1000 seating only	
	Freight Ro-Ro										
91	Krasnograd	14308	161	23.05	9	18.8	242	83		Quarter loading; 3 cranes	
90	Intrepido	13150	157.4	26.3	7.2	15.2			3010	Side bow and stern; 3 cranes	
91	Helena	11843	157.2	25.6	6.7	14.6			2278		
90	Ahlers Baltic	9515	148	25	7.2	19.47			2100	Stern loading	
93	Hokuren Maru	4674	142.8	21.4	6.6	23.5			1300	Milk delivery	
92	Via Ligure	6200	137.32	23.4	5.6	19.2			1850		
90	Shinka Maru	5043	130	21.2	6.6	17.75	50	80		Quarter loading; newsprint rolls	
95	Island Commodore	5238	118.5	21	6	18			1810	Short haul for trailers	
95	MN Toucan	4250	105.6	20	3.65	15.5			–	Ro-Ro/Lo-Lo for Ariane rockets	
90	Bore Sea	4000	98	17	5.8	15.3			910	Stern loading	
	Train Ferries										
95	Polonia	7250	159	28	5.9	20			2200	920	204
91	Tycho Brahe	2500	106	27.6	5.5	13.5	240		815	1250	–

Table 16.11

Passenger ships. Cruise liners in order of passenger numbers. Passenger/cargo ships in order of deadweight

Year	Name	Dwt	Main dimensions			Speed	Capacity passengers	Remarks
			L ×	B ×	T			
Cruise Liners								
91	Monarch of the Seas	6000	236	32.2	7.53	22	2744	Geared diesel
90	Fantasy	7000	224	31.5	7.85	21	2604	Diesel electric
95	Legend of the Seas	4800	221.5	32	7.55	24.1	2066	Diesel electric
95	Oriana	7270	224.05	32.2	8.2	24	1975	Geared diesel
90	Crown Princess	5400	204	32.25	7.85	19.5	1900	Diesel electric
93	Statendam	5500	185	30.8	7.5	20	1629	Diesel electric
93	Windward	4800	160	28.5	6.8	21	1500	Geared diesel
90	Horizon	4300	175	29	7.2	21.4	1354	Geared diesel
90	Crystal Harmony	5039	205	29.6	7.3	22	960	Diesel electric
95	Crystal Symphony	8000	203	30.2	7.6	22	960	Super-luxury
92	Crown Jewel	1800	139.83	22.5	5.4	19.5	916	Geared diesel
91	Asuka	3596	160	24.7	6.6	21	584	Geared diesel
92	Radisson Diamond	1300	115.78	32	8	12.5	354	SWATH design
92	Royal Viking Queen	820	112.4	19	5	19.3	212	Mini cruise liner
91	Society Adventurer	1100	105.5	18	4.7	17	188	Geared diesel
90	Renaissance II	523	74.85	15.2	3.7	15.5	100	Geared diesel
Sail Cruise Ships								
92	Club Med 2	1600	156	20	5.09	16	410	Cruising under sail
92	Star Clipper	620	70.2	15	5.73	11	194	16 knots under sail

Passenger/Cargo							Capacity TEU (T) Bale (B)	Passengers (P) Cabins (C)
95	Zi Yu Lan	5881	137.5	24	6.9	20	2930(T)	392(P) 122(C)
94	Xin Jian Zhen	4321	143	23	6.2	21	2181(T)	355(P) 55(C)
90	St Helena	3130	96	19.2	6	14.5	52(T) 3750(B)	132(P) – (C)
91	Sirimau	1397	90.5	18	4.2	15	– (T) 492(B)	969(P) 54(C)

Table 16.12

Specialist service ships in order of deadweight

Year	Name	Dwt	Main dimensions			Speed	Capacity	Remarks
			L ×	B ×	T			
Dredgers					Dredging		Hopper	
94	Pearl River	25000	135.86	28	10.44	16	17000	Trailing suction to 30 m
92	J F J De Nul	17150	127	25.5	9.2	15.2	11750	Trailing suction to 45 m
90	Antigoon	12948	109.6	22.4	8.68	14	8300	Trailing suction to 33 m
92	Camdijk	9945	104.6	19.6	7.65	13.33	5110	Trailing suction to 50 m
95	Hang Jun 5001	8105	106.5	17.95	7.25	13.9	5018	Trailing suction to 30 m
90	Sand Heron	5715	94.5	16.3	6.4	12.5	2500	Trailing suction to 33 m
Cablelayers								
95	Cable Innovator	10500	131.9	24	8.3	14.5	–	Stern working
92	KDD Ocean Link	5464	121	19.6	7	15	2650	Bow and stern working
91	CS Sovereign	5060	106.94	21	5.9	14	–	Bow working
94	Asean Restorer	4800	117.5	21.8	6.3	16	–	Stern working
Factory-freezer stern trawler							Refrig	
95	Johanna Maria	5675	112.31	17.5	7.16	17	7200	Worldwide trawling pelagic fish
93	Kapitan Nazin	4895	89.5	20	9.2	14.5	5932	
Heavy lift carrier							Bale	
95	Jumbo Spirit	5199	87.55	17.75	6.8	15.25	7280	500 tonnes lift capacity
Research vessels							Grain+liquid	
95	Marion Dufresne	4871	108.33	20.6	6.95	15.7	5600+1170	Carries supplies to sub Antarctic
91	James Clark Ross	2589	90	18.85	6.3	12		Antarctic and hydrographic reserach;
Icebreaker/Offshore Supply Ship								
93	Fennica	1650	96.7	26	7	16	–	4800dwt at 8.4 m draft

Chapter 17

Specification and Tender Package

17.1 SPECIFICATION PRINCIPLES

Shipowners and shipbuilders tend to take fundamentally different approaches to specification writing. A shipowner will usually write a "performance" type specification which states the performance required from the ship and from its equipment and systems. It need not, and indeed from a legal point of view it is possibly better if it does not, give any guidance on how the performance is to be obtained, leaving this to be entirely the responsibility of the shipbuilder. This is to take a somewhat academic view of a very practical piece of prose, however, and most shipowners believing their experience to be valuable include a lot of guidance, particularly about trade peculiarities, in their specification as well as stating the required performance.

A shipbuilder's specification will usually concentrate on describing the equipment and systems which will be provided and say as little as possible about the performance, recognising that the more that is said about performance the more opportunities there are for the shipowner to claim that the specification is not being met. Two examples of this are:

1. A shipowner will specify the speed which he wants the ship to achieve in service; a shipbuilder on the other hand will prefer to specify the speed that the ship is to achieve on trial, rightly regarding service conditions as outside his control.

2. A shipowner will specify that the ship should have a turning circle of a specific number of ships lengths; a shipbuilder will prefer to specify that the ship is to be equipped with a rudder of a particular area and a steering gear of a particular power.

Before a contract is agreed the draft specification, whether this was originally written by a shipowner or a shipbuilder, should be modified to include both the performance characteristics required by the shipowner and a description agreed by both shipbuilder and shipowner of how that performance is to be obtained. Such an agreement is particularly important where the required performance is in any way unusual.

It is important to remember that the specification forms part of the contract and its precise wording may become important in law in the event of a dispute.

17.2 ADVANTAGES OF STANDARDISING THE FORMAT

It is highly desirable for both shipowners and shipbuilders to write all their specifications in a standard format. Shipowners will from time to time find themselves having to accept a shipbuilders specification and *vice versa*. When either party is writing a specification they will, however, find it time-saving and the best way of avoiding omissions if they use a standard format. The advantages of a standard format continue during building and indeed, as far as the shipowner is concerned, they continue throughout the ship's life when it enables any desired information to be found quickly.

The best English syntax in which to write a specification is a matter for debate. Some alternatives are:

(i)   Note form with no verbs — "classification Lloyds Register"
(ii)   Verbs in the present tense — "The ship is classed to Lloyds Register"
(iii) Verbs in the future tense — "The ship will be classed to Lloyds Register"
(iv)   Verbs in the imperative tense — "The ship shall be classed to Lloyds register"
(v)   Verbs in the infinitive tense — "Ship to be classed to Lloyds Register" or "Ship is to be classed to Lloyds Register".

The author's personal preference is for the note form (i) for outline specifications and for one of the two infinitive tenses (v) for detailed specifications. But the only real rule is to choose one of the syntaxes and stick to it.

A reason for stressing the importance of using one tense throughout is the fact that a change in syntax may be considered to represent a change in emphasis in what is written which could become important in a legal sense should the contract result in litigation.

The way in which working to a standard format helps to ensure nothing is omitted has been mentioned. Almost equally important is the avoidance of repetition, not only within the specification itself, but also between the specification and the contract.

The danger of repetition lies in the fact that when there are two statements there is quite likely to be some difference between them, either *ab initio* or because one has been altered at some time to accommodate some change and the other has been forgotten. This can lead to unnecessary disagreements.

One cause of repetition is the division of most merchant ship specifications into hull, machinery and electrical sections, written by different disciplines. As will be seen from the specification format in the next section, there are a number of items which are dealt with in more than one of these sections. For example:

– Plumberwork in H6 and M4
– Fire extinguishing in H5.7 and M5.5

A way of eliminating this problem is to have an additional systems specification which deals with all systems that cross the hull, machinery divide.

17.3 THINGS TO AVOID WHEN WRITING, OR ACCEPTING A SPECIFICATION

This section is written on the premise that at the contract date both shipbuilder and shipowner wish the specification to be a document that gives clear guidance on what the former is to supply and the latter wishes to receive and is not loaded in favour of either party. This is by no means always the case.

Some shipowner's specifications have a statement along the lines of "everything to be to the owner's complete satisfaction". This is a very one-sided clause, the acceptance of which in good faith took a reputable shipbuilder to bankruptcy some years ago when the company had to accept quite unreasonable demands made on its strength.

Shipbuilders accepting a contract based on a shipowner's specification should make quite sure that they fully understand what is being specified and insist that clauses of this sort are modified to incorporate the word "reasonable" rather than "complete" satisfaction. They should also ensure that there is an agreed arbiter such as the Classification Society to adjudge on what is "reasonable".

By the same token a shipowner should remember the Latin tag *"caveat emptor"* and make sure they fully understand the provisions of a shipbuilder's specification and that any special requirements are incorporated prior to signing a contract. Shipbuilder's standard specifications are designed to enable them to quote a competitive price and the corollary of this may be a long list of expensive extras for an owner who has special needs and has failed to negotiate the inclusion of these prior to contract.

Some specifications have a clause with words to the effect "anything specified twice shall only be supplied once" but this is surely only an unsatisfactory remedy for slipshod drafting.

The amount of writing necessary in a specification can be reduced by quoting appropriate standards such as Classification Society Rules, International Maritime Organisation rules, National Standards etc. Having invoked such rules it is important not to appear to qualify them by further specification statements on the subject unless some significant difference from the standard is in fact required.

Trade names should not be used, nor should subcontractors be specified in the main text of the specification. This applies whether the specification is written by a shipowner or a shipbuilder. The reason is, of course, the possibility (some would say probability) that if the subcontractors concerned learn of such a mention they will be tempted to increase their quotation. A list of nominated subcontractors of main equipment can however, with advantage, be included at the end of the specification. To improve the comparability of shipbuilder's quotations, it is quite usual practice for shipowners to ask that bids be made on the basis of nominated suppliers. If they do this, they will usually ask that alternative quotations be provided on the basis of the shipbuilders preferred suppliers, with the cost savings and/or technical advantages for changing to these being itemised.

In addition to forming a firm basis for the shipbuilder's quotation, the specification should identify a number of alternatives which may provide enhanced performance or result in operational cost savings at some additional capital cost and get the shipbuilder to quote for these options. At a time when the shipbuilder is still trying to win a contract, the extra costs quoted for these alternatives can be expected to be realistic and not loaded as they may be if quoted later as a variation to contract.

17.4 A SPECIFICATION FORMAT

A specification format, which with minor changes has served the author well for many years for a wide variety of merchant ship types is written in four parts: General, Hull, Machinery and Electrical.

For specialist vessels another part may be added to deal with special equipment if this is too extensive to insert in the hull specification: Section 2 "Specialist Hull".

Section headings of this standard specification are:

G   *General Specification*
1.0   General Description
1.1   Principal Dimensions,
1.2   Deadweight and Capacities
1.3   Machinery, Power and Speed
1.4   Hull Form and Model Testing
1.5   Stability and Trim Requirements
1.6   Ship Operating Conditions
1.7   Vibration and Noise Levels
1.8   Alterations and Additions
1.9

2.0   Classification, Regulations
2.1   Certificates
2.2   Working Plans
2.3   As Fitted Plans and Instruction Manuals
2.4   Inspections, Tests and Trials
2.5   Materials and Workmanship, Cleanliness
2.6   Standardisation
2.7   Spares
2.8   Contract and Completion
2.9

H   *Hull Specification*
H1   *Structure*
1.0   General, Materials and Methods
1.1   Shell Plating and Framing
1.2   Stem, Sternframe, Rudder and Stock
1.3   Bottom Construction and Machinery Seats
1.4   Bulkheads and Tanks
1.5   Decks, Pillars and Girders
1.6   Casings and Superstructure
1.7   Stairs and Ladders
1.8   Access Hatches and Doors
1.9   Masts and Miscellaneous Steelwork

H2   *Specialist Hull*

(Dry Cargo Ships)		(Tankers)	
2.0	Cargo hatches and Doors	2.0	Cargo Hatches
2.1	Cargo Handling Gear	2.1	Cargo Pumps
2.2	Cargo Stowage	2.2	Pump Room
2.3	Cargo Ventilation	2.3	Tank Piping
2.4	Cargo and Stores Refrig. Mc.	2.4	Deck Piping
2.5	Cargo and Stores Insulation	2.5	Cargo Heating
2.6		2.6	

Other modified versions of this section can be devised for different ship types

H3   *Accommodation*
3.0   Joinerwork General
3.1   Furniture and Fittings
3.2   Upholstery
3.3   Sidelight and Windows

The grouping of items under some of the headings can be criticised, and more rational arrangements can probably be developed. The important thing, however, is to have a system and stick to it with the minimum of changes.

One possible improvement would be to use an identical system for the grouping of weights, costs and specification headings. Each of these groupings is, however, arranged for different reasons and attempting to rationalise in this way may not be entirely successful.

A warship specification using this basic layout would require at least one extra section dealing with weapons and command and control on the following lines:

W   *Weapons, Command and Control*
1.0   Surface to Surface Weapons, Ammunition Handling and Stowage
1.1   Surface to Air Weapons, Ammunition Handling and Stowage
1.2   Anti- Submarine Weapons, Ammunition Handling and Stowage
1.3   Helicopter, Handling, Hangar, Weapons
1.4   Minelaying, Minehunting
1.5   Decoy Launchers, Small Arms
1.6   Action Information Room and Weapon Control Systems
1.7   Weapon Radars
1.8   Sonars
1.9   Electronic Warfare systems

An alternative to this layout for a warship specification would be to model it (other than the general section, which would need to be much expanded from the headings suggested for merchant ships) on the warship weight classification shown in Fig. 4.14. This system, originally developed by the American Bureau of Ships for the United States Navy and adopted with minor modifications by many other navies including the British Navy, is already used by both the U.S. and British navies for costing, so extending its use to the specification would complete a loop.

Because of the complexity of warships the demarcation common in merchant ship practice between the hull and the engine room which has led to separate hull and machinery specifications does not apply and the eight sections used in warship practice for both weight and cost estimates and weight and cost records apply to the whole ship.

To recap, the eight sections are:
   1. Structure
   2. Propulsion
   3. Electrical
   4. Control and Communications
   5. Auxiliary Systems
   6. Outfit and Furnishing
   7. Armament
   8. Variable loads

It may be worth commenting that in offshore oil industry projects it is also usual to try to specify the whole of a system in one section and with the advantages of this becoming apparent, it is possible that the traditional merchant ship demarcation into separate hull and machinery specifications should be seen to have outlived its usefulness.

A demarcation devised by the Norwegian Shipping and Offshore Service and now quite widely used is reproduced as Fig. 17.1. This system, which avoids a hull/machinery division, is used not only for specification writing, but as a basis for both weight and cost estimating and recording.

	1 SHIP GENERAL	2 HULL	3 EQUIPMENT FOR CARGO	4 SHIP EQUIPMENT	5 EQUIPMENT FOR CREW AND PASSENGERS	6 MACHINERY MAIN COMPONENTS	7 SYSTEMS FOR MACHINERY MAIN COMPONENTS	8 SHIP COMMON SYSTEMS
	10 Specification, estimating, model test, drawing, ordering	20 Hull materials, general hull work	30 Hatches and ports	40 Manoeuvring machinery and equipment	50 Lifesaving protection and medical equipment	60 Diesel engines for propulsion	70 Fuel systems	80 Ballast & bilge systems, gutter pipes, outside accommo.
	11 Insurance, fees, general expenses, representation	21 Afterbody	31 Equipment for cargo in holds and on deck	41 Navigation and searching equipment	51 Insulation, panels, bulkheads, doors, sidescuttles, skylights	61 Steam machinery for propulsion	71 Lube oil systems	81 Fire & lifeboat alarm-, fire fighting-, wash down systems
G	12 General work and models	22 Engine area	32 Special cargo handling equipment	42 Communication equipment	52 Internal deck covering, ladders, steps, railings etc.	62 Other types of propulsion machinery	72 Cooling systems	82 Air and sounding systems from tanks to deck
R	13 Provisional rigging during construction (staging etc.)	23 Cargo areA	33 Deck cranes with rigging etc. for cargo	43 Anchoring, mooring and towing equipment	53 Ext. deck covering, step and ladders, fore & aft gangway	63 Transmissions and foils	73 Compressed air systems	83 Special common hydraulic oil systems
O U	14 Work in connection with ways, launching and docking	24 Forebody	34 Masts, posts with derrick booms, rigging & winches for cargo	44 Repair/maint./cleaning equipment, outfitting in workshop/store	54 Furniture, inventory and entertainment equipment	64 Boilers, steam and gas generators	74 Exhaust systems and air intakes	84 Central heat transfer systems with chemical liquids
P S	15 Inspection, measurements, tests and trials	25 Deck houses and super-structures	35 Loading and discharging systems for liquid cargo	45 Lifting and transport equipment for machinery components	55 Galley and pantry equip. arr. for prov. ironing, laundry	65 Motor aggregates for main electric power production	75 Steam, condensate and feed water systems	85 Electrical systems general part
	16 Guarantee and mending work	26 Hull outfitting	36 Freezing, refrigerating and heating systems for cargo	46 Hunting, fishing and processing equipment	56 Transport equipment for crew passengers and provisions	66 Other agg. and gen. for main and emergency elec. power prdduction	76 Distilled and make-up water systems	86 Electrical power supply
	17 Services to ship during repair	27 Material protection external	37 Gas/ventilation systems for cargo holds/tanks	47 Armament, weapons and weapon counter-measures	57 Ventilation, air conditioning and heating system	67 Nuclear reactor plants	77	87 Electrical distribution common systems
	18	28 Material protection internal	38 Auxiliary systems and equipment for cargo	48 Special equipment	58 Sanitary systems with discharges, drainage systems for accommo.	68	78	88 Electrical cable installation
	19 General consumption articles	29	39	49	59 (Reserved for passenger ships)	69	79 Automation systems for machinery	89 Electrical consumers

Fig. 17.1. Specification and cost sections used in the S.F.I. system.

17.5 THE CONTRACT

The contract is a legal document. The initial preparation of this is best handled by the respective technical staffs of the contracting parties, but there should always be a final careful vetting by lawyers, which will usually involve redrafting into legal language, before an agreement is concluded.

The following paragraphs appear in the majority of contracts, with of course considerable expansion from the brief kernel statements given here.

1. *Introduction.* This names the contracting parties and outlines the purpose of the contract.
2. *Interpretation.* This gives the meaning of various words used in the contract such as "Purchaser", "Builder", "Components", "Specification", and details which document rules if there is any inconsistency — Contract over Specification over Plans being usual.
3. *Particulars of Vessel.* Lists such items as deadweight, engine M.C.R., speed and associated power, cubic capacity. (These particulars will be given in the specification and repetition in the contract is a potential source of conflict; however some such statement is desirable for identification purposes.)
4. *Regulations.* Lists the rules to which the ship is to be built, with pertinent dates (see also note under paragraph 3 above).
5. *Inspection.* Requires submission of plans, lays down right of access, responsibility for payment of Classification and other fees.
6. *Access.* Provides for access of subcontractors employed by purchaser.
7. *Transfer of Title.* Provides for progressive transfer as security for instalment payments.
8. *Risk and Insurance.* Defines that prior to delivery the vessel is the responsibility of the builder, who must arrange suitable insurance.
9. *Trials.* States that trials are at builders expense and under his control. (Trial performances are generally stated in the specification.)
10. *Trial Deficiencies.* Details the liquidated damages (which for some legal reason must not be called penalties) payable for failure to provide the specified performance.

Usually there is a clause relating to speed with damages for each 0.1 of a knot (or similar) below that specified and the right of rejection if the speed is say one complete knot less.

A second clause relating to specific fuel consumption may provide for damages if this is more than 5% above that specified

A third clause will provide for damages for failure to provide the specified deadweight and here again there will be a point at which the ship may be rejected.

Cubic capacity and/or container numbers may also be the subject of a damages clause.

11. *Delivery.* A contract delivery date will be specified along with liquidated damages *per diem* if this is late by more than a limited "grace period" which is usually set at 15 days; there should be provision for rejection if the delivery is delayed beyond a stated period. There should be a statement requiring force majeure claims to be lodged on the contract delivery date.

12. *Guarantee.* A guarantee period (usually of one year) will be stated, together with conditions for making good defects.

13. *Alterations and Additions.* The right of the purchaser to make alterations and conditions attached thereto.

14. *Cancellation by Purchaser, Cancellation by Builder.* The conditions applying thereto.

15. *Arbitration.* Conditions relating to the appointment of an Arbitrator should be stated.

16. *Default of Contractor*

17. *Patents.* Indemnification of purchaser against any patent infringement.

18. *Assignment.* Permitting the Purchaser to assign to another company.

19. *Taxation.* Each party to bear taxes imposed by its own government.

20. *Law.* A statement defining which countries law applies to the contract.

21. *Price and Payment.* The price is stated in an agreed currency together with a statement either that it is a fixed price or that it is a provisional price. In the latter case the basis on which the final price is to be determined should be most carefully spelt out.

Payment terms can vary widely. In general these should aim to match income to the outgoings. Milestones at which payments are to be made should be easily identified. A fairly typical schedule of payments might be:

    15% with order,

    25% on commencement of steelworking,

    20% when one third of steelwork erected at berth,

    15% when two thirds of steelwork erected,

    15% when vessel launched,

    10% on delivery and satisfactory trials.

Although the one-third and two-third steelwork erection milestones are quite good as divisions of the building cost it has to be admitted that they are not very easy milestones to define. for example, does "erected" mean erected only or erected, faired and welded? Some owners would limit the last payment to 5%, retaining 5% until the end of the guarantee period to ensure satisfactory service.

22. *Finance.* There may be a clause dealing with any loan provided to the Purchaser by Banks operating at the behest of the Builder. Sometimes these extend to several years after the delivery of the ship.

17.6 TENDER PACKAGE

For merchant ships the usual tender package sent to shipyards by a shipowner seeking tenders consists of:
- Draft Contract
- Specification
- Ship General Arrangement Plan(s)
- Outline Machinery Arrangement
- Outline Midship Section
- Plans of any special features to be incorporated in the ship.

A shipyard seeking to interest a shipowner in a standard design would probably send a somewhat similar package.

The tender package normally sent with a warship enquiry contains a very much larger number of plans (hundreds rather than tens) and many documents detailing naval standards. The intent is to avoid any claim by the tendering shipyard that they were not fully informed about the precise requirements, but those familiar with merchant practice may be excused for believing there is often a substantial overkill in the inclusion of plans of familiar and comparatively low cost systems and/or fittings. Neither the production of these nor their examination by tendering shipyards seems cost-effective.

Tender packages for offshore oil projects are also usually accompanied by quite a large package of plans and specifications. The reason for this is the novelty of many of these projects which quite often is such as to require the engineering design to be completed as a separate contract before construction bids are sought. This practice, of course, raises the question of design responsibility — a matter that should be clearly defined in the contract.

Chapter 18

Cost Estimating

18.1 COST AND PRICE

The words cost and price are used colloquially as though they had the same meaning, but as used here they are fundamentally different.

The cost of a ship is the sum needed to pay for all the materials and labour involved in its construction plus the overhead costs incurred. The material and labour costs attributable to a particular contract are easily identifiable, but making a correct and equitable allocation of overheads is not an easy matter depending as it does not only on the ship being costed but on the general level of activity in the shipyard at the time.

Costs can be divided into two categories — estimated and actual. The estimated cost is that calculated when the shipyard is tendering; the actual cost is that ascertained to have been incurred at the end of the contract. The price is the sum of money which the shipyard quotes to, and eventually receives, from his customer. The tender price is that given in the quotation, the contract price that agreed in subsequent negotiations whilst the final price is the sum for which the contract is concluded. The tender and contract prices are based on the estimated cost and on the state of the market.

The difference between the cost and price will take account of any allowances necessary for cash flow finance, for any anticipated inflation, for the shipyard's profit with these additions being reduced by any Government subsidy which can be claimed. If the price has to be quoted in a foreign currency it may also be wise for it to include some provision for possible exchange rate fluctuations.

The price may be a fixed price or there may be provision for it to vary with inflation. In some circumstances where it is impossible to specify exactly what is required, the tender price may be little more than an indication with the contract agreed on an ascertained cost basis. Needless to say this is not very desirable for the buyer, but may be the only way to get work under way on a novel project.

The final price depends very much on the contract. If this has stipulated a "fixed" price, then the final price will be based on the tender price adjusted as necessary by the agreed variations to contract.

When there have been difficulties with the contract or the specification/design have been inadequately defined, there may be considerable extra costs due to changes required by the owner or due to delays that the shipyard can claim were the responsibility of the owner. With inflation and possibly with a variety of different currency exchange rates entering the equation there can be fierce arguments before a final price is agreed.

18.2 TYPES OF APPROXIMATE ESTIMATE

18.2.1 Approximate or budget prices

A shipowner contemplating new tonnage will want to know budget prices for various alternatives he may be considering but is not usually very interested in the costs. Unless the ship in question is of an unusual type a scan of trade journals will probably provide prices for a number of reasonably similar ships, from which prices per "unit" can be derived. The unit will vary with ship type and may be deadweight tonnes, cubic metres of capacity, number of containers, number of vehicles, number of passengers, or tonnes of lightship weight if known.

There will usually be merit in plotting the unit prices against a unit of size such as the length BP to see whether there is an obvious scaling effect. This is particularly worth doing if the ship for which a price is required differs significantly in "size" from the data points. It will generally also be helpful to note against each data point the country of build to see which country appears to quote the keenest prices.

If the ship is of an unusual type, only a few recent prices may be available and it may become necessary to use information relating to ships built some years ago. This information, although dated, can still be useful provided the quoted prices are updated using a suitable inflation index.

The need to convert prices from a variety of currencies is likely to pose a problem. Should such conversions be made at the current exchange rate, at that in force at the time of contract or at that applying at the date of settlement? Assuming the price is a contract one, it is probably best to update this to the required date staying in the currency of the shipbuilder and using an inflation index applicable to the country of build and then convert to the required currency.

Fortunately, a high degree of accuracy is not necessary for budget prices. Noting the variations in prices that occur even when detailed quotations are obtained it should be accepted that ±15% is about as accurate as can be expected.

18.2.2 Approximate or budget costs

Although shipbuilders may also make use of the price estimating method described above, they are of course much more interested in estimating approximate costs. This is needed to show whether the shipyard is likely to be competitive for a particular order, so that the considerable cost involved in preparing a detailed estimate or detailed ship design is only undertaken when there is an acceptable chance of success.

It may be noted that even if a shipbuilder was able to arrive at a sufficiently accurate cost figure for tender purposes by the use of an approximate method, a detailed estimate would still be required for cost control purposes during building.

The overall configuration of an approximate cost estimate should be much the same as that of a detailed estimate. The difference between these estimates lies in the provenance of the individual figures with which they are made up.

The figures in a detailed estimate will come from quotations for materials and detailed work assessment for labour costs, whilst those in an approximate estimate are generally derived by the use of costs per tonne or manhours per tonne from records of recent construction or from figures used in a recent detailed estimate for a similar ship. The following sections describe detailed estimate methods first and then try to show how approximate estimates can be made using much the same methods. Some data for use in approximate estimates is given in the last section of the chapter.

18.3 DETAILED ESTIMATES — BASIS

It is useful to define a basis for estimates, to apply unless otherwise stated. In general, cost estimates should be prepared, in the first instance, on the basis of material costs, labour rates, productivity indices and overhead rates prevailing at the date the estimate is made.

Before such a basis estimate is built up into a price it has of course to be amended to allow for any change in costs, labour rates, etc. that seem likely to apply during the planned building timetable.

Cost estimates for merchant ships are generally made in the first instance on the basis of a single ship against which all the first-off costs are charged. If more than one ship is to be tendered, a second estimate for a repeat ship, excluding first off costs is then made.

Warship builders follow a more logical practice by separating first of class (F.O.C.) costs completely from production costs — a practice merchant shipbuilders could adopt with advantage, even though F.O.C. costs are very much smaller proportionally for merchant ships than they are for warships.

F.O.C. costs include design and drawing office costs, mould loft or equivalent costs, tank test and similar costs. Depending on the overhead structure of the firm, they may also include buying dept and similar non repeating costs.

Other factors which may reduce the costs of repeat ships include discounts offered by material suppliers for bulk buying and the improvement in labour productivity resulting from experience gained on earlier ships in a series. Shipyards should conduct an ongoing analysis of all contracts to determine productivity indices and repeat ship savings.

18.4 DEMARCATIONS AND SUBDIVISIONS OF COSTS

18.4.1 Between material, labour and overheads

Each shipyard has its own demarcation between materials, labour and overheads, and when making inter-firm cost comparisons it is essential to make sure the same demarcation applies or that suitable corrections are made.

18.4.2 Material costs

As well as the obvious items of steel, outfit and machinery, the materials cost includes the cost of work carried out by subcontractors working on the ship. This can introduce a difficulty when making inter-firm comparisons or even when comparing the performance of the same shipyard on two different contracts if the work put out to subcontractors differs.

Electrical and plumbing work are the two main activities for which some yards have their own departments whilst others use subcontractors, but there are others as will be mentioned later. Where a difference of this sort occurs, the material cost in one yard will include costs which in another yard are included in labour costs.

If a shipyard which normally does its own electrical or plumbing work uses a subcontractor or contract labour on a particular ship it is wise to synthesise cost records corrected to the shipyard's standard methods for future estimating use.

It is worth noting that all material costs include labour costs and the distinction is whether or not these labour costs are, or are not, incurred in the shipyard. As an example, the steel used in a shipyard starts as iron ore in the ground and needs labour to mine it, more labour to smelt and roll it into the plates and sections as which it enters the shipyard, at which point all the costs involved become a material cost.

18.4.3 Labour costs

Labour costs by definition include the cost of the time charged to the ship contract by the labour force, including contract labour, employed by the shipyard. This includes all the tradesmen and charge-hands, but shipyards differ in their treatment of the cost of foremen and managers. Both of these are likely to divide their time between several contracts and would find it difficult to record this, so many shipbuilders let them charge to an overhead number. Whilst this is an easy answer, such additional charging to overheads makes cost control more difficult.

18.4.4 Overhead costs

Overhead costs include a wide variety of costs incurred in the operation of the shipyard which are not directly chargeable to particular ship contracts. They include such items as interest on bank loans, rates and taxes, insurance, electricity, telephone and postage, salary costs of managers and office staff, etc. That there are a number of items which could be charged to either labour or overheads is obvious, but there are a number of items which could be charged as either materials or overheads. For example, it may be more convenient to issue stock items such as acetylene and oxygen without charging them to a ship contract, although once again this practice is not to be recommended as it reduces cost control.

The total overhead expenditure is usually calculated at the end of the financial year and expressed as a percentage of the labour cost. The rate to be used for estimates in the future is based on this analysis with adjustments for anticipated changes in costs and/or any expected change in the workload. An increase in throughput can bring a most pleasing reduction in the overhead rate, but a recession in demand forces an increase at a most unwelcome time.

Although the attribution of overheads as a percentage of labour cost is normal practice, it should be noted that quite a lot of overhead costs are more strictly time related. An appreciation of this leads to an understanding of the desirability of reducing building time even if this is not accompanied by a reduction in labour manhours.

It is usual for one standard overhead rate to be applied to all labour costs. This is certainly convenient because it is easy to calculate and easy to use. It may not give the right answer however if used in a comparison between the cost of using in-house labour or subcontracting a particular task. The true overhead rate of an outfit department such as painting which requires little by way of buildings, machine tools or management must be less than that attributable to steelworkers who require a great deal of each of these, but in most shipyards they are assumed to be identical.

18.4.5 Subdivision of costs — merchant shipbuilding practice

As well as the "horizontal" divisions into materials, labour and overheads an estimate is usually divided "vertically" — in merchant shipbuilding practice into at

Fig. 18.1. Typical cost breakdown for a frigate.

least three cost groups —structure, outfit, and machinery. As approximate estimates can be more accurate if they are based on a larger number of cost sections, it is useful to divide detailed estimates into a number of sub sections which can then become a data base for approximate estimates.

18.4.6 Subdivision of costs — warship practice

The eight sections used by the British M.O.D. for costing warships are also used for weight estimates and are shown in Fig. 4.14.

A typical breakdown of the cost of a warship is shown in Fig. 18.1 abstracted from Admiral Sir Lindsay Bryson's 1984 R.I.N.A. paper "The Procurement of a Warship". In this the cost is divided into six, rather than eight sections with the costs of the usual Controls and Communications section presumably split between Weapon Equipments and Propulsion and Auxiliary systems with the Variable loads section omitted.

The high proportion of cost attributable to weapons and the low proportion attributable to the hull should be noted as very significant factors in the design of warships. Another point to note is the fact that warship cost sections are not divided into Outfit and Machinery as is usual for merchant ships, for the very good reason that warship machinery and systems are spread throughout the ship.

18.4.7 The estimate sheet

An estimate sheet with eight "vertical" sections that can be used for either merchant ship or warship cost estimates is given in Fig. 18.9. It will be noted that this estimate sheet uses the unit production cost concept and makes provision for different overheads rates being used for each cost section.

Definitions of the items included in each of the merchant ship outfit sections are given in §18.6, whilst definitions of the items included in the machinery sections are given in §18.7.

Definitions of the items in the warship sections are given in Fig. 4.14.

18.5 STRUCTURAL COSTS

18.5.1 Structural material

Methods of calculating the net structural weight have been described in Chapter 4. For costing purposes the net weight must be divided into mild steel; higher grades of steel D; E; higher tensile steels AH; DH; EH plus aluminium if any of this material is to be used. The net weight of each material must be grossed up to an invoiced weight by the use of a factor which allows for a suitable scrap percentage and the weight thus obtained then divided into plates and sections as these have different prices per tonne.

Depending on the supplier of the steel there may be extras over the base cost per tonne for unusually large or small plates, for particularly thick or thin plates, for delivery, for testing, for flanging quality, etc. There may be discounts for large quantities or on the other hand extras for small quantities. Some special sections, such as Admiralty T bars, must be costed separately as they carry a very considerable premium.

The cost of welding rods and gases is normally added to the structural material cost as a percentage of this based on an analysis of completed ships.

A number of items which form an integral part of the structure but are nevertheless generally included in outfit are noted in §18.6.

18.5.2 Structural labour

The structural labour cost is the product of the manhours required multiplied by the labour cost per manhour. The manhours can be estimated in several ways, depending on the information available.

If sufficient structural drawings are available the estimate can be made by detailed work assessment, which may in turn be broken down into shop manhours; berth manhours and afloat manhours.

Shipyards with well organised planning departments can make very accurate estimates by this method. In preparing such estimates, attention should be paid to the fact that productivity often varies considerably between "shop", "berth" and "afloat", with the first of these being the best and the last the poorest. Organising the work to maximise shop work and minimise afloat work can significantly reduce the manhours.

With modern modular construction methods this reduction in manhours applies also to outfit and machinery installation manhours. Walker and McCluskey in their 1998 IESIS paper "Restructuring the Manufacturing Process of Modern Warships" quote some striking figures for the improvement in productivity attainable by advancing works in this way.

As well as the split into shop, berth and afloat manhours, a detailed manhour assessment will calculate manhours separately for shell plating, double bottom, decks, superstructure, and sundry steel, as each of these requires different manhours per tonne. It should also look into the manhours associated with special owners' requirements such as shot blasting (when carried out after construction to suit the use of special paint systems) or zinc spraying both of which can severely disrupt normal production rates.

Although there is a high degree of flexibility nowadays in the work that individual steelworkers can do and all steel workers in a shipyard are generally paid at one standard hourly rate, it may still be convenient to divide the total steelwork manhours into the four main trades of platers, caulkers and burners, welders and shipwrights or indeed to synthesise the total from these components.

When only a total steelweight is available, as is generally the case when a shipyard is tendering, the estimate must be made by the use of a plot of manhours derived from completed ships against the total steel weight (either net or invoiced) as shown in Fig. 18.2 or of manhours per tonne as shown in Fig. 18.3. Both plots have advantages but that of manhours per tonne is the more usual. As both these figures show, the manhours per tonne vary greatly with ship type and to a lesser extent with the steelweight itself.

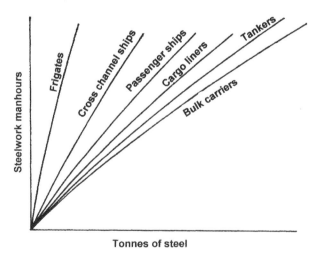

Fig. 18.2. Steelwork manhours versus steel weight.

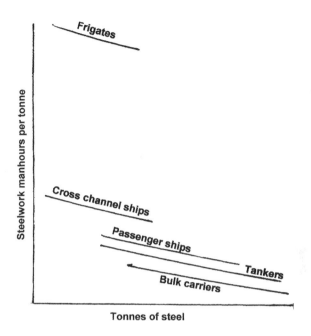

Fig. 18.3. Manhours per tonne versus steel weight.

The data on which these graphs were based dates back many years and for this reason no figures are shown on the axes as productivity increases have made these out of date. The comparative differences between ship types and with increasing steel tonnage still give some useful guidance although the differences are not now as great as they used to be.

A disadvantage of both these plots is the fact that they fail to distinguish between two ships of the same steelweight, one of which is smaller but heavily constructed and the other larger but with lighter scantlings. The latter ought, of course, to require more manhours.

As an alternative to a base of steelweight, manhours or manhours per tonne can be plotted against a numeral such as Lloyds old equipment numeral which gives an approximate indication of ship size (see Chapter 4). Both of these plots avoid the disadvantage noted above but the manhour plot to this base, as shown in Fig. 18.4 fails to show a proper allowance for the steelweight and a plot of manhours per tonne may be better against this base.

After much thought about the problems posed by these alternatives, a plot was made of manhours against the total area of plate used as shown in Fig. 18.5. Remarkably, this was found to give a straight line relationship irrespective of ship type, suggesting that area rather than weight was the best criterion for labour manhours.

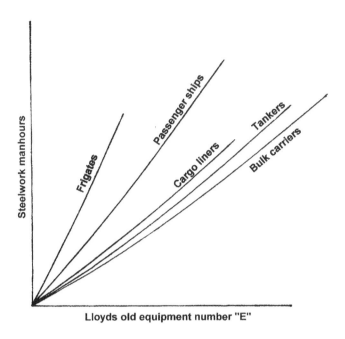

Fig. 18.4. Steelwork manhours versus ship size.

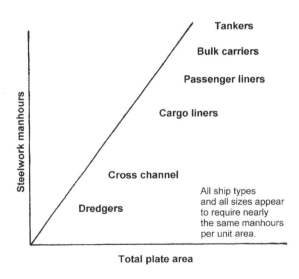

Fig. 18.5. Steelwork manhours versus total plate area.

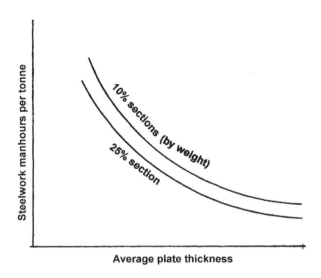

Fig. 18.6. Manhours per tonne versus average plate thickness.

Thicker and heavier plates require more work by way of edge preparation for welding and more work to form them if they are to be shaped; on the other hand, the stiffeners on heavier plates are more widely spaced and need less work per unit of area. On balance, therefore, it seemed not unreasonable that the labour content per unit area might be constant. Following this thesis, Fig. 18.6 was prepared on the basis of a single fixed figure of manhours per unit of plate area derived from the original of Fig. 18.5. Two assumptions of the percentage of stiffeners (sections) by weight were made and the manhours per unit weight calculated for a number of plate thicknesses.

As with the previous manhour graphs no actual figures are given on this graph because the author's data is now out of date due to major improvements in productivity. The method is however a very useful one, which can be recommended.

Finally it is worth emphasising that it is vitally important when discussing manhours per tonne to be sure whether the base is gross or net tonnes.

18.6 OUTFIT COSTS

Some outfit material costs are obtained from sub-contractor's quotations, others by costing items on a cost per unit or per unit weight. Where greater accuracy is required more sub-contractors quotations should be used, but where speed in making up a price is essential the cost/tonne method is used for more items.

18.6.1 Outfit material — merchant ships

To make the estimate format helpful for approximate estimates it seems sensible to construct groupings which bring together items whose costs per tonne do not differ significantly and which are related to some fairly easily assessed ship parameter. The following groupings already given in Chapter 4 for weight estimation seem to meet this criterion reasonably well and are repeated here for convenience:

Group 2. Structure related (steelweight)
    Structural castings or fabrications (rudder and sternframe)
    Small castings (bollards, fairleads etc.)
    Steel hatch covers
    W.T. doors

Group 3. Cargo related (cargo capacity or ship size)
    Cargo space insulation and refrigeration machinery
    Cargo ventilation
    Firefighting
    Paint
    3(a) Plumberwork

Group 4. Accommodation related (complement)
    Joinerwork
    Upholstery
    Deck coverings
    Sidelights and windows
    Galley gear
    Lifts, HVAC, LSA
    Nautical instruments
    Stores and sundries
    4(a) Electrical work

Group 5. Deck machinery related (by units or by ship type × size)
    Steering gear
    Bow and stern thrusters
    Stabilisers
    Anchoring and mooring M/C
    Anchors, cables and mooring ropes
    Cargo winches, derricks and rigging
    Cranes

Plumberwork 3(a), and electrical work 4(a), are given their subheadings because they are such major items that it is frequently advisable to treat them separately.

SECTION	ITEM	WEIGHT	COST PER TONNE	COST £
2	QUOTED ITEMS WEIGHT COSTED			
	TOTAL			
3	QUOTED ITEMS WEIGHT COSTED			
	TOTAL			
4	QUOTED ITEMS WEIGHT COSTED			
	TOTAL			
5	QUOTED ITEMS WEIGHT COSTED			
	TOTAL			
	TOTAL WEIGHT		COST	

Fig. 18.7. Outfit material cost calculations.

The estimating parameter which it is suggested should be used for each group is shown in brackets.

A convenient format for estimating outfit material costs for a merchant ship is shown in Fig. 18.7 — the correlation with the weight estimate should be noted.

18.6.2 Outfit material — warships

Although the term outfit is not used in warship design practice, it has been found convenient to use it in this book defining it as consisting of the warship weight and cost groups 4, 5, 6 and 7. (see Fig. 4.14). It should be noted that these groups also include items in the machinery spaces, whereas group 3 which is taken as part of the machinery weight includes some items located outside the machinery spaces.

18.6.3 Outfit labour costs

Outfit labour costs can be calculated in two ways, both of which require an assessment of the manhours and the multiplication of this by an average wage rate per manhour.

The most accurate method is based on detailed work assessment for each of the trades, work areas or systems involved. This is a lengthy task, especially when the ship being costed differs significantly from those for which the shipyard has records, but it is in just those cases that its use is most desirable.

An alternative method used in approximate estimates arrives at the outfit manhours by proportioning from the manhours used on some similar ship. The parameter used for proportioning is usually the ratio of the respective outfit weights (W_o) preferably in the form of this raised to the two third power — $(W_o)^{2/3}$ — as suggested by Jack Carreyette in his 1978 R.I.N.A. paper "Preliminary ship cost estimation".

When making a calculation by this method it is vitally important to check that the demarcation between work carried out by the shipyard and that subcontracted is to be the same on the ship for which the estimate is being made as it was on the ship used as the basis; or, if not, a suitable correction must be made.

The accuracy of an approximate outfit labour estimate can be significantly increased by breaking the total labour manhours up into the four sections suggested for the material estimate — if the basis ship data is available in this format, and then proportioning these separately using relevant parameters of weight, area or power.

Although in most shipyards there is now a considerable degree of flexibility in working practices, it is still convenient to divide the manhours into a number of entities corresponding to the old trades. Some of these trades work is entirely for

Table 18.1

Trade/work area	Group	Remarks
Stagers	3	Painters are the main users
Painters	3	
Shot blast	3	
Iron workers	3	Mainly hold ventilation
Carpenters	3	
Plumbers	3(a)	
Joiners	4	
Labourers	4	Mainly in accommodation
Cleaners	4	
Temp light	4	
Electricians	4(a)	
Deck engineers	5	
Riggers	5	

one of the four groups suggested under outfit material but others work for more than one of these. The number of different trades, and indeed their names, differ from yard to yard, but the list given in Table 18.1 is fairly representative. The table is annotated with a distribution to the groups, which — whilst to some extent arbitrary — may be convenient for approximate estimate methods such as that given at the end of this chapter.

18.7 MACHINERY COSTS

18.7.1 Machinery material costs — general

Machinery material costs are obtained mainly from subcontractor's quotations but partly by costing items either on unit, unit power or unit weight basis. Where greater accuracy is required more subcontractor's prices should be used, but where speed is essential the cost per tonne basis is necessarily used for most items.

18.7.2 Machinery material costs — merchant ships

For merchant ships a split into three groups seems to provide a way of bringing together items whose costs per unit weight are fairly similar and which can be related to an easily assessed parameter.

It could be argued that it would be better to make controls and switchboards into a separate group because of their high cost and low weight, but they are so closely connected with propulsion and generation respectively that groups 6 and 7 seem the best homes for them. The groups already suggested in Chapter 4 are repeated here for convenience.

Group 6. Propulsion
   Main engine(s)
   Gearbox
   Shafting
   Propeller(s)
   Main engine controls

Group 7. Auxiliary machinery
   Generators and switchboard
   Pumps
   Compressors, etc.

Group 8. Structure related
   Funnel and uptakes
   Ladders and gratings
   Pipework and ventilation trunking within engine room.

SECTION	ITEM	WEIGHT	COST PER TONNE	COST £
6	QUOTED ITEMS WEIGHT COSTED			
	TOTAL			
7	QUOTED ITEMS WEIGHT COSTED			
	TOTAL			
8	QUOTED ITEMS WEIGHT COSTED			
	TOTAL			
	TOTAL WEIGHT		COST	

Fig. 18.8. Machinery material cost calculations.

The division into groups 6 and 7 is, as discussed in Chapter 4, not suitable for diesel electric installations for which these groups should be combined.

A machinery materials estimate summary sheet is given in Fig. 18.8.

18.7.3 Machinery material costs — warships

In the arbitrary division of the warship weight groups already discussed under outfit, Group 2 propulsion and Group 3 electrical have been defined as the warship machinery weight and cost group.

Some of the items in Group 3 are located outside the machinery space whilst some machinery related weights and costs are included in Group 4 Controls and communications.

18.7.4 Machinery labour costs

Machinery labour costs are estimated as the product of the manhours required and the average wage rate applicable. The manhours can be obtained either by a detailed work assessment — the most accurate way but a lengthy process — or for approximate estimates by proportioning from available data on the manhours and total machinery power (P) of a suitable reference ship using this in the ratio (P) to the power 0.82, again as recommended by Carreyette.

18.8 UNIT PRODUCTION AND FIRST OF CLASS COSTS

18.8.1 Unit production cost

A calculation sheet for the total unit production cost is given as Fig. 18.9. This can be used for either warships or merchant ships and for either detailed or approximate estimates. The presence of the weights and costs per tonne on this sheet is not strictly necessary when it is being used for a detailed estimate but makes it useful as a quick reference when preparing an approximate estimate.

All the entries on the table have been discussed in the preceding sections with one exception — the column headed "services and miscellaneous". These items which do not involve weight include:
- classification and DTI or similar fees
- launch party expenses
- drydocking
- port dues
- trials costs
- delivery voyage costs
- insurance and provision for guarantee repairs, etc.

18.8.2 First of class cost

One item remains to be added to arrive at the total cost and this is the "First of class cost". Costs included in this are design and drawing office costs, planning department costs, tank tests or similar investigative work.

If the costing is being prepared for a multi-ship tender, the first of class cost can be added once as a package cost. If, however, the price is being quoted on a per ship basis, the first of class costs must be divided by the number of ships before being added to the unit production cost.

18.9 FROM COST TO PRICE

The step from cost to price is a complicated one requiring answers to a number of questions.
1. How many ships are being tendered?
2. What are the building periods of each of the ships?
3. What instalment payment dates are being linked to the price?
4. Do these result in a positive or negative cash flow?
5. What interest is payable on borrowed money?
6. What cost reductions can be obtained as a result of bulk buying?
7. What productivity improvements can be expected in later ships?
8. What inflation rates are expected to apply in each year from the base date to delivery of each ship to materials, labour and overheads?

SECTION	1	2	3	4	5	6	7	8	SERVICES & MISC.	TOTAL
WARSHIPS	STRUCTURE	PROPULSION	ELECTRICAL	CONTROL & COMMS	AUXILIARY SYSTEMS	OUTFIT & FURNISH	ARMAMENT	VARIABLE LOADS		
			OUTFIT			MACHINERY				
MERCHANT SHIPS	STRUCTURE	STRUCTURE RELATED	ACCOMMOD RELATED	OUTFIT CARGO RELATED	DECK M/C RELATED	PROPULSION	MACHINERY AUXILIARIES	RELATED		
WEIGHT TONNES										
MATERIAL COST £ PER TONNE										
MANHOURS PER TONNE										
MANHOURS										
LABOUR RATE £ PER MANHOUR										
OVERHEADS										
MATERIAL COST										
LABOUR COST										
OVERHEAD COST										
UNIT PROD COST										

Fig. 18.9. Cost estimate summary sheet for both merchant ships and warships.

Based on the answers to these questions, a corrected unit production cost is calculated for each ship based on its proposed production schedule. These are then summed and the first of class costs added to give the cost of the total ship package.

The next item to be considered is the sum to be added for contingencies and profit to arrive at the gross package price.

The contingency part requires careful consideration of the risks involved in each step of the estimate whilst the profit element requires an equally careful assessment of the likely competition.

The last step consists of making a deduction for any government subsidy, Intervention funding or the like which may be available to reach the selling price for the package, or dividing by the number of ships the average selling price per ship. Alternatively separate and different prices can be quoted for each ship.

18.10 APPROXIMATE COST DATA

18.10.1 General discussion

This chapter has so far been restricted to describing the theory of cost estimating, but this section now attempts to live up to the book's title *Practical Ship Design* and provide data that will enable readers to make a reasonably accurate estimate of the cost of any type of merchant ship to a 1993 cost base.

Obtaining cost and related data is not easy, as shipyards regard this information as something to be keep secret from their competitors. In spite of this the author managed to persuade a few shipyards to give him information, having assured them that this would be presented in a form which whilst providing useful data for making approximate cost estimates would at the same time preserve the confidentiality of the original data.

The original intention was to give data in much the same form as is used in traditional shipyard cost estimating methods — labour manhours and rates per manhour plus costs for materials used, each with some appropriate estimating parameter. It quickly became apparent however that the shipyards supplying the data all had different practices in relation to work for which in-house labour (manhours) was used and that for which they employed sub-contractors (material).

Items for which shipyard practice was found to vary in this way included paintwork and joinerwork as well as the anticipated electrical and plumber work. In addition shipyards demarcation between labour and overheads differed. Obviously there could be no consistency in either labour manhours or material costs on the basis of such ill conditioned data and the decision was therefore taken to convert each item to a total cost which included material, relevant labour and overhead costs as it would do as a sub contract item.

The next decision was to use weight as the estimating parameter throughout. Weight has the advantage that it applies to almost all components of ship cost. It is true that some individual items can be more accurately estimated on the base of another parameter, for example the cost of main engines on a base of horsepower, that of electrical work on a base of total installed kilowatts and that of joinerwork on a base of accommodation area, but weight is the one parameter that can be used for almost all items and methods of estimating the components of structural, outfit and machinery weight have already been given in Chapter 4.

18.10.2 Structural cost

The use of a total cost per tonne method for structural costs is made difficult by the large differences that exist in the labour manhours required for structural work of different types and sizes of ships. At one extreme a small fine lined specialist ship constructed of light plating may require something of the order of 130 manhours per gross tonne whilst at the other extreme a large full lined tanker constructed of heavy plating may be constructed at about 25 manhours per gross tonne.

Quite apart from the different manhours per tonne which are associated with different types of construction, manhour figures from different shipyards are likely to be affected by different operating efficiencies. When converting from manhours to cost considerable differences were noted in the labour cost per manhour and in the overheads of different shipyards. Whilst these differences generally to some extent cancel themselves out, that this will be the case is by no means certain, but pains have been taken to ensure that all the figures used in calculating cost per tonne figures relate to the particular ship concerned.

The invoiced cost of structural material was found to differ with ship type and size with large ships getting lower basic costs but generally requiring a higher percentage of the more expensive special quality steels. Different ship types were also found to require differing proportions of plates and sections, quite an important consideration as the latter are more expensive.

The costs per tonne quoted are based on net steel weights but the basic material costs allow for the gross steel ordered and it was noted that the scrap percentages which the different yards regard as normal also differed quite widely.

Finally, an additional sum was added to the steel material cost to allow for the cost of electrodes and miscellaneous steelworkers stores.

The resulting costs in US dollars per tonne of net steel are shown in Fig 18.10. Like the other graphs in this section, this graph is based on limited data and should only be used for approximate cost estimates, but the author knows of no other published information which even ventures figures on this subject.

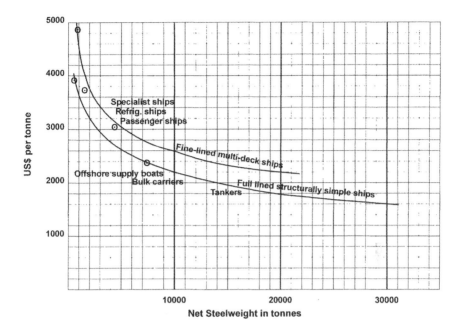

Fig. 18.10. Approximate costs of structural steelwork per tonne. Costs are on a 1993 basis and include materials, labour and overheads.

18.10.3 Outfit costs

Whilst the costs per tonne of different items of outfit were found, as expected, to differ quite widely, the costs per tonne of the total outfit (of an admittedly not very large data sample) were found to be much closer even though the ships were of quite different types and sizes. This appears to indicate that the proportions of high and low cost per tonne items are not too dissimilar whatever the ship type and, if this can be relied on, the task of making a reasonably accurate first level cost estimate is certainly eased.

After some experiments it was found that a closer convergence of the data could be obtained if the cost per tonne was calculated for a "normalised" outfit which excluded any items which would have a major influence on the average cost per tonne either because it had a very high or a very low cost per unit weight — combined in the latter case with a sufficiently large weight to make its influence felt.

An example of the former might be the high cost, with very low weight, of special scientific equipment in a research vessel and of the latter the relatively modest cost associated with the considerable weight of the refrigerating machinery and insulation of a reefer ship.

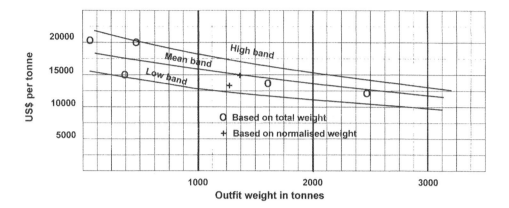

Fig. 18.11. Approximate costs of outfit per tonne. Costs are on a 1993 basis and include materials, labour and overheads.

Even with this normalisation, costs per tonne still seemed to vary quite a lot and it was without much hope of finding a pattern, that the plot given in Fig. 18.11 was prepared. The result turned out to be quite surprising as there appears to be a pattern with a higher cost per tonne for ships which have a lot of specialist equipment and a relatively high specification as compared with ships which have a fairly basic specification. There was also a clear tendency for the cost per tonne to be higher for small weights with this reducing as the weight increases.

The difference between the cost per tonne of the high and low bands would be sufficient to result in a considerable error in an estimate if the graph is used without discretion, but if users make an educated guess as to where the ship they are interested is likely to lie in the band the accuracy of their approximate cost estimating can be considerably increased.

Costs per tonne for the four outfit groups suggested in §18.6 and for a number of individual outfit items are given in Table 18.2.

It is perhaps worth mentioning that the costs of some of the subcontract items included in the calculations leading to the data have been adjusted from the figures originally quoted by suppliers to allow for the price reductions that keen sellers offer when a sale looks imminent.

The relative costs per tonne of some of the groups and of some of the individual items were as expected but some came as a surprise and it may be worth making some comments on these:

– Group 2: this is, as expected, the lowest cost per tonne, higher than ship structural work but in the same general bracket.

Table 18.2

Costs per tonne of outfit groups and of some individual items of outfit

Item	$US per tonne
Group 2	5,000
Group 3	28,500
Group 3(a)	41,000
Group 4	18,500
Group 4(a)	23,000
Group 5	10,500
Some individual items	
steel hatch covers	5,000
sternframe, rudder	5,000
cargo refrig and insulation	7,000
anchors and cables	2,000
anchoring and mooring machinery steering gear	8,500
cranes	12,500
nautical instruments on a specialist ship	1,000,000 (total cost - weight only a tonne or so)

- Group 3: much of this cost is painting and the associated staging. It was a surprise to find how costly this is on a weight basis and indeed as a total cost as the weight is not inconsiderable.

- Group 3(a): whilst a high cost per tonne was anticipated, it was a surprise to find that plumberwork is very nearly twice as expensive as electrical work, but on reflection bending and fitting pipes is a lot more difficult than leading cables!

- Group 4: - this group is made up of more different items than any of the others groups and this may explain why the cost per tonne is nearest to the total normalised cost per tonne than that of any other group.

- Group 4(a): see 3(a).

- Group 5: the average cost of this group is brought down sharply by the low unit cost and high weight of anchors and cables (see individual items).

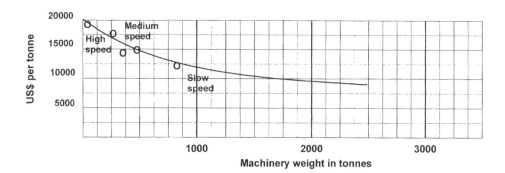

Fig. 18.12. Approximate costs of machinery per tonne. Costs are on a 1993 basis and include
materials, labour and overheads.

18.10.4 Machinery costs

Figure 18.12 shows a graph of the cost per tonne of machinery against the total
machinery weight. The trend towards reduced specific cost as the weight increases
with a flattening out to a figure of just under $10000 per tonne at machinery
weights in excess of 1500 tonnes may be noted.

The high cost per tonne end of the line can be identified with lighter high speed
machinery, the middle with medium speed and weight and the low cost per tonne
end with the heavier slow speed diesels.

18.10.5 Non weight costs

Finally, there are a number of costs which do not have an associated weight, or
cannot easily be divided into costs associated with the three groups of structure,
outfit and machinery. For example, although drawing office labour and overheads
are usually recorded separately for Ship and Machinery it is rare for the first of
these to be divided into structure and outfit. Other costs to which similar considerations apply include Classification and Department of Transport or similar fees,
consultancy, tank test, model costs, launch expenses, drydock, pilotage, towage,
trials costs, insurance, provision for guarantee repairs and miscellaneous similar
expenses.

For approximate estimate purposes it seems best to express the cost of all these
as a percentage of the total ship and machinery cost. From the analysis made a
suitable percentage seems to lie between 7.5% and 12.5% with the lower figure
applying to smaller ships/shipyards and the higher to larger ship/shipyards with
10% as a most probable general figure.

18.10.6 *Cost of sister ships*

For a sister ship built in the same shipyard there will be savings on expenditure on such items as tank tests, drawing office costs, loft or CAD work and, provided the sister ship is built in reasonably close succession to the lead ship, some general improvements in manhours. In total the costs of a second ship can be expected to be reduced by about 8%. On further ships of a class there should be additional saving but this is unlikely to be of more than 1%.

18.10.6 *The price*

The total cost arrived at after the addition of these percentages still needs a final correction to convert it to a price that a shipyard might quote. This further correction consists of:

   (i)   the addition of a profit and/or contingency margin which may vary from zero to about 10% with 5% being possibly the best figure to use in this approximate estimating method;
   (ii) the addition of an allowance for the effect of anticipated inflation on costs during the building period which might be 2% at present;
   (iii) the deduction of any government support such as Intervention funding of 9% currently available to shipyards in the E.C.

Taken together these items would mean that the selling price might be about 2% less that the cost arrived at above.

All the cost figures in this section have been given in US$ as this is the currency in which worldwide tendering is most commonly carried out. As most of the data used in this section was originally in sterling it may be worth recording that the conversions to dollars were based on a round figure exchange rate of £1 = $1.50.

Both costs and price relate to a 1993 quotation for an early delivery. If the data given is used for later years the figures should be corrected for inflation, altered exchange rates, etc.

Readers with access to cost data should make their own checks on the data given and build up additional and updated data on the lines suggested.

Chapter 19

Operational Economics

19.1 SHIPOWNERS AND OPERATORS

The operational economics of a ship can be looked at in a number of different ways depending on the type of trade in which it is used and how it is employed.

19.1.1 Types of trade

Whilst there is an enormous diversity in the type and size of ships, all are generally employed in one of five principal ways, namely as liners, cruise ships, industrial carriers, service vessels or as tramps. The first four of these categories can be classed as owner-operated ships, whilst the last category consists mainly of ships let out on charter.

19.1.1.1 Liners

To be designated as a liner, a vessel must ply on a regular advertised service; examples are container ships and ferries. Because ships providing this sort of service sail on scheduled dates and, when passengers are carried, at scheduled times, departing whether the ships are fully loaded or not, the cost of running a service of this type can be high and freight rates and ticket prices must be set to achieve a satisfactory return over a period of time against the anticipated demand.

19.1.1.2 Cruise ships

The first cruises were offered by passenger liner companies using their liners either in their normal country to country service or on special voyages. These cruises were usually arranged at a time of year when passenger numbers in their normal services were likely to be on the low side.

With the decline of passenger services caused by the growth of air travel, passenger liners ceased to be available for use in this way and purpose built cruise liners started to make their appearance. These are now becoming more like floating hotels or holiday camps and the cruise business is currently one of the fastest growing areas of shipping.

Typically, cruise ships undertake trips of one or two weeks duration generally steaming at night and with arrangements made for passengers to go ashore and see the sights and enjoy a new locality each day.

Although each cruise is a scheduled service, the fact that cruise schedules and itineraries can be changed at relatively short notice gives these ships an operational flexibility which liner services do not have.

19.1.1.3 Industrial carriers

A number of large companies with a substantial shipping requirement either for the import of their raw materials or for the export of their finished products or both own a number of ships to cover at least a baseload part of their shipping requirement.

Typical examples of this are the tanker fleets owned by oil companies; ships specially designed to carry iron ore and/or coal owned by steelmakers; and ships designed to carry cars in bulk owned by major car manufacturers.

The owners of these ships generally assume total responsibility for all aspects of cost when the vessel is employed in their own trade. The object of such an ownership is to minimise the costs of an overall industrial process, but the lack of flexibility which has often been a characteristic of such operators has sometimes been found to do the opposite and this type of shipowner has been diminishing in recent years.

The U.S. anti-pollution laws have had a severe impact on some of the major oil companies who now refuse to trade with their own vessels in U.S. waters because of the virtually unlimited liability that applies there and instead charter in from traditional shipowners.

19.1.1.4 Service vessels

Very few, if any, service vessels carry cargo, their function being to supply services to other vessels or installations at sea. Examples of service vessels are tugs, dredgers, navigational service vessels, offshore safety vessels, etc. These services may be paid for directly as in the case of tugs or indirectly through port dues or taxation in some other cases. But the owners of all these ships need to calculate ship operating expenses on an owner operator basis.

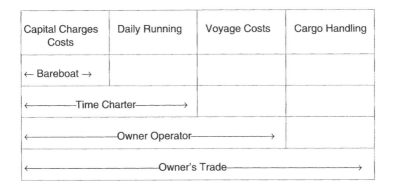

Capital Charges Costs	Daily Running	Voyage Costs	Cargo Handling
← Bareboat →			
←————————Time Charter————————→			
←————————————Owner Operator————————————→			
←————————————————Owner's Trade————————————————→			

Fig 19.1. Changing responsibilities of the owner from bareboat to owner's trade.

19.1.1.5 Tramps

A ship can be said to be tramping when it is prepared to go wherever a suitable cargo is available. Tramp ships can be employed in various ways under different types of charter which are explained in §19.1.2. Most bulk carriers and oil tankers, together with many small container ships and coasters operate as tramps, making this the method of employment of the majority of ships.

19.1.2 Methods of employment

An owner will generally employ a ship in one of four ways, namely: in his own trade, in tramp trades as an operator, or in tramp trades by time chartering or bareboat chartering the ship to another party. The extent to which an owner bears the costs of operations under each of these situations is discussed in the following paragraphs and is illustrated in Fig. 19.1 which is a slightly modified version of a figure originally given in Dr. Buxton's 1972 R.I.N.A. paper "Engineering economics applied to ship design" — a paper which, along with Dr. Buxton's earlier B.S.R.A. report "Engineering economics and ship design", contributed substantially to this chapter.

19.1.2.1 Ships used by an owner in his own trade

The types of trade in which ships are used by owners in their own trade have been outlined in §§19.1.1.1–4. When ships are used in this way, the owner will generally assume total responsibility for all aspects of cost incurred.

19.1.2.2 Ships used by an owner as operator

An owner operator can arrange for the employment of a ship in a number of different ways, viz:

(i)   by taking on Contracts of Affreightment to move a large volume of cargo in regular shipments of a set size, based on a set rate per tonne moved;

(ii) by letting the ship on Voyage Charter to carry a single cargo on a set rate per tonne; or

(iii) by letting the ship for a single voyage on Time Charter for a set rate per day.

Under Contracts of Affreightment and Voyage Charters the owner will meet the capital cost, running costs and voyage costs (comprising port charges and bunkers). The terms of the charter will determine who pays the cargo handling costs as follows:

Gross terms (Gross)	Shipowner pays for loading and discharge
Free on board (FOB)	Charterer pays for loading
Free discharge (FD)	Charterer pays for discharge
Free in and out (FIO)	Charterer pays for loading and discharge

Under a single voyage time charter the charterer will meet the voyage costs as well as the cargo handling costs.

19.1.2.3 Tramping — let out on time charter

In a time charter, the shipowner undertakes to provide a ship for the charterer to use either for a fixed time of anything from a few months to 20 years or for a single round voyage.

The charterer is responsible for arranging cargoes and voyages during the charter and also for paying all voyage expenses including fuel, port and canal dues, cargo handling charges.

The shipowner provides the ship and crew and is responsible for the capital charges and daily running costs. Hire is only payable for time in service and ceases during breakdown and repair, although it continues if the ship is delayed in port or sails empty for reasons not attributable to the ship.

19.1.2.4 Tramping — let out on bareboat charter

In this case the charterer provides the crew and is responsible for maintenance with the shipowner's sole responsibility being the provision of the ship and meeting the capital charges. In effect the charterer uses the ship as if he owned it.

19.2 ECONOMIC CRITERIA

19.2.1 The basis of these criteria

There are a number of different economic criteria which may be used to assess the likely success of a shipping investment or to compare the profitability of alternatives. These criteria should take account of:
- the time value of money,
- the full life of the investment,
- changes in items of income and expenditure which can be expected over the life,
- the economic facts of life such as interest rates;taxes; loans and investment grants.

The time value of money represents the fact that a sum of money available now is of much more value than the same sum not available for a number of years.

Interest is fundamental to the calculations whether there is a need to borrow or not. This takes account of the fact that if available cash is used the interest it might have earned is being foregone.

19.2.2 Interest

This may be simple or compound and the following relationships apply:
- Simple interest
  Total repayment after N years: $F = P (1 + N \cdot i)$
- Compound interest
  Total repayment after N years: $F = P (1 + i)^N$
  In this case the factor $(1 + i)^N$ is called the compound amount factor (CA), and P = original investment.

19.2.3 Present worth

The reciprocal of CA is called the present worth (PW) factor.
  $PW = 1/(CA) = (1 + i)^{-N}$
  $P = (PW) F$
The present worth of F, which includes all the accumulated interest is the same as the present sum of money P.

19.2.4 Repayment of principal

If the loan is repaid by annual instalments of principal plus interest, this may take two forms:

(i)   principal repaid in equal instalments with interest being paid on the reducing balance; or

(ii) equal annual payments with interest predominating in the early years and capital repayments in the later years.

The concept of equal annual payments enables a present sum of money to be converted into an annual repayment sum spread over a number of years with the annual sum A being linked to the sum invested — the "present sum P" by the capital recovery factor (CR)

$$A = (CR)P; \text{ and } CR = \frac{i(1+i)^{N}}{(1+i)^{N}-(1)} \text{ or } \frac{i}{1-(1+i)^{-N}}$$

The reciprocal of (CR) is Series Present Worth factor (SPW). This is the multiplier required to convert a number of regular annual payments into a present sum.

19.2.5 Sinking fund factor

To find the annual sum (A) which accumulates to provide a future sum (F), this is multiplied by the sinking fund factor (SF)

$$A = F(SF); \text{ and } (SF) = \frac{i}{(1+i)^{N}-(1)}$$

The reciprocal of (SF) is the series compound amount factor (SCA)

$$SCA = 1/SF \text{ and } F = (SCA)A$$

With this brief introduction to, or refresher on, economics, the economic criteria commonly used in shipping can now be introduced.

19.2.6 Net present value

In this type of calculation the net present values (NPV) of income and expenditure are calculated over the assumed life of the ship (N) years. The final sum should be positive for the investment to be profitable at the assumed discount rate — or where alternatives are being compared it should be the larger sum.

$$NPV = \sum_{1}^{N} [PW \text{ (cargo tonnage} \times \text{freight rate)}$$
$$- PW \text{ (operating costs)} - PW \text{ (ship acquisition costs)}]$$

19.2.7 Required freight rate

The required freight rate (*RFR*) is that which will produce a zero *NPV*, i.e. the break-even rate. Transposing the equation above gives:

$$RFR = \sum_{1}^{N} \left[\frac{PW(\text{Operating costs}) + PW(\text{Ship acquisition costs})}{\text{Cargo tonnage}} \right]$$

19.2.8 Yield

In the above calculations a rate of interest must be assumed. If the freight rate is known or at least assumed, the rate at which money can be borrowed with *NPV* = 0, can be made the criterion.

19.2.9 Inflation and exchange rates

It is perhaps worth pointing out that economic forecasts of the sort described in the foregoing paragraphs are made on fixed money values. Inflation and the consequent reduction in the future value of money together with changes in exchange rates do not enter into these calculations although both of these must be estimated and taken into account in more detailed projections such as those made when fixing rates which are intended to apply over more than a limited period of time and/or when payments are to be made in a currency other than that in which the costs are incurred.

19.3 OPERATING COSTS

The next three sections as well as describing the components of operating costs try to suggest some ways of minimising these.

19.3.1 Capital charges

As Fig 19.1 shows, capital charges are included in the costing of all the different modes of ship operation and are in fact the only cost component in Bareboat chartering. Included in capital charges are:
- loan repayment
- loan interest
- profit
- taxes

19.3.2 Capital amortisation

Loan interest and loan repayment can conveniently be taken together as capital amortisation.

The biggest component of capital charges is the repayment of the loan used to pay the shipbuilder. Payments to shipbuilders are almost invariably made in a number of instalments during the building period with a final instalment at the end of the guarantee period (usually a year after delivery).

Before the ship starts earning, its total cost will have increased above the tender price due to the interest payments on the sums paid out together with such other costs as those incurred in supervising construction, engaging the crew and in providing owners' supply items and initial stores.

Moreover, it will be an exceptional contract that does not result in some additional payments for changes in specification during building.

One obvious way to minimise capital charges is to keep the capital cost low, which may be achieved by good buying in relation to shipbuilding prices.

The initial building cost can, in principle, be kept down by building to a lower standard, although if this involves accepting that the ship will have a shorter than normal life this may not be a cost effective thing to do.

When considering capital economy measures, care must be taken to ensure that any lower standards adopted do not lead to higher operating costs that will negate any savings made.

The second largest component of capital charges is the sum paid in interest on the money borrowed to meet the costs incurred in building the ship and getting it into service.

Consequently, another way — and probably in the long term one of the most important ways — of minimising capital charges, is by obtaining the most advantageous interest rates available.

Finally, at the end of whatever operating life is being assumed in the financial costing, the ship will still have a value, even if this is only as scrap, and an allowance for this should be made when assessing the cost of capital amortisation.

The general assumption made in most financial assessments is that ships will have an operating life of 20 years. Although many continue in service for much longer periods others become obsolete much earlier either as a result of changes in technology and/or in trading patterns and a 20 year period is probably a reasonable compromise.

19.3.3 Profit and taxes

The profit which the shipowner plans to make together with the taxes which this profit will incur forms the second part of capital charges.

19.3.4 Depreciation

Although depreciation does not enter into a operating cost calculations, it seems desirable to include a short paragraph on the subject at this point as it does have a very significant effect on shipping company accounts, the tax paid and the profit made in particular years.

Depreciation is the process of writing off capital costs in company accounts. There are two classical methods of treating depreciation, namely:

(i) Straight line depreciation. If a 20-year life is assumed, the depreciation would be 5% per annum.

(ii) Declining balance depreciation. If a 15% per annum basis is assumed, then:
Year 1: 15% × 100   = 15%
Year 2: 15% × (100–15)   = 12.75%
Year 3: 15% × (100–15–12.75) = 10.84%
Year 10: 3.52%
Year 20: 0.94%

In most countries there are special provisions for the treatment of shipping depreciation from a taxation point of view. These treatments vary from country to country as do the rates of tax imposed.

Most of these treatments permit the writing off of a ship's capital cost at rather faster rates than the classical treatments. In general it pays a shipowner to depreciate as fast as the profits permit thus reducing or at least deferring tax payments.

19.3.5 Ship values

Although the book value of a ship at any time will be its original cost plus the cost of any repairs or alterations and minus the accumulated depreciation, the value of a ship as measured by its possible selling price is likely to fluctuate dramatically during its lifetime. This does not enter into operating cost calculations although some owners significantly improve their profits by playing the market in this way!

19.4 DAILY RUNNING COSTS

Included in daily running costs are:
– crew costs
– provisions and stores
– maintenance and repairs
– insurance
– administration and general charges

These costs are added for time charter calculations and of course also apply to voyage charters and owner operation. These are costs incurred whether the ship is at sea or in port.

19.4.1 Crew costs

The two major factors which determine crew costs today are crew numbers and the nationality of different sections of the officers and crew.

The effect of numbers is offset to some extent by the fact that a smaller crew will generally tend to have more "chiefs" and fewer "indians" and the fact that all the members of a reduced crew will (or certainly ought to) have a higher standard of training and as a consequence will (or ought to) be paid more *per capita*.

The automation and higher quality materials required to reduce watch-keeping and maintenance and thus enable the reduced crew to work the ship satisfactorily will increase the capital cost, whilst there is also likely to be a demand for higher class accommodation although this will be offset by the reduced number of cabins required.

19.4.2 Provisions and stores

Provisions are usually bought locally at the ship's trading ports and the annual cost is calculated on a per person per day basis.

Ships consume an extraordinary variety and quite considerable quantity of miscellaneous stores with the three most important items being chandlery, paint, chemicals and gases but with smaller sums being expended on such items as fresh water, laundry and charts.

Lubricating oil is sometimes included with this item, but it seems more logical to include it with bunkers.

19.4.3 Maintenance and repair

With today's small crews, maintenance at sea is necessarily limited, but careful planning by the ship's staff whilst at sea can greatly speed work carried out when in port and minimise its cost.

A big item under this heading is drydocking but, as discussed in §7.7.3, this is no longer an annual event with three or even five year intervals becoming usual.

Budgets for maintenance will generally include sums for work on the hull and superstructure, cargo spaces and systems, the main and auxiliary machinery, the electrical installation and the safety equipment plus survey fees.

Also included under this heading is the cost of riding squads which are now used to carry out maintenance and repairs which would have formerly been done by the crew but which is beyond the capability of the reduced crews of today.

19.4.4 Insurance

Insurance can be subdivided into Hull and P & I. The cost of Hull insurance is directly related to the capital cost of the ship with the insurance history of the managing company exercising a secondary effect. Costs have escalated significantly in recent years due to the number of major casualties and a generally ageing tonnage. Policies now provide for more deductibles and in the event of a claim these can increase running costs considerably.

P & I premiums have also increased greatly because of the U.S. Oil Pollution act and worries about crew standards.

19.4.5 Administration and general charges

Administration costs are a contribution to the office expenses of a shipping company or the fees payable to a management company plus a not inconsiderable sum for communications and sundries, together with flag charges.

Amongst the items included in general charges can be the cost of hiring items of ship's equipment such as the radio installation which are sometimes hired rather than bought as part of the ship.

The charge for the hire can be reduced by making a bulk deal for several ships with one company. The decision between buying and hiring demands reconsideration from time to time as prices, interest rates and tax measures change. At present the use of hired equipment is reducing.

It is also wise to allow in this heading a sum for exceptional items when preparing a cost estimate as regrettably only too often there will be something which cannot be foreseen.

19.5 VOYAGE COSTS

Included in voyage costs are:
  – bunkers
  – port and canal dues
  – tugs, pilotage
  – miscellaneous port expenses
These items are added when moving from a time charter to a voyage charter calculation and of course apply to owner operation.

19.5.1 Bunkers

19.5.1.1 Oil fuel

The factors affecting oil fuel costs are the distance travelled, the average power used, the specific fuel consumption and the cost per tonne of fuel. The first of these can be minimised by good navigation which must also take into account favourable and adverse currents.

The second can be minimised by steaming at as slow a speed as enables the required schedule to be kept; by keeping the hull finish to a high standard of smoothness (a task that is much easier than it used to be with the latest long life and self polishing antifouling paints); and at an earlier stage, by good design of the ship's lines and the propeller.

Specific fuel consumption can be minimised at the design stage by a good choice of machinery and at the operating stage by keeping the engine well maintained.

The cost of fuel can be minimised by a careful choice of bunkering port, although any cost saving thus obtained must first meet any additional costs if a diversion is required or there is any reduction in cargo carrying capacity or increase in average voyage displacement increasing power and consumption. The fuel cost can also be reduced by the use of a poorer quality of fuel, although any saving must be assessed against any extra costs for purifiers, etc. needed for the fuel to be used and any increases in maintenance and repair costs that may result from its use. Bulk buying is yet another way of getting fuel at an advantageous price.

19.5.1.2 Diesel oil

Here the factors involved are the number of days, as generators are kept running in port as well as at sea, and the average electrical load. Because the cost of diesel oil is much higher than that of oil fuel it is advantageous to meet as much as possible of the electrical load by the use of shaft driven alternators.

19.5.1.3 Lubricating oil

Although the quantity of lubricating oil consumed is relatively small its high unit cost results in it being a considerable item of expenditure. This item is sometimes included with stores, but as the usage depends on the distance travelled it seems better grouped with bunkers.

19.5.2 Port and canal dues, pilotage, towage etc.

19.5.2.1 Port and canal dues

Port and canal dues depend on the tonnage of the vessel and on the trading pattern. Low gross and/or net tonnages are particularly important on some routes, such as those using the Suez or Panama canals or The St. Lawrence Seaway.

Booklets giving canal dues can be obtained from:
– Panama Canal Commission, Balboa, Republic of Panama (Fax: 507-272-2122)
– Suez Canal Authority, Ismailia, Arab Republic of Egypt (Fax: 064-320-784)
– St. Lawrence Seaway Authority, 360 Albert St, Ottawa, Canada (Fax: 613-598-4620)

19.5.2.2 Pilotage costs

Pilotage costs are usually also assessed on gross tonnage but can be reduced in certain trades by having a ship's officer with a pilotage certificate where this procedure is followed.

19.5.2.3 Towage and mooring costs

Tug charges can be eliminated or reduced if the ship is fitted with a bow thruster or approved high performance steering equipment.

The time spent in mooring can be reduced by fitting special deck machinery such as self-tensioning winches.

19.6 CARGO HANDLING COSTS

Cargo handling costs include the costs arising from both loading and unloading cargo together with any claims that may arise relating to the cargo.

Cargo handling costs are excluded from voyage charter costs but have to be met in owner operation.

Cargo handling time can be reduced and with it the costs of this operation, by the provision of good cargo handling features such as:
1. large hatches giving good access;
2. shipside doors where appropriate;
3. hatch covers which can be speedily opened and closed;
4. fork lift trucks to speed stowage;
5. cargo handling cranes or derricks on the ship with a lift capacity optimised to the cargo carried and a speedy cycle time;
6. in appropriate cases by providing the ships with self discharging facilities.

Where the trade is based on a small number of specific ports there is the alternative of minimising the ship cost and using shoreside cargo handling facilities.

Containerisation or palletisation of the cargo can make a step change in cargo handling time and cost.

19.7 SOME COST FIGURES

The previous sections of this chapter have tried to introduce the more important factors involved in operational economics. Following the practice of the other chapters of this book, this last section now tries to give some figures which will enable readers to make their own approximate calculations.

The treatment given calculates a typical annual budget by amortising the capital costs on an equal annual payment basis. It should be emphasised that this is a broad brush treatment suitable for a naval architect's quick look at the subject.

A shipowner making a similar calculation would calculate the costs for each year of the assumed life of the ship and aggregate these to a net present value in the way described in §19.2.6.

As many as possible of the cost items have been dealt with in a way which enables approximate figures to be estimated for any type and size of ship, but this has not been possible for all items. No allowance has been made for inflation in this example (see §19.2.9) and caution should be exercised in the use of the figures given.

19.7.1 Operational cost example

The ship chosen as an example is a 40,000 tonne deadweight products tanker with the following main characteristics:
- main engine power (MCR) 9000 kW;
- fuel consumption 30 tonnes per day;
- diesel consumption 2.7 tonnes per day;
- gross tonnage 23000.

19.7.2 Ship cost

The shipyard tender price for this ship is:

$30,000,000

This price is given in US dollars following general practice in this market and the same unit is used for all the subsequent figures.

Although the figures are approximate they should provide a useful outline picture of the relative importance of the various items.

In addition to the shipyard tender cost, the total capital cost at the time that the ship enters service and starts to earn should also include allowances for the following.

19.7.2.1 Extras likely to be claimed by the shipyard

The minimisation of extra costs forms an important part of specification writing and contract negotiation. In the case of specialist ships it may be difficult to avoid extras, but for fairly standard ship types, extras should not exceed 0.5% of the shipyard contract price

= $150,000.

19.7.2.2 The costs of owner supply items

The costs of owner supply items and of owner's costs during construction including supervision and visits to the shipyard could be of the order of 1% of the contract price

= $300,000.

19.7.2.3 Interest

The cost of interest on the instalments paid prior to delivery. Assuming a building period of two years from order to delivery and a reasonably evenly spread schedule of payments the average sum on which interest would need to be paid might be about half the contract price, making the interest paid prior to delivery:

2(years) × 9% × $15 million = $2,700,000.

The total capital invested in the ship when it enters service is therefore:

$33,150,000.

19.7.2.4 Ship value at end of 20 years

It is estimated that the scrap value at the end of 20 years (at today's money value) will be $750,000.

The present value of this sum using the interest rate discussed in the next paragraph is:

$750,000/(1.09)^{20} = $134,000.$

The sum to be dealt with in the capital amortisation calculation is therefore reduced to:

$33,016,000.

19.7.3 Capital charges

19.7.3.1 Capital amortisation

Capital amortisation is made on the assumption of a 20-year life, an interest rate of 9% and on the basis of equal annual payments based on the formula given in §19.2.4.

The interest rate suggested reflects the fact that in order to stimulate their shipbuilding industries, most shipbuilding countries have introduced loans for ship purchases at reduced rates of interest.

From a variety of figures these rates have now mainly become standardised as OECD terms which cover 80% of the contract price for eight and a half (8½) years at 8% interest.

For the other 20% the shipowner will either use his own funds or borrow elsewhere. Borrowing elsewhere will also apply if funds to repay the shipbuilding loan at the end of the 8½ year period are not available. The interest payable on such borrowing is likely to be higher than 8% and for a broad brush treatment it may be not unreasonable to use a rate of 9% as applicable to the whole capital sum.

$$\text{Annual payment} = \$33,016,000 \times \frac{0.09(1.09)^{20}}{(1.09)^{20} - (1)}$$

$$= \$3,620,000.$$

19.7.3.2 Profit and taxes

For an approximate calculation of this sort it seems reasonable to base the profit on the investment in the ship although clearly other items will be taken into account when this calculation is done "in anger". Using the ships capital cost as the basis a net profit of 6% might be looked for which would increase to 8% if allowance is made for a tax of 33%.

The combined figure would then be:

$2,650,000.

19.7.4 Daily running costs

19.7.4.1 Crew costs

General guidance on crew numbers has been given in Chapter 5. The ship used as the example has a crew of 10 officers and 15 ratings.

Rates of pay and supplementary payments to both officers and ratings vary considerably with nationality, and it is outside the scope of this book to give more than one example. With much of the world's tonnage "flagged out", the example is reasonably typical, consisting as it does of Indian officers and Phillipino ratings.

Both pay and supplementary payments for leave pay, overtime etc., scale with rank with a Master receiving rather more than twice the pay of a Fourth engineer or Third officer. For officers an average basic annual pay is $16,000, which doubles to about $32,000 with supplements and increases to about $40,000 when travel and manning expenses are added.

For ratings the average basic pay is about $6,000, increasing to $15,000 with supplements and to $18,000 with travel, etc.

The annual crew cost for the example is therefore:

$10 \times \$40,000 = 15 \times \$18,000 = \$670,000.$

19.7.4.2 Provisions and stores

Provisions can be based on what seems a remarkably low daily rate of $7 per person giving a figure of $25 \times 365 \times \$7$ for the example

$= \$64,000.$

Stores were estimated to amount to $85,000 for the example ship. There does not seem to be any easy way of scaling this to other ship types or sizes, although such items as paint must scale with size, whilst chandlery will vary with ship type, making a total of:

$149,000.

19.7.4.3 Maintenance and repair

These costs vary greatly with ship type and shipping company policy and only the roughest guidance can be given.

Costs should be relatively low in the early years increasing in the later ones. In principle, investment in high quality machinery and outfit should reduce these costs. For a ship very similar to that used as the example a detailed estimate after an inspection gave a figure of $100,000 for 5-year-old ship, and this figure is carried forward to the example:

$= \$100,000.$

19.7.4.4 Insurance

Here again (to the indignation of at least one very well known naval architect who thinks that increased investment in equipment that will improve safety, such as a second radar, should reduce the insurance premium), the only guidance that can be given is to suggest an allowance of about 1% of the capital value, remembering that this will be reducing annually. The P&I insurance is based on the Gross tonnage and a rate of about $7 per G.R.T. can be expected

For the first years operation the insurance for the example will thus be:

$1\% \times \$33.15 \text{ million} + \$7 \times 23000$

$= \$493,000.$

19.7.4.5 Administration and general charges

A management fee of $100,000 constitutes the largest part of this item with communications and port charges other than those associated with the cargo, plus flag expenses and sundries aggregating about $80,000 and a provision for exceptional items of $50,000 to give a total of:

$230,000.

19.7.5 Voyage costs

19.7.5.1 Oil fuel

Oil fuel costs for the product tanker used as the example are estimated on the premise that the ship will be at sea for 280 days per year and will operate for this period at 85% MCR. The time at sea for other ship types/other trades will differ with bulk carriers spending rather less time at sea, possibly about 260 days.

Bunker prices are quoted in a number of shipping publications, the following figures being taken from the August 1993 *Marine Engineers Review*. This shows the cost of oil fuel IFO 380 varying from $59 per tonne in Rotterdam or Houston to $119 per tonne in Buenos Aires or Dakar. IFO 180 being $5 to $10 per tonne more.

For the example, say 280 days × 30 tonnes × $65

= $546,000.

19.7.5.2 Diesel oil

Diesel oil prices in the same ports range from about $148 to about $200 per tonne.

For the example, say 365 days × 2.7 tonnes × $160

= $158,000.

19.7.5.3 Lubricating oil

Lubricating oil usage for the main engine crankcase and cylinders plus that used in generators and other machinery can be approximated to 35 litres per day per 1000 KW of main engine power.

The cost per litre is about $1 and the annual bill for the example is 280 days × 9000 / 1000 × 35 litres × $1

= $88,000

The relatively high costs of diesel oil and particularly of lubricating oil are worthy of notice.

19.7.5.4 Canal dues

Panama: These are based on Panama net tonnage and for a one-way trip are:
- loaded $2.21 per net tonne(PC/UMS)
- in ballast $1.76 per net tonne

Suez: These are again based on the net tonnage (SCNT), but are more complicated than the Panama ones, varying with ship type and reducing per tonne with ship size. Table 19.1 gives figures for three types of ship.

St. Lawrence Seaway: These are calculated partly as a charge per gross ton and partly as a charge per metric tonne of cargo carried with rates varying with the type of cargo. The dues are split between a fee for transit to or from Montreal to Lake Ontario and from Lake Ontario to or from Lake Erie (Table 19.2).

Table 19.1

Suez Canal dues

Type	First $5,000	Next $5,000	Next $10,000	Next $20,000	Next $30,000	Remainder
Crude oil tanker	6.49	3.62	3.25	1.40	1.40	1.21
Bulk Carrier	7.21	4.14	2.97	1.05	1.00	1.00
Container ship	7.21	4.10	3.37	2.42	2.42	2.42

There are different dues for other ship types, but generally in the same ballpark.
Dues in ballast are about 85% of dues loaded.

Table 19.2

St. Lawrence Seaway dues

	Montreal to Ontario $(Can)	Ontario to Erie $(Can)
Per Gross ton	0.08	0.13
Per metric tonne of cargo		
bulk cargo and containers	0.83	0.55
general cargo	2.00	0.88

There are special rates for some other types of cargo, the figures given are illustrative only.

Table 19.3

Annual operating budget

Item	$	$	$	%
Capital charges				
Capital amortisation	3,620,000			41
Profit and taxes	2,650,000			30
Basis for bareboat charter		6,270,000		
Daily running costs				
Crew	670,000			8
Provisions and stores	149,000			2
Maintenance and repairs	100,000			1
Insurance	493,000			6
Administration and general	230,000			3
Total		1,642,000		
Basis for time charter			7,912,000	
Voyage costs				
Oil fuel	546,000			6
Diesel oil	158,000			2
Lub. oil	88,000			1
Port charges etc.	?			
		792,000		
Basis for voyage charter			8,704,000	100
Cargo handling	?			
Basis for owner operation			?	

Because they vary so widely no figures can be given for the other items of voyage expenses port dues, pilotage and towage costs, cargo handling costs.

19.7.7 Summary

The complete budget is given in Table 19.2. Each of the cost headings has also been expressed as a percentage of the total budget to highlight its relative importance.

Chapter 20

Conversions

20.1 GENERAL

Most of the methods which are used in the design of new ships apply equally to a major conversion to an existing ship. There are, however, some special difficulties in conversions and some established methods of overcoming these will be discussed in this chapter.

First of all it may help to define what is meant by a conversion in this context. This may vary from an extreme case where a ship's whole purpose is changed as when T2 tankers were converted to become the first container ships or wartime corvettes to be the first factory trawlers, to jumboisations where the role is unchanged but the ship is increased in size to enable the quantity of cargo or the number of passengers carried to be greatly increased. In addition to conversions such as these which are usually made after a ship has been in service for a number of years and the trade basis for which it was designed has altered, major modifications are sometimes found necessary before a ship can enter service because the design has turned out to be defective in some way. Some of these remedial modifications are similar to those required for role modifications and will be discussed in this chapter.

20.2 ANTICIPATING THE NEED AT THE DESIGN STAGE

If the possibility of a conversion being required is thought likely at the ship design stage, there are a number of things that can be done to facilitate this. Probably warships are the ship type where this is most worthwhile. The hull and machinery of a warship normally has a much longer life than the weapons and weapon systems installed in it, and a mid life refit is common. Warships have traditionally been very tightly packed with equipment, making access for repairs difficult and necessitating a "piecee small" approach to any new item being installed. Providing good access paths and arranging the complete kit of a major assembly as a module can greatly reduce both the time and cost of a refit. Modularity can be greatly helped by such relatively minor design changes as the elimination of sheer and camber whose presence demand tailor-made fittings.

As well as their use on warships, these concepts are often designed into oceanographic and fishery research vessels where the next job is likely to require different equipment from the last one and on offshore working ships for the oil industry where much the same applies.

20.3 COST AND TIME OF CONVERSIONS

Two major factors which seem to apply to all conversions are a need to minimise cost and the time taken. These factors are not strange to designers of new ships, but they are significantly stronger as applied to conversions. The reasons for this are not very difficult to identify.

The cost sensitivity stems from the fact that the remaining life over which the conversion cost must be amortised is almost invariably less than applies to a new ship; and where the work is being done to rectify a design fault someone — either shipowner or shipbuilder — is already out of pocket and is trying to limit losses.

Time sensitivity in relation to the conversion of an existing ship is bound up with cost as the ship is a capital asset failing to earn its keep whilst under conversion.

20.4 NEED TO MEET NEW RULES

When planning any major modification, it should be borne in mind that the modified ship will almost certainly have to meet the latest IMO rules which may be much more severe than the rules to which the ship was originally built and changes may be required in all sorts of items not in any way affected by the desired modifications.

20.5 MODIFICATIONS TO INCREASE DEADWEIGHT OR CAPACITY

Quite a large proportion of conversions, whether involving a change in role or not, require an increase in the displacement or in the internal volume of the hull.

Most jumboisations involve adding a length of parallel middle body by splitting the existing ship in two in dry dock. One part is then floated out of the dock, the new section floated or lifted in, the part which was removed floated back in and the three sections joined up to give a new ship of increased length but with the same beam and depth as the original vessel. The split need not be at amidships but if possible should be within the parallel middle body of the original ship.

In some cases the much more difficult task of adding a section of which has a greater beam has been achieved. This is bound to involve some unfairness in the waterlines which will offend a purist, but the powering penalty may be acceptable against the gain made in cargo carrying ability.

Some ships have been split horizontally with all the structure above a certain level being raised to permit the insertion of units which will increase the depth to a desired figure. The boldness of some conversions in recent years has been striking and they have on occasion given their entrepreneurs a virtually new ship in a time scale much shorter than could be obtained by new building.

The addition of a length of parallel middle body to a ship poses an unusual and quite interesting problem in naval architecture. This is the calculation of the positions of the longitudinal centre of gravity and longitudinal centre of buoyancy of the jumboised ship and an algebraic solution to this problem is shown in Fig. 20.1.

The effect of an addition of parallel middle body on speed and power can be quite surprising as the reduction in the specific resistance stemming from the reduced Froude Number quite frequently more than offsets the increased displacement and wetted surface with the surprising result that the ship gains in speed or requires less power for the same speed.

The effect on the structure of an addition to the length is not so happy as the new structure requires heavier scantlings than the original ship and doubler plates may be required on the strength deck of much of the original structure to bring the section modulus up to that required for the longer ship.

20.6 MODIFICATIONS TO IMPROVE STABILITY

Modifications to improve stability are the next most common requirement, and once again these may be required either to permit the addition of top weight in a conversion or as a way of rectifying inadequate stability found when a ship is inclined on completion.

Improved stability may be required either at an operating draft, or at a draft resulting from damage. The required improvement may consist of an increase in KM, an increase in large angle righting lever or an increase in the range of positive stability or all of these may be required. It is not unusual for a need for an improvement in stability to be linked with a need for increased displacement and fortunately most solutions to stability problems do help with the other deficiency, if only to a minor extent.

There are two main approaches to improving stability — one which must be used if a major improvement is necessary and another approach which may be adopted if a modest improvement will suffice.

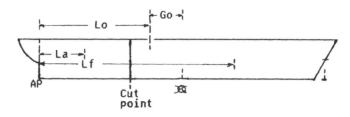

Aft portion		Original ship		Ford portion	
Length	$= X_a$	Length B.P.	$= X_o$		
Weight	$= W_a$	Weight of ship	$= W_o$	Weight	$= W_f$
LCG from AP	$= L_a$	LCG from AP	$= L_o$	LCG from AP	$= L_f$
		LCG from $\text{C\!\!\!\!L}$	$= G_o$		

$W_o = W_a + W_f$ $\qquad W_o \cdot L_o = W_a \cdot L_a + W_f \cdot L_f$ $\qquad G_o = X_o/2 - L_o$

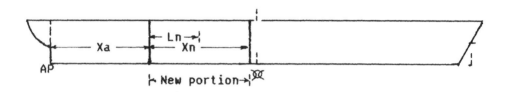

New portion		Jumboised ship	
Length	$= X_n$	Length	$= X_j = (X_o + X_n)$
Weight	$= W_n$	Weight	$= W_j = (W_o + W_n)$
LCG from aft cut	$= L_n$	LCG from AP	$= L_j$
LCG from AP	$= X_a + L_n$		

$$(W_o + W_n) L_j = W_a \cdot L_a + W_n (X_a + L_n) + W_f(L_f + X_n)$$
$$= W_a \cdot L_a + W_f \cdot L_f + W_n(X_a + L_n) + W_f \cdot X_n$$
$$L_j = \frac{W_o \cdot L_o + W_n (X_a + L_n) + W_f \cdot X_n}{W_o + W_n}$$

Fig. 20.1. Finding the longitudinal centre of gravity of a jumboised ship. (The method applies equally to the longitudinal centre of buoyancy).

20.6.1 Sponsons

Sponsons can provide a major improvement in stability and are an expedient which has been used quite frequently by naval architects. The lower edge of a sponson should be some distance below the lowest operational waterline at which the gain in stability is required in order to avoid a sudden loss of stability when a sponson comes out of the water as the ship rolls or heels under wind or other forces. A possible design criterion would be to ensure that it remains immersed when the ship at its minimum operating draft rolls to at least 5 or possibly 10 degrees.

The upper edge of the sponson will be determined by whether the sponsons are being provided to increase the GM at intact operational drafts or at the draft after damage and whether it is required to increase GZ and/or range in either of these conditions. In the latter case the upper edge must be above the heeled waterline after damage.

The fore and aft length of a sponson on a fine lined ship must be determined to a large extent by the length of parallel middle body at the height covered by the sponson as there is a diminishing return in increased stability as the lines move away from maximum beam and a danger that the forward end of the sponsons will cause a measurable increase in powering resistance if it is positioned forward of the shoulder.

Unless there is a limitation on the overall width of the ship to suit berthing or some similar consideration or the required increase in stability demands a wider sponson, a width of about 800 mm is commonly used as lesser widths tend to pose constructional access problems and greater widths require more extensive internal stiffening.

The ends of sponsons should be arranged on bulkheads and the sponsons should have watertight divisions in line with each of the ship's watertight bulkheads.

The addition of sponsons can pose a multitude of problems in relation to the outreach of cranes and lifeboat davits.

Figure 20.2 shows a possible sponson on a passenger ferry which was investigated as a design exercise (the ship in question having no need for improved stability). It was found that the sponson shown would give an increase in KM at the load draft of 0.90 m or 6.4% and 0.78 m at a "margin line" draft with an increase in the range of stability after damage, for three different damage conditions, of between 3 and 4°.

The sponson arrangement shown in Fig. 20.3 was fitted to an oilfield support vessel in which the need to operate equipment over the ship's side was a major constraint. In the event, two almost separate sponsons were fitted — a long lower one to give the required GM at the service draft and a shorter upper one to improve the range of stability and the GZ values at large angles.

20.6.2 Stern wedges

High stability lines have been discussed in Chapter 8. Something along the lines discussed there can be done — and indeed has been done — as a modification to an existing ship. Again the same passenger ferry is used to illustrate the effect of adding stern wedges.

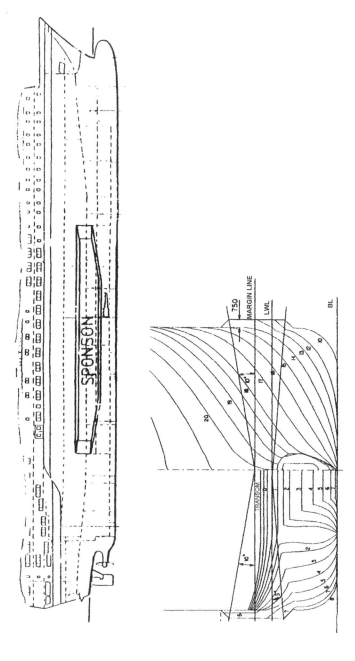

Fig. 20.2. Sponsons fitted to a Ro-Ro ferry. In worst damage condition these increase GZ from about 0.08 m to 0.23 m and the range from 9.5 to 13°.

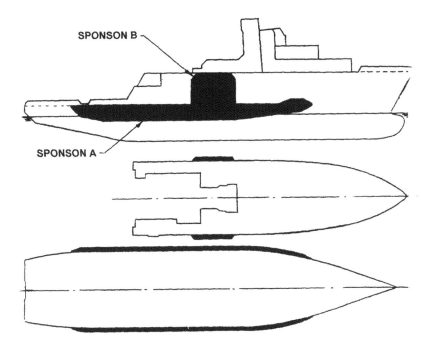

Fig. 20.3. Sponsons fitted to an oilfield support vessel. The unusual configuration is necessary to keep these clear of overside gear. Sponsons "A" increase KM, sponsons "B" increase GZ and the range.

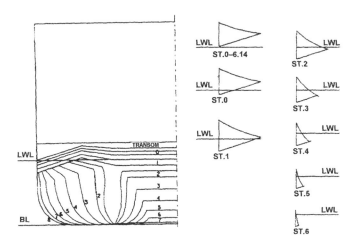

Fig. 20.4. Added stern wedges on a Ro-Ro ferry to increase KM at load draft. These increase KM at load draft by 1.28 m.

The original stern lines together with modified lines with added stern wedges are shown in Fig. 20.4.

The KM at the load draft is increased by 1.28 m or 9% and at a "margin line" draft by 0.12 m with an increase in range of stability of between 1.5 and 4° for three damage conditions.

The biggest effect is attained at the load draft, but the modest improvement in the damaged condition may be equally or more valuable.

Bibliography

The date given for most papers is that of the year of presentation; these generally appear in the transactions of the following year.

CHAPTERS 2–5

Books:
Basic Naval Architecture. K.C. Barnaby, 1953.
Basic Ship Theory — Rawson and Tupper, 1983.
Introduction to Naval Architecture — Gillmer and Johnson, 1982.
Manual of Naval Architecture — White, 1984.
Merchant Ship Types — Munro Smith, 1975.
Modern Ship Design — Gillmer, 1975.
Muckle's Naval Architecture (revised by Taylor).
Principles of Naval Architecture — S.N.A.M.E. revised 1988.
Ship Design and Construction — D'Arcangelo (S.N.A.M.E.) 1969.
Ship Design and Construction — Taggart, 1980.
Standard Ship Designs — Vols. of Fairplay, 1984–5.
The Design of Merchant Ships — Van Lammeren (ed), 1959.
Theory and Technique of Ship Design — Manning, 1959.
Theory of Naval Architecture — Robb.

Papers:
Andrew, 1986. An integrated approach to ship synthesis. R.I.N.A.
Brown and Tupper, 1988. The naval architecture of surface warships. R.I.N.A.
Buxton, 1980–81. Matching merchant ship designs to markets. N.E.C.I.E.S.
Eames, 1980. Advances in naval architecture for future surface warships. R.I.N.A.
Eames and Drummond, 1976. Concept exploration — an approach to small warship design. R.I.N.A.
Gallin, 1977–78. Inventiveness in ship design. N.E.C.I.E.S.
Gilfillan, 1968. The economic design of bulk cargo carriers. R.I.N.A.

Hills and Buxton, 1989. Integrating ship design and production considerations during the pre-contract phase. R.I.N.A.

Griethuysen, 1992. On the variety of monohull warship geometry. R.I.N.A.

Griethuysen, 1993. On the choice of monohull warship geometry. R.I.N.A.

Katsoulis, 1975. Optimising block coefficient by an exponential formula. Shipping World and Shipbuilder.

Meek, 1980. Developments in merchant ships over the last decade. Thomas Low Gray Lecture, IMechE.

Nethercote and Schmitke, 1981. A concept exploration model for SWATH ships. R.I.N.A.

Watson, 1958. A note on the distribution of steel weight in ship's hulls. B.S.R.A. Report 266.

Watson, 1962. Estimating preliminary dimensions in ship design. I.E.S.I.S.

Watson and Gilfillan, 1976. Some ship design methods. R.I.N.A.

Welsh, 1987–88. Preliminary ship design using a micro-based system (CAD). N.E.C.I.E.S.

CHAPTERS 6–8

Books:

Manual for the Use of the 1978 ITTC Performance Prediction Method as Modified in 1984 and 1987 — ITTC '87.

Ship Design for Efficiency and Economy — Schneekluth, 1987.

Ship Performance — Hughes, 1987.

Papers:

Burcher, 1980. The influence of hull shape on transverse stability. R.I.N.A.

Buxton and Logan, 1986. The ballast performance of ships with particular reference to bulk carriers. R.I.N.A.

Cox, 1976. Fishing vessel safety. R.I.N.A.

Grigson, 1981 and 1983. The drag coefficients of a range of ship surfaces. R.I.N.A.

Grigson, 1982. Propeller roughness, its nature and its effect upon the drag coefficients and ship power. R.I.N.A.

Grigson, 1987. The full-scale viscous drag of actual ship surfaces and the effect of quality of roughness on predicted power. Journal of Ship Research.

Grigson, 1989. Note on a rational method for approximating the full-scale viscous drag from roughness geometry. R.I.N.A.

Grigson, 1992. Drag losses of new ships caused by hull finish. Journal of Ship Research.

Grigson, 1993. An accurate smooth friction line for use in power prediction. R.I.N.A.

Guldhammer and Harvald, 1974. Ship resistance: effect of form and principal dimensions. Akademisk Forlag, Copenhagen.

Hagen, Comstock and Slager, 1986. Investigation of design power margin and correlation allowance for surface ships. Marine Technology.

Holtrop and Mennen, 1982. An approximate power prediction method. International Shipbuilding Progress.

Holtrop, 1984. A statistical analysis of resistance and propulsion data. International Shipbuilding Progress.

Lackenby and Parker, 1966. The B.S.R.A. methodical series — an overall presentation. R.I.N.A.

Lloyd, Salsach and Zseleczky, 1986. The effect of bow shape on deck wetness in head seas. R.I.N.A.

Lloyd, 1991. The seakeeping design package. R.I.N.A.

Lyster, 1978–79. Prediction equations for ship's turning circles. N.E.C.I.E.S.

Moor, 1971. The effective horsepower of single screw ships: average and optimum standards of attainment 1969. B.S.R.A. Report 317.

Moor and Patullo, 1968. The effective horsepower of twin screw ships. BSRA Report 192.

Moor and Small, 1960. The effective horsepower of single screw ships. R.I.N.A.

Morrall, 1980. The capsizing of small trawlers. R.I.N.A.

Morrall, 1983. A study of the safety of fishing. R.I.N.A.

Nethercote, Murdey and Datta, 1987. Some warship bulbous bow and stern wedge investigations. R.I.N.A.

Robson, 1987. Systematic series of high speed displacement hull forms for naval combatants. R.I.N.A.

Townsin, Byrne, Milne and Svensen, 1980. Speed, power and roughness: The economics of outer bottom maintenance. R.I.N.A.

Townsin and Kwon, 1983. Approximate formulae for the speed loss due to added resistance in wind and waves. R.I.N.A.

Turner, Harper and Moor, 1963. Some aspects of passenger liner design. R.I.N.A.

Watson, 1981. Designing ships for fuel economy. Parsons Memorial Lecture. R.I.N.A.

Winters, 1997. Application of a large propeller to a container ship with keel drag. R.I.N.A.

CHAPTER 9

Books:

Introduction to Marine Engineering — Taylor, 1983.

Marine Diesel Engines — Calder, 1988.

Ships and their Propulsion Systems — Gallin, Hiersig and Heiderich.

Slow Speed Marine Diesels — Woodward, 1981.

Warship Propulsion System Selection — Plumb, 1987.

Papers:

Bonney and Walker, 1986–87. Walker Wingsail operating experience on the M.V. Ashington. N.E.C.I.E.S.

Hieda and Kusano, 1985–86. The application and fuel economy of gas turbine combined cycle for LNG carriers. N.E.C.I.E.S.

Hopkins, 1991. Fantasy and reality. IMarE.

Ikeda, Itoh and Someya, 1982–83. Development of marine coal fired boilers with special reference to the fluidised bed boiler. N.E.C.I.E.S.

Spears, 1983–84. Advanced systems for the gas turbine, steam and combined cycle propulsion of ships. N.E.C.I.E.S.

Watson and King, 1980. Machinery for the future — back to coal? W.E.M.T.

YARD and GEC authors, 1981. Trends in propulsion machinery. IMarE.

CHAPTER 10

Books:

Merchant Ship Construction — D.A. Taylor, 1980.

Structural Design of Sea-going Ships — N. Barabanov, 1970.

Ship Construction — Eyre.

Papers:

Chalmers and Price, 1980. On the effective shear area of ship sections. R.I.N.A.

Chalmers, 1982. Preliminary structural design of warships. R.I.N.A.

Chalmers, 1988. Hull structural design using stiffness as a criterion. R.I.N.A.

Ferguson, 1992. Safety of bulk carriers. I.E.S.I.S.

Ferguson, 1994. Detail design of double hull tankers — no room for complacency. I.E.S.I.S.

Fransman, 1988. The influence of passenger ship superstructures on the response of the hull girder. R.I.N.A.

Grigson, 1997. On the contention that bulkers and tankers are too weak in storms. R.I.N.A.

Kuo and McCallum, 1984. An effective approach to structural design for production. R.I.N.A.

Mikelis, Miller and Taylor, 1984. Sloshing in partially filled liquid tanks and the effect on ship motions. R.I.N.A.

Smith et al., 1991. Strength of stiffened plating under combined compression and lateral pressure. R.I.N.A.

Smith and Chalmers, 1987. Design of superstructures in fibre reinforced plastic. R.I.N.A.

Winkle and Baird, 1985. Towards more effective structural design through synthesis and optimisation of relative fabrication costs. R.I.N.A.

CHAPTERS 11–13

Papers:

Allan, 1997. The 1995 SOLAS Diplomatic Conference on Ro-Ro Passenger Ferries. R.I.N.A.

Burcher, 1979. The influence of hull shape on transverse stability. R.I.N.A.

Buxton and Logan, 1987. The ballast performance of ships with particular reference to bulk carriers. R.I.N.A.

Kuo, Pryke, Sodahl and Craufurd, 1997. A Safety Case for Stena Line's High Speed Ferry HSS1500. R.I.N.A.

Sarchin and Goldberg, 1962. Stability and buoyancy criteria for U.S. surface ships. S.N.A.M.E.

CHAPTERS 14–16

Books:
Chemical and Parcel Tankers — M..Grey, 1984.
Caldwell's Screw Tug Design — J.N. Wood, 1969.
Natural Gas by Sea — Roger Fooks, 1979.
Tanker Practice — G.A.B. King, 1969.

Papers on Merchant Ships:
The Clyde Class. Motor Ship, June 1970.
Baxter, 1973. Hydrographic survey or research ship. R.I.N.A.
Beattie and Robson, 1978–79. A review of cargo handling systems for dry cargo vessels. N.E.C.I.E.S.
Bell and Brow, 1984. Design and operation of a dynamically positioned drillship. R.I.N.A.
Bengtsen and Walker, 1980. Modern car ferry design. R.I.N.A.
Burrows, 1983. Clyde 252: a new class of economy offshore supply ships. I.E.S.I.S.
Cox and Wilson, 1977. The development of sand and gravel dredges. R.I.N.A.
Dick and Corlett, 1976. The pan type post office cable repair ship. R.I.N.A.
Kanerva and Lunnberg, 1985. Icebreaking cargo ships. R.I.N.A.
Kay, Jones and Mitson, 1991. FRV Corystes: a purpose built fisheries research vessel. R.I.N.A.
Meek and Ward, 1973. Accommodation in ships. R.I.N.A.
Parker and Woolveridge, 1988. B.P. Swops an operator's and shipbuilder's perspective. R.I.N.A.
Payne, 1989. The evolution of the modern cruise liner. R.I.N.A.
Payne, 1992. From Tropicale to Fantasy: a decade of cruiseship development. R.I.N.A.
Rorly, Hansen and Bowes, 1978. Corvette KV 72. R.I.N.A.

Papers on Warships:
Andrews, 1981. Creative ship design. R.I.N.A.
Brown, 1995. Advanced warship design — limited resources. R.I.N.A.
Brown and Tupper, 1988. The naval architecture of surface warships. R.I.N.A.
Bryson, 1984. Procurement of a warship. R.I.N.A.
Gates, 1986. Cellularity — an advanced weapon electronics technique. R.I.N.A.
Gates and Rusling, 1982. The impact of weapons electronics on surface warships design. R.I.N.A.
Harris, 1980. The Hunt Class mine countermeasures vessels. R.I.N.A.
Honnor and Andrews, 1982. H.M.S. Invincible: the first of a new genus of aircraft carrying ships. R.I.N.A.
Marsh and Gilchrist, 1985–86. Building aircraft carriers. N.E.C.I.E.S.
Rawson, 1989. Ethics and fashions in design. R.I.N.A.
Robson, 1983. Development of the Royal Australian Navy GRP minehunter design. R.I.N.A.
Sadden and McComas, 1992. Modern corvette design and construction. I.Mar.E. (Conference).

Symons and Sadden, 1982. Design of seabed operations vessel. R.I.N.A.
Thomas and Easton, 1991. The Type 23 Duke Class Frigate. R.I.N.A.
Usher and Dorey, 1981. A family of warships. R.I.N.A.
Ware, 1988. Habitability in surface warships. R.I.N.A.
Watson and Friis, 1991. A new Danish fishery inspection ship type. R.I.N.A.

CHAPTERS 17–19

Books:
Design and Cost Estimating of Merchant and Passenger Ships — Kari, 1938.
Fundamentals of Ship Design Economics — Benford, 1965.
Liner Shipping Economics — Jansson and Shneerson, 1987.
Ship Maintenance: A Quantitative Approach — Shields et al., 1982.

Papers:
Benford, 1972. Optimal life and replacement analysis for ships and shipyards. R.I.N.A.
Brown, 1987–88. An aid to steel cost estimating and structural design optimisation. N.E.C.I.E.S.
Buxton, 1972. Engineering economics applied to ship design. R.I.N.A. (also B.S.R.A. 1976).
Carreyette, 1977. Preliminary ship cost estimation. R.I.N.A.
Fisher, 1973. Relative costs of ship design parameters. R.I.N.A.
Meek and Ward, 1983–84. Ship performance and contractual requirements. N.E.C.I.E.S.
Rawson, 1972. Towards economic warship acquisition and ownership. R.I.N.A.
Sato, 1967. Effect of principal dimensions on weight and cost of large ships. S.N.A.M.E.
Southern, 1980. Work content estimating from a ship steelwork data base. R.I.N.A.
Walker and McCluskey, 1998. Restructuring the manufacturing rocess of modern warships. I.E.S.I.S.

Subject Index

Printed and bound by CPI Group (UK) Ltd, Croydon, CR0 4YY

13/05/2025

01869555-0002